Lecture Notes in Physics

W0227329

The Editorial Policy for Proceedings

The series Lecture Notes in Physics reports new developments in physical research and teaching – quickly, informally, and at a high level. The proceedings to be considered for publication in this series should be limited to only a few areas of research, and these should be closely related to each other. The contributions should be of a high standard and should avoid lengthy redraftings of papers already published or about to be published elsewhere. As a whole, the proceedings should aim for a balanced presentation of the theme of the conference including a description of the techniques used and enough motivation for a broad readership. It should not be assumed that the published proceedings must reflect the conference in its entirety. (A listing or abstracts of papers presented at the meeting but not included in the proceedings could be added as an appendix.)

When applying for publication in the series Lecture Notes in Physics the volume's editor(s) should submit sufficient material to enable the series editors and their referees to make a fairly accurate evaluation (e.g. a complete list of speakers and titles of papers to be presented and abstracts). If, based on this information, the proceedings are (tentatively) accepted, the volume's editor(s), whose name(s) will appear on the title pages, should select the papers suitable for publication and have them refereed (as for a journal) when appropriate. As a rule discussions will not be accepted. The series editors and Springer-Verlag will normally not interfere with the detailed editing except in fairly obvious cases or on technical matters.

Final acceptance is expressed by the series editor in charge, in consultation with Springer-Verlag only after receiving the complete manuscript. It might help to send a copy of the authors' manuscripts in advance to the editor in charge to discuss possible revisions with him. As a general rule, the series editor will confirm his tentative acceptance if the final manuscript corresponds to the original concept discussed, if the quality of the contribution meets the requirements of the series, and if the final size of the manuscript does not greatly exceed the number of pages originally agreed upon. The manuscript should be forwarded to Springer-Verlag shortly after the meeting. In cases of extreme delay (more than six months after the conference) the series editors will check once more the timeliness of the papers. Therefore, the volume's editor(s) should establish strict deadlines, or collect the articles during the conference and have them revised on the spot. If a delay is unavoidable, one should encourage the authors to update their contributions if appropriate. The editors of proceedings are strongly advised to inform contributors about these points at an early stage.

The final manuscript should contain a table of contents and an informative introduction accessible also to readers not particularly familiar with the topic of the conference. The contributions should be in English. The volume's editor(s) should check the contributions for the correct use of language. At Springer-Verlag only the prefaces will be checked by a copy-editor for language and style. Grave linguistic or technical shortcomings may lead to the rejection of contributions by the series editors. A conference report should not exceed a total of 500 pages. Keeping the size within this bound should be achieved by a stricter selection of articles and not by imposing an upper limit to the length of the individual papers. Editors receive jointly 30 complimentary copies of their book. They are entitled to purchase further copies of their book at a reduced rate. As a rule no reprints of individual contributions can be supplied. No royalty is paid on Lecture Notes in Physics volumes. Commitment to publish is made by letter of interest rather than by signing a formal contract. Springer-Verlag secures the copyright for each volume.

The Production Process

The books are hardbound, and the publisher will select quality paper appropriate to the needs of the author(s). Publication time is about ten weeks. More than twenty years of experience guarantee authors the best possible service. To reach the goal of rapid publication at a low price the technique of photographic reproduction from a camera-ready manuscript was chosen. This process shifts the main responsibility for the technical quality considerably from the publisher to the authors. We therefore urge all authors and editors of proceedings to observe very carefully the essentials for the preparation of camera-ready manuscripts, which we will supply on request. This applies especially to the quality of figures and halftones submitted for publication. In addition, it might be useful to look at some of the volumes already published. As a special service, we offer free of charge LATEX and TEX macro packages to format the text according to Springer-Verlag's quality requirements. We strongly recommend that you make use of this offer, since the result will be a book of considerably improved technical quality. To avoid mistakes and time-consuming correspondence during the production period the conference editors should request special instructions from the publisher well before the beginning of the conference. Manuscripts not meeting the technical standard of the series will have to be returned for improvement.

For further information please contact Springer-Verlag, Physics Editorial Department II, Tiergartenstrasse 17, D-69121 Heidelberg, Germany

Arnold O. Benz   Albrecht Krüger (Eds.)

Coronal Magnetic Energy Releases

Proceedings of the CESRA Workshop
Held in Caputh/Potsdam, Germany
16-20 May 1994

 Springer

Editors

Arnold O. Benz
Institut für Astronomie, ETH-Zentrum
CH-8092 Zürich, Switzerland

Albrecht Krüger
Astrophysikalisches Institut
D-14482 Potsdam, Germany

Frontispiece:

Dynamic evolution of a current sheet following the localized occurrence of anomalous resistivity (here applied at $x = 13, 27, 53$, $y = 0$, and $0 < t < 2$ in units of the current sheet half width and the Alfvén time, respectively). Current density maxima (bright) are localized at the X-points formed by the magnetic field (curves). Induced tearing leads to the formation of three current filaments by $t = 10$. Subsequently, complete coalescence occurs, connected with the induction of an electric field. Both processes are supposed to lead to rapid energy release in the corona (courtesy B. Kliem).

ISBN 978-3-662-14002-4   ISBN 978-3-540-49189-7 (eBook)
DOI 10.1007/978-3-540-49189-7

CIP data applied for

Originally published by Springer-Verlag Berlin Heidelberg New York in 1995
Softcover reprint of the hardcover 1st edition 1995

Typesetting: Camera-ready by the authors
SPIN: 10481062   55/3142-543210 - Printed on acid-free paper

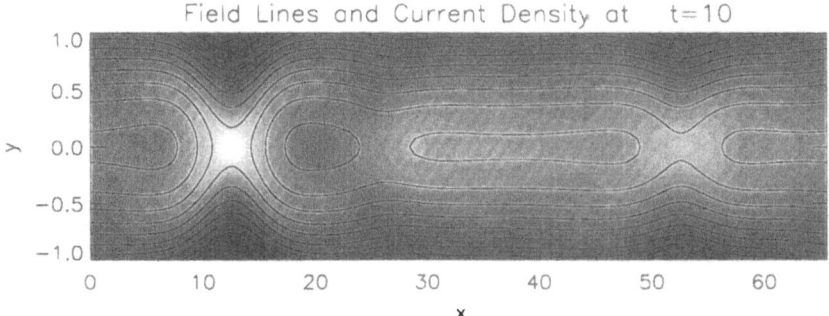

Field Lines and Current Density at   t=10

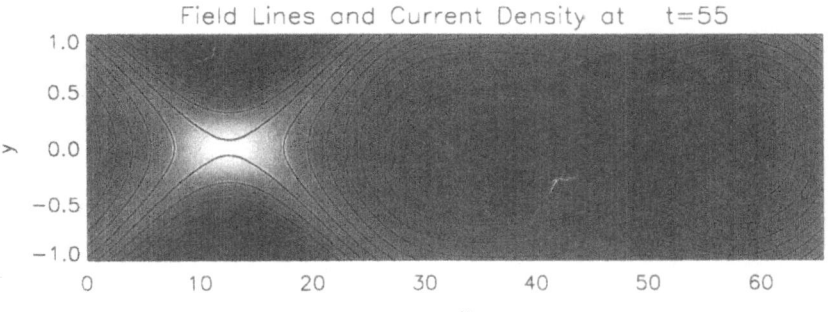

Field Lines and Current Density at   t=55

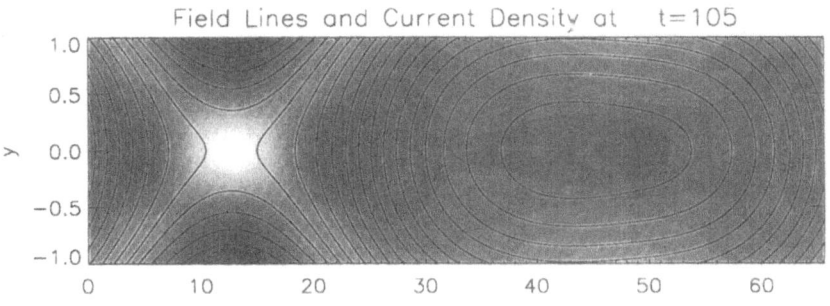

Field Lines and Current Density at   t=105

Preface

In the last ten years it has become increasingly clear that more dynamical phenomena occur in the solar corona than just flares. Not only are there different kinds of flares, as classified for example in hard X-rays, but magnetic energy is released in several different forms that are not necessarily related to some brightening in Hα, the traditional definition of flares. Radio emissions, such as S-component and noise storms, transient brightenings of active regions in radio and X-rays, X-ray bright points, filament eruptions, coronal mass ejections, and others have been recognized as non-flare releases of magnetic energy stored in the corona. New instruments, too many to be listed, with higher sensitivity, better resolution or longer observing time have been put into operation during the present solar cycle. They have recorded not only smaller and smaller flares, but also a wealth of information on various kinds of coronal dynamics. The coronae of more rapidly rotating stars exhibit even more activity and seem to be in a state of very frequent or continuous flaring.

The Board of CESRA, the Community (formerly named "Commission") of European Solar Radio Astronomers, decided to organize a workshop with a unifying scope on magnetic energy releases. Not excluding that coronae may also be fed by other forms of energy, focusing on magnetic energy limits the range of topics and observations to phenomena of possibly similar physics. These exciting, but poorly understood, events comprise instabilities of electric currents, acceleration of nonthermal particles, shock waves, and heating processes.

The generation and maintenance of hot coronae around central gravitating astrophysical bodies, the processes of their heating and activity are intimately related to the question of energy release and also pose a burning problem of contemporary cosmic physics. One of the key questions is expected to be in the understanding of the role which magnetic fields are playing in this respect.

The experimental base to explore coronal magnetic energy releases includes the methods of radio astronomy and space research. Thus the combined discussion of ground-based and space-borne projects, complementing each other, appears useful. For this reason special attention was paid at the workshop to the presentation of actual satellite experiments, in particular the SOHO project and the CORONAS-I mission. Other sessions were devoted to the time, space, and energy scaling of flares, radio observations, theory, time-series analyses and statistics, coronal and interplanetary shock waves, new radio instrumentation, software, and data bases. The Scientific Organizing Committee consisted of A.O. Benz (Zürich, chair), B. Fleck (Noordwijk), A. Krüger (Potsdam), A. Magun (Bern), M. Pick (Meudon), G. Trottet (Meudon), and P. Zlobec (Trieste).

This volume contains invited reviews and invited contributions of the workshop. There were 82 participants from 14 European countries, and also from the USA, China, and Japan. The program comprised 78 contributions, 59 of which orally presented, among them were 14 invited reviews and 10 specific invited contributions, most of them contained in this issue. (The other contributed papers and posters have been recommended to be published in a volume of "Solar Physics".)

CESRA, which took the initiative for the present workshop, was founded in the beginnings of the seventies and one of its major activities is the stimulation of scientific meetings devoted to actual topics of solar radio physics and neighbouring disciplines. The present meeting was organized by the Astrophysical Institute Potsdam and the University of Potsdam under the auspices of the European Physical Society. We are grateful that the workshop could be sponsored by the "Deutsche Forschungsgemeinschaft", the Commission of European Communities, the International Science Foundation, the Astrophysical Institute Potsdam, and the University of Potsdam. We also thank the members of the Local Organizing Committee (H. Auraß, J. Hildebrandt, B. Kliem, J. Kurths, G. Mann) and many assistants (G. Haase, L. Kurth, D. Lehmann, A. Marks, and A. Trettin) for their untiring help in making the workshop a success. Moreover the editors of this volume are indebted to a number of referees for evaluating the contributions. Also the assistance of Ljudmila Kurth in editing the volume is gratefully acknowledged.

Zürich and Potsdam, December 1994   Arnold O. Benz
                                                                     Albrecht Krüger

Contents

Flares and Coronal Heating in the Sun and Stars

Arnold O. Benz

Institute of Astronomy, ETH-Zentrum, CH–8092 Zürich, Switzerland

Abstract. Many forms of energy input into coronae have been proposed as the dominant heating mechanism. Here I review topical aspects of impulsive releases of magnetic energy. Several solar phenomena from bright points to coronal mass ejections are attributed to free magnetic energy apparently available in the corona. The possibility that magnetic energy release is the dominant energy input into the corona is discussed for the Sun with special emphasis on small radio events, with negative results. The evidence is better, however, for active stars where a correlation between thermal radiation and gyrosynchrotron emission by energetic electrons has been found recently. It suggests that a flare-like release of magnetic energy is the dominant coronal heating process of active, rapidly rotating stars. However, the required cadence of flares has not (yet) been found. The link between stellar coronal heating and magnetic energy release is not clear as long as the various flare-like phenomena in the solar corona are not better understood.

1 What Heats the Corona?

At about the same time — in the middle of this century — when astrophysicists started to understand the nuclear heating process in the interior of the Sun, W. Grotrian and B. Edlén suggested that the corona has a temperature of some million degrees. It raises a new, major problem. What heats the solar corona? Recent reviews are contained in a book edited by Ulmschneider et al. (1991). For an overview (but not a review), the many ideas may be grouped into six classes:

1. MHD waves driven by the convection below the photosphere.
2. Body waves in dense flux tubes or surface waves in boundary layers of the inhomogeneous corona.
3. Gradual dissipation of currents in magnetic flux emerged into the corona from below.
4. Impulsive release of magnetic energy in flares, microflares, or flare-like events.
5. Bulk motion (spicules etc.) in the transition layer and into the lower corona.

1

6. Evaporation of particles in the tail of the velocity distribution, resulting in nonlinear heat transfer in the inhomogeneous atmosphere.

None of these processes can be rejected off hands, they probably all contribute to the heating. The much more difficult question is: Which kind of heat input dominates? That we still cannot answer this question may have several reasons: (i) Every new generation of instruments has revealed more pieces of ignorance about the physics of the coronal plasma. (ii) The leading actors may not have been observed yet (e.g. low-frequency collisionless waves, nonthermal ions below 1 MeV). (iii) There may be different inputs dominating in different coronal regions.

Here I review some new aspects on the energy input by impulsive release of magnetic energy and, in particular, the contribution of radio observations. In the following section, new evidence for flare-like heating of stellar coronae is reviewed. Section 3 discusses the evidence against heating of the solar and stellar coronae by regular flares, and Section 4 presents some new observations of solar microbursts in decimetric radio waves. Concluding remarks on the present state and future possibilities are given in Section 5.

2   Flare-like Heating in the Solar and Stellar Coronae

The energy input into the solar corona that can be best observed and that we know best (but still do not understand) is the solar flare. In Figure 1 the microwave and soft X-rays of a typical (impulsive) solar flare are shown. As usual, the microwave emission peaks before the soft X-rays. Neupert (1968) was the first to observe that the hard X-rays (correlating usually well with the microwave emission) peak before the soft X-rays. The broadband microwave radiation is compatible with gyrosynchrotron emission of mildly relativistic electrons. The spectral peak of the microwave emission of Fig. 1 was near 10 GHz.

The temperature of the soft X-ray emitting plasma peaks after the emission of the energetic electrons in agreement with a scenario where the acceleration and thermalization of energetic electrons cause the heating. As the heat is distributed and the plasma is heated, the emission measure still increases (e.g. by evaporation) for some time after maximum temperature.

The classical 'Neupert effect' for the impulsive release of magnetic energy in solar flares suggests that acceleration and heating are related. A linear correlation between the peak value of the 5 GHz flux and the maximum GOES flux has been suggested by Drake et al. (1989). The correlation increases considerably when (i) a higher microwave frequency is used and its flux is corrected for optical thickness, and (ii) the bulk part of the soft X-ray flux in the 0.1–2.4 keV range is compared (Benz & Güdel 1994).

In Fig. 2 the ratios of soft X-ray to microwave fluxes of sixteen publicly available flares are presented. They comprise the impulsive and gradual flares used by Benz & Güdel (1994) to show correlation. The microwave optical thickness was estimated from the spectrum. For the gradual flares the 8.4 GHz emission was

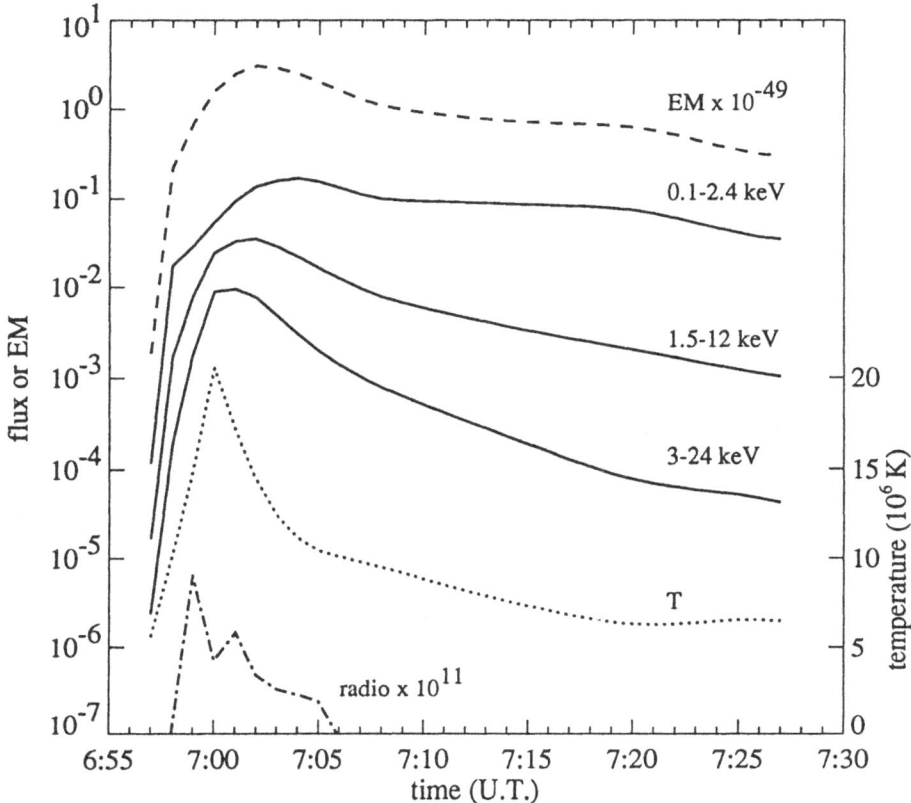

Fig. 1. A typical impulsive flare on the Sun. The observed radio flux density at 8.4 GHz (Bern Radio Observatory) and the soft X-ray flux at 1.5–12 keV and 3–24 keV observed by the GOES satellite are presented on a logarithmic scale. The temperature (on linear scale to the right), the 0.1–2.4 keV flux representing the main thermal emission of the flare plasma, and the emission measure have been calculated from the originally observed GOES data, using model calculations for the thermal line emissions according to Raymond & Smith (Benz & Güdel 1994)

clearly on the high-frequency side of the spectral maximum, thus assumed to be caused by optically thin emission. For the impulsive flares with spectral peaks above 8.4 GHz we used the extrapolation from a power-law fitted at higher frequencies, where the spectrum indicates optically thin emission. Only flares with a single maximum in the 5–12 GHz range have been included in this study. The peak microwave luminosities corrected to represent optically thin emission at 8.4 GHz then range from $3.78 \cdot 10^{10}$ to $2.3 \cdot 10^{12}$ erg/s Hz. The soft X-ray peak luminosities of these flares span a range from $1.4 \cdot 10^{26}$ to $3.2 \cdot 10^{27}$ erg/s.

Figure 2 shows that the ratio of soft X-ray to microwave peak luminosities of regular flares is distributed over an order of magnitude. The scatter is not reduced when the luminosities are integrated over the duration of the flares.

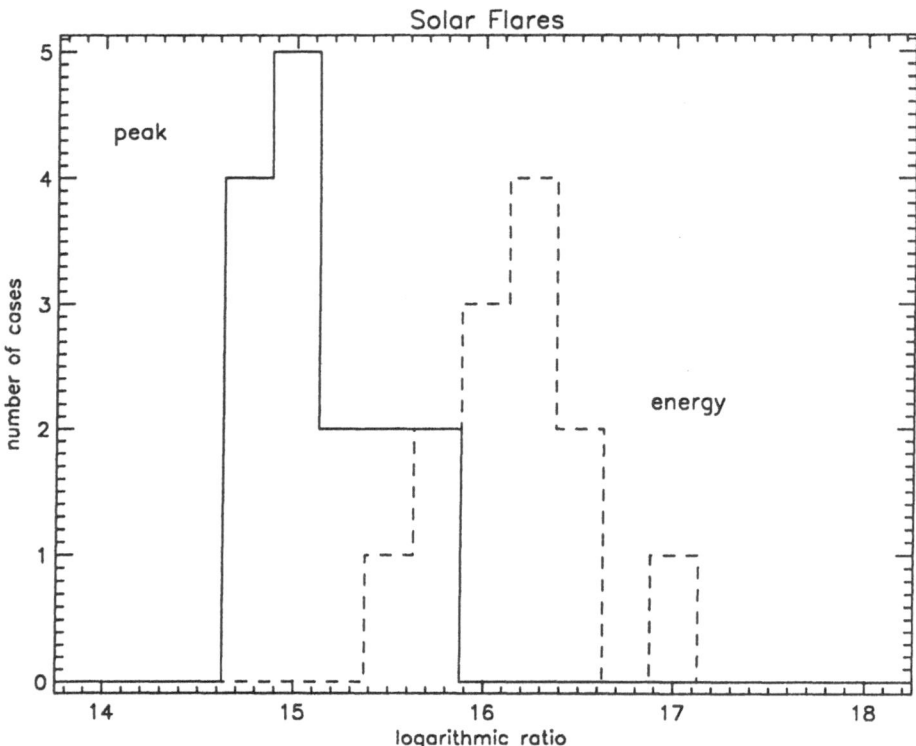

Fig. 2. The ratio of the 0.1–2.4 keV soft X-ray flux and the microwave flux at 8.4 GHz corrected for optical thickness. The distribution of sixteen solar flares is presented in logarithmic scale. The ratio of the peak values is shown with a solid histogram, the ratio of the time integrated fluxes (energy) with a dashed histogram

The ratio is shifted, however, to larger values since the microwave flare duration is considerably smaller than the soft X-ray duration.

The scatter in the soft X-ray to microwave ratio becomes larger if non-regular flares are included. Figure 3 compares the flares of Fig. 2 (peak values) with microflares, 2–3 orders of magnitude smaller than regular impulsive flares (Fürst et al. 1982). The average soft X-ray to microwave ratio is an order of magnitude larger than the mean value for regular flares. As the microflares have all been observed in the same small active region this may be interpreted as energy release in a weak magnetic field leading to less efficient acceleration or in higher density reducing the lifetimes (and thus emission times) of energetic electrons. More observations are clearly needed.

Homologous flares that occur at the same place within a few minutes of a previous flare have even larger ratios (Strong et al. 1984). The second flare of a double takes place in a hotter and much denser environment, and this change in plasma parameters seems to reduce the acceleration efficiency or particle lifetime, and hence increase the ratio of thermal to nonthermal emissions.

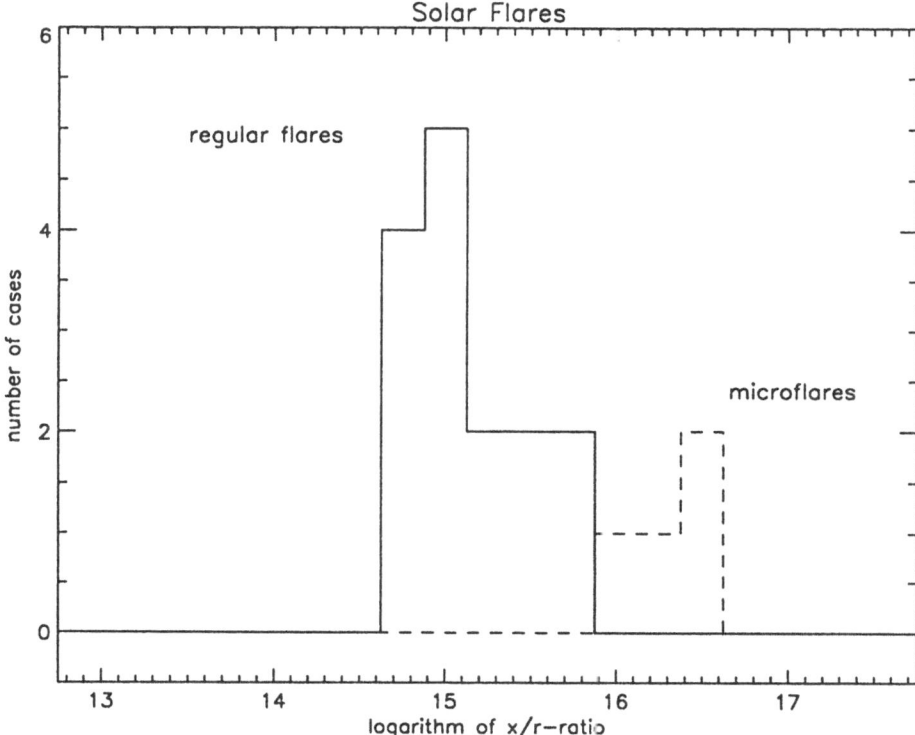

Fig. 3. Comparison of the soft X-ray to microwave ratios of regular flares and microflares

The more surprising then is the relatively small scatter observed in the ratio of quiescent soft X-ray and microwave emissions of active dwarf M and K stars in Fig. 4, presented previously by Güdel & Benz (1993) in a different way. With quiescent emission we mean the stellar radiation at low level. Its characteristics are (i) small variability, (ii) with time scale longer than 10 minutes, and (iii) a small degree of polarization for radio emission.

Except for the value at 14.42 of UV Cet, the stars scatter in Fig. 4 only a factor of 2 around the mean soft X-ray to microwave ratio (15.47 ± 0.3). All these stars are rapidly rotating, frequently flaring, and have chromospheric emission lines indicating magnetic activity. As the soft X-ray and microwave emissions are generally interpreted like in flares by thermal bremsstrahlung and, respectively, optically thin gyrosynchrotron radiation, Fig. 4 suggests that the thermal and nonthermal electron populations in quiescent stellar coronae exist in an astonishingly similar admixture as in an average solar flare.

The radio and soft X-ray luminosities of solar flares and stellar coronae are distributed over more than 8 orders of magnitude (Benz & Güdel 1994). The correlation between the two emissions noted in various objects has been critized to be exaggerated by the different volumes (Bastian 1994). The source size is elim-

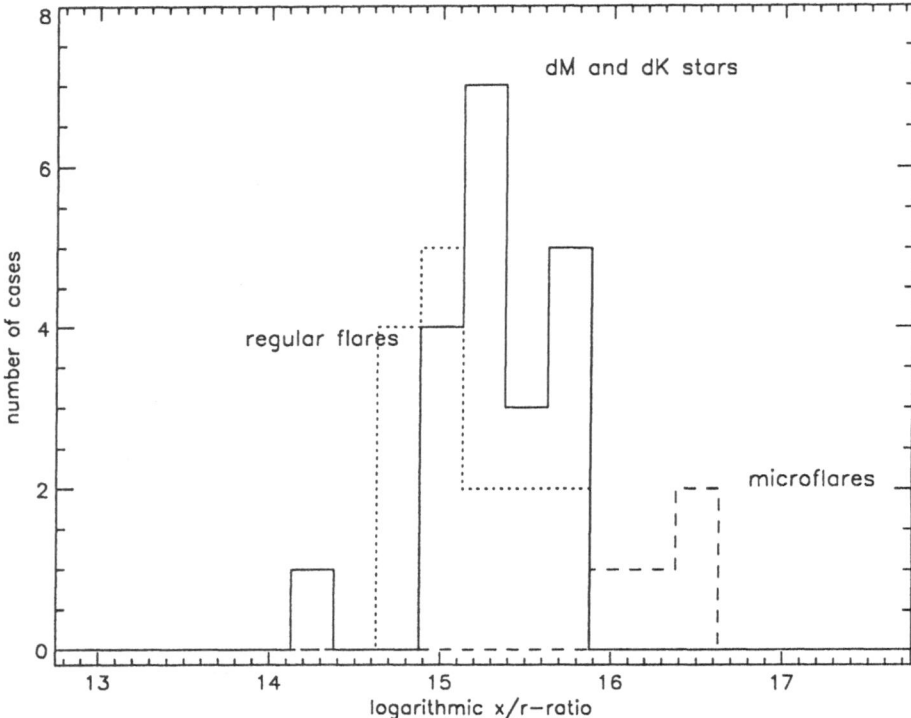

Fig. 4. The ratio of quiescent emissions of twenty active dwarf M and K stars in soft X-rays and microwaves is compared to the ratio found in solar flares. The same soft X-ray range and similar microwave frequencies are used in the solar and stellar observations

inated here by the representation of Fig. 4, which just displays the distribution of the ratios. It shows that the correlation is clearly not a volume effect.

Figure 4 reveals a correlation between thermal and nonthermal emissions in solar flares and active stellar coronae. It indicates that the dominant heating process of active stellar coronae has some of the characteristics of solar flares: it accelerates electrons to similar energies with similar efficiency. Secondly, the similarity of the solar flare and the stellar coronae ratios seems to suggest that similar physical processes are at work, i.e. that the stellar coronal heating process is a flare-like phenomenon.

3 The Energy Input of Flares into the Coronae

Are the observed flares energetic enough to heat the solar corona? The answer is clearly negative. Let the energy of a flare be E. The rate N of flares per second and per emitted energy has been reported to follow a power-law distribution

$$N(E) = A \ E^{-\alpha} \quad [(\text{erg s})^{-1}] \tag{1}$$

over several orders of magnitude (e.g. for soft X-rays: Drake 1971, and for hard X-rays: Lin et al. 1984; Dennis 1985; Crosby et al. 1993). The coefficient A varies in yearly averages by more than an order of magnitude during the 11 year solar cycle (e.g. Crosby et al. 1993). If some thick target model is used to convert from hard to soft X-rays, the results in the two energy bands agree in the value of A. The exponent α has been found around 1.8 in all cases.

Equation (1) can easily be integrated to yield the total power

$$P = \int_{E_{\min}}^{E_{\max}} N(E)\, E\, \mathrm{d}E = \frac{A}{2-\alpha} E^{-\alpha+2} \Big|_{E_{\min}}^{E_{\max}} \quad [\mathrm{erg/s}] \qquad (2)$$

If most of the flare energy is radiated away in soft X-rays, P is representative for the power injected into the corona by flares.

Since α is smaller than 2, Eq. (2) indicates that the energy input is dominated by the most energetic flares. Lacking evidence for superflares, we may assume that there is a cutoff near the largest observed flares radiating soft X-rays with a power of a few 10^{30} erg. The dominance of energetic, but rare flares does not fit at all the idea of a 'quiet' corona heated by flares.

The integration of Eq. (2) for solar flare energies between 10^{26} and 10^{32} erg yields a power of about $2 \cdot 10^{25}$ erg/s (Hudson 1991). This is a small fraction even of the 'quiet' Sun's soft X-ray luminosity of some 10^{27} erg/s (Vaiana & Rosner 1978). Both the dominance of energy from large flares and the limited energy in flares contradict the hypothesis of coronal heating by regular flares.

In most characteristics that can be compared, stellar flares appear similar to solar flares. The soft X-ray variability of dMe stars has a similar power-law exponent, $\alpha = 1.52 \pm 0.08$, as shown by Collura et al. (1988) from Exosat observations. The most remarkable difference is their energy that can surpass solar flares by more than 3 orders of magnitude. An important difference is also the scarcity of long soft X-ray observations. Thus the upper limit of the flare energy is not known. Again, the heating of stellar coronae by rare huge events is not supported by the observations.

4 Microbursts

If coronae are heated by flare-like processes as suggested for active stars in Section 2, the previous section has shown that these processes cannot simply be identified with the observed flares. There must be a different population at much larger or much lower energies. In the first possibility, this could be a quasi-permanent current-sheet associated with an active region. The current may be driven by the fast rotation of active stars in the presence of finite conductivity.

More popular is the alternative, microflares having a 'softer' distribution ($\alpha > 2$). This idea has been around for many years. Small impulsive events have been detected at EUV wavelengths in the transition region (e.g. Brueckner and Bartoe 1983), and weak soft X-ray brightenings have been reported by Schadee et al. (1983). Observations of a large-area hard X-ray detector by Lin et al. (1984) have given new impetus. Note, however, that these hard X-ray observations

appear to have about the same sensitivity as optical observations, as many of
the microflares were also detected as small events in $H\alpha$ (Canfield and Metcalf
1987). Even worse, the one microflare overlapping in time with the Phoenix radio
spectrometer was associated with a group of regular type III bursts (Benz 1987).

If microflares heat the solar corona, they have to be smaller than the smallest
regular flare by several orders of magnitude. *Regular* means here to belong to
the set of flares distributed with a power-low in energy with index 1.8. The lower
boundary of this set is a few 10^{26} erg.

Imaging soft X-ray telescopes such as Yohkoh can detect brightening far
below this limit. Also, imaging radio telescopes may see microflares. If nonthermal, the microflares may emit *coherent* radio emission. Since such emission
is very efficient, but unpredictable in frequency, it may be observable with a
sensitive full-Sun broadband spectrometer.

The non-thermal radio emission of microflares may be easier to detect than
their thermal emission in soft X-rays. Microbursts could thus be a guide to the
microflares. In the following I review four different kinds of microbursts and
report on new observations in Zurich.

4.1 Transition Region Activity

Radio emission at 8.5 GHz correlates extremely well with photospheric structures
in quiet solar regions. Enhanced emission originates in the transition region
above the magnetic network boundary of supergranular cells (Kundu et al. 1987,
Gary et al. 1990). The radio emission is highly variable at time scales of a few
minutes and only the integration over hours yields the close correspondance to
the network. The emission is generally believed to be thermal bremsstrahlung.
However, the short time scales have led Gary and Zirin (1988) to leave open the
question of possible nonthermal contributions.

4.2 Type III Microbursts

Extremely weak fast drifting bursts have been reported at long metric wavelengths by Kundu et al. (1986) from Clark Lake observations. They have the
characteristics of small electron beams propagating on open field lines in the upper corona and producing type III bursts. The microbursts occurred at a rate of
about five bursts per hour over an extended period in sunspot minimum. Their
flux density was typically a few sfu (10^{-19} erg s^{-1} Hz^{-1} cm^{-2}). The energy
carried by these beams may be estimated from scaled down regular bursts. It
may be as low as 10^{23} erg/beam yielding an average power of some 10^{20} erg/s.
If the accelerator of such a beam is not very efficient, the primary energy release
may be much larger. Nevertheless, the energy input into the corona seems to be
much smaller than required for heating the minimum corona (some 10^{27} erg/s).

4.3 Weak Gyrosynchrotron Bursts

Small radio bursts of presumably gyrosynchrotron origin have been observed at
11 GHz with the 100m telescope at Effelsberg (Benz et al. 1981, Fürst et al.

1982). The microbursts of the order of 1 sfu have been observed in a weak active region. They were not associated with $H\alpha$ flares, but coincided with coherent radio emissions between 0.6 and 1.3 GHz. The emission also was accompanied by soft X-ray emission which indicated an excess thermal energy of 10^{26} erg, similar to the weakest hard X-ray flares. The total average power in these microflares was 10^{23} erg/s.

4.4 Noise Storms

Noise storms at metric radio wavelengths may be considered as a third group of microbursts. The storms (also called type I radio emission) are associated with emerging magnetic flux and developing active regions. The bursts have been suggested by Benz and Wentzel (1981) to be signatures of small reconnection events caused by the interaction of rising field lines with preexisting flux. In their model a single burst has been estimated to require the release of at least a few 10^{21} erg. With a burst rate of about 1 Hz the average power for this process in an active region may then be estimated to be at least of the order of 10^{21} erg/s.

4.5 Decimetric Microbursts

Figure 5 displays a short interval of full-Sun spectrometer data recorded during a 30 minute test run. The receiver and antenna pointing were newly adjusted for optimum sensitivity. The full observing band surveys the solar radio emission from 200 MHz to 3 GHz. Some small parts of the spectrum at low frequencies show terrestrial interference and have been omitted in the figure. Enhanced emission was recorded only around 1.4 GHz in three isolated small events. No $H\alpha$ flare, microwave burst, nor GOES soft X-ray event was reported by the observing patrol instruments (Solar and Geophysical Data 1994) throughout the observing period. Magnetograms show several small active regions (the international sunspot number was 54).

The enlargement of a small event in Fig. 5 (bottom) shows two bursts each with a total duration of about 0.5 s and a bandwidth of about 60 MHz. The peak flux density is 2.0 sfu. The small bandwidth, 4% of the center frequency, clearly indicates a coherent emission process. The observed degree of polarization was 0±3% circular. The microbursts do not easily fit into any class of regular decimetric bursts with larger flux. The observed bandwidth and polarization agree with the characteristics of decimetric and microwave spikes. Their duration, however, is an order of magnitude too long at their center frequency (Güdel and Benz 1990). More likely is the classification as 'blips', reported in association with a weak gyrosynchrotron burst at 600–900 MHz (e.g. Fürst et al. 1982) and later identified as narrowband type III bursts (Benz et al. 1983). The observed time series is too short to be representative. More observations are needed to establish a new class of microbursts.

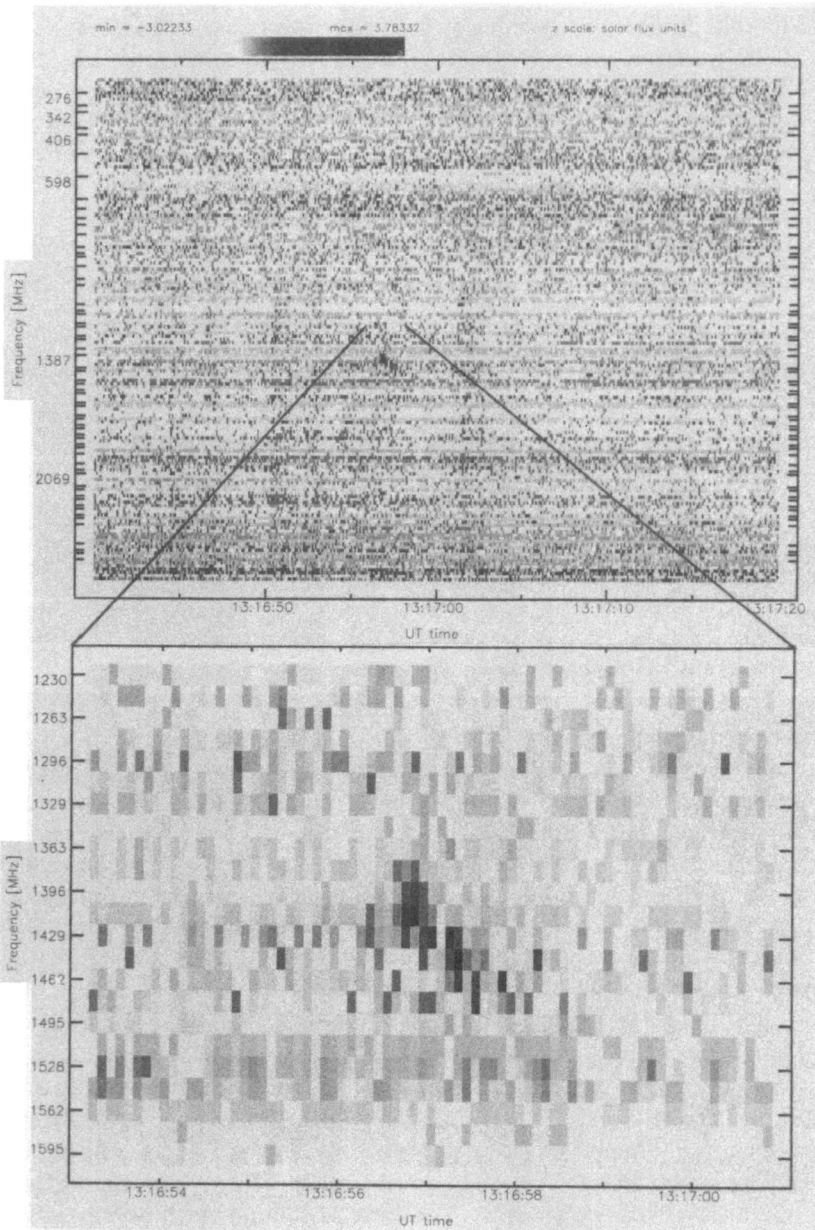

Fig. 5. Spectrogram of solar radio emission observed by the Phoenix spectrometer of ETH Zürich on 1994 March 4. Enhanced flux density is shown dark in the frequency-time plane. The gray scale is linear and ranges from −3.0 sfu (white) to 3.8 sfu (black). **Top**: The full observing band is presented in a 40 second interval. **Bottom**: Enlargement of the figure at the top showing some decimetric microbursts.

5 Conclusions

A universal ratio of soft X-ray thermal emission and microwave gyrosynchrotron emission in active stellar coronae and solar flares suggest a flare-like heating process for active stellar coronae. The heating does not seem to be caused, however, by the observed flares both in stars and the Sun.

The solar corona is different from active stars. Outside of flares the thermal coronal emission in soft X-rays does not correlate with gyrosynchrotron emission. Nevertheless, it is extremely important to study the small-scale dynamics of the solar corona. Solar radio astronomers have known for a long time that magnetic energy release in the corona occurs in many different forms including

- flares
- post-flares
- coronal mass ejection
- type III emitting electron beams
- gyrosynchrotron microflares
- noise storms
- coronal bright points
- transition region activity

The investigation of small radio events has not yet been systematic. It is clear, however, that some of the reported microbursts (such as single type I bursts, type III microbursts, decimetric microbursts) involve extremely small amounts of energy even compatible with the theoretical ideas of 'nanoflares'. It is not clear at all what role they play in the heating of the corona. None of them seem capable to satisfy the need for heating the corona or even an active coronal region. Definite conclusions are not yet in place before these processes are studied in a more quantitative way.

To substantiate the model of magnetic heating of coronae by flare-like processes, non-thermal processes have to be found that are (i) capable of particle acceleration and (ii) that consist of an average power release of about 10^{27} erg/s for the 'quiet' corona and of a similar value for each large active region. Solar observations may clarify the principle of reconnection in small events and assess the importance of different heating phenomena.

Acknowledgements

It is a pleasure to thank the Radio Astronomy Group at ETH Zurich for cooperation in observation and data reduction and M. Güdel for helpful discussions on the soft X-ray emission of the Sun and stars. This work is financially supported by the Swiss National Science Foundation Grant Nr. 20-040336.94.

References

Bastian T.S., 1994, Sp.Sci.Rev. 68, 261
Benz A.O., 1987, ESA Sp-275, 105

Benz A.O., Bernold T.E.X., Dennis B.R., 1983, ApJ 271, 355

Benz A.O., Fürst E., Hirth W., Perrenoud M.R., 1981, Nature 291, 210

Benz A.O., Güdel M., 1994, A&A 285, 621

Benz A.O., Wentzel D.G., 1981, A&A 94, 100

Brueckner G.E., Bartoe J.-D.F., 1983, ApJ 272, 329

Canfield R.C., Metcalf T., 1987, ApJ 321, 586

Collura A., Pasquini L., Schmitt J.H.M.W., 1988, A&A 205, 197

Crosby N.B., Aschwanden M.J., Dennis B.R., 1993, Solar Phys. 143, 275

Dennis B.R., 1985, Solar Phys. 100, 465

Drake J.F., 1971, Solar Phys. 16, 152

Drake S.A., Simon T., Linsky J.L., 1989, ApJ 71, 905

Fürst E., Benz A.O., Hirth W., 1982, A&A 107, 178

Güdel M., Benz A.O., 1990, A&A 231, 202

Güdel M., Benz A.O., 1993, ApJ 405, L63

Hudson H.S., 1991, Solar Phys. 133, 357

Lin R.P., Schwartz R.A., Kane S.R., Pelling R.M., Hurley K.G., 1984, ApJ 283, 421

Neupert W.M., 1968, ApJ 153, L59

Schadee A., de Jager C., Svestka Z., 1983, Solar Phys. 89, 287

Solar and Geophysical Data, 1994, NOAA Boulder, Nr. 596, 31

Strong K.T., Benz A.O., Dennis B.R., Leibacher J.W., Mewe R., Poland A.I., Schrijver J., Simnett G., Smith J.B., Sylvester J., 1984, Solar Phys. 91, 325

Ulmschneider P., Priest E.A., Rosner R. (eds.) 1991, Mechanisms of Chromospheric and Coronal Heating, Springer Verlag, Berlin

Vaiana G., Rosner R., 1978, Ann. Revs. Astron. Astrophys. 16, 393

Imaging, Stereoscopy, and Tomography of the Solar Corona in Soft X-Rays and Radio

Markus J. Aschwanden

University of Maryland, Astronomy Department, College Park, MD 20742, USA
(e-mail: markus@astro.umd.edu)

Abstract. We review simultaneous imaging observations of the (quiet) solar corona in soft X-rays (or EUV) and radio, and outline recent developments that involve three-dimensional (3D) reconstruction techniques, such as stereoscopy and tomography. The 3D reconstruction of coronal structures involves not only accurate measurements of geometric parameters (position, altitude, rotation rate), but also the deconvolution of physical parameters (density, temperature, magnetic field) along the line-of-sight, which is most feasible with simultaneous observations in complementary wavelengths, e.g. in soft X-rays, EUV, and radio.

Keywords. Solar Corona, Soft X-rays, EUV, radio emission

Soft X-ray (SXR), (extreme) ultraviolet (EUV, UV), and radio wavelengths provide a powerful "color palette" of wavelengths that complement each other in an ideal way to draw a contrastful, physical picture of the solar corona. SXR images represent a sensitive visualization of the high-temperature ($\gtrsim 1$ MK) plasma, while UV and EUV images trace also the cooler plasma ($\lesssim 1$ MK), confined in the transition region and near the footpoints of cool active region loops. Radio images convey important complementary information: (1) free-free emission from a hot plasma is very sensitive to absorption by cool plasma in the line-of-sight, and (2) gyroresonance emission represents the only method to directly measure the coronal magnetic field strength. Thus, simultaneous observations in SXR, EUV, and radio provide complementary diagnostics of the physical properties of the coronal plasma, which is of fundamental importance for 3D models with stereoscopic and tomographic methods. Reviews on active regions observed in SXR, EUV, and radio can also be found in Schmahl (1980) and Holman (1995).

1 Imaging Observations

An overview of primary instrumentation used in joint SXR (or EUV) and radio observations is compiled in Table 1, containing instrumental characteristics such as the wavelengths, frequency (or energy) ranges used, temperature ranges and magnetic field strengths to which the instruments are sensitive at a particular wavelength, field of views, and spatial resolutions.

1.1 Soft X-ray Imaging

Soft X-ray images of the Sun can only be obtained from space-borne detectors, either on rocket flights (with typical durations of $\approx$ 7 minutes) or on spacecraft missions. SXR images provide primarily information on hot coronal plasmas with temperatures of $\gtrsim$ 1 MK (see peak formation temperatures of SXR lines in Table 1). Extensive documentation on imaging observations in SXR and EUV can be found in Bray et al. (1991).

The first crude X-ray photograph of the Sun was obtained from a pinhole camera on a *Aerobee* rocket by the *Naval Research Laboratory (NRL)* in 1960. In 1966, a grazing-incidence telescope was built by *Goddard Space Flight Center (GSFC)* that produced the first good-quality X-ray images. Since 1968, a series of *American Science & Engineering (AS&E)* rocket flights have been flown, often scheduled during solar eclipses. Improved spatial resolution (1″) and less diffractive scattering was achieved with the development of *Normal Incidence X-Ray Telescopes (NIXT)*, using multi-layered coated X-ray optics, as demonstrated by Golub et al. (1990) and Walker et al. (1988). Similar instrumentation has been flown since, such as the *Multi Spectral Solar Telescope Array (MSSTA)* by Stanford (Walker et al. 1993), and the *Normal Incidence X-ray Imager (NIXI)* in 1994 by Lockheed (LPARL; M.Bruner).

Spacecraft missions with solar-dedicated soft X-ray imagers were flown since 1973: *Skylab, P78-1, OSO-8, Hinotori, SMM, Yohkoh,* and *Coronas.* The *Skylab* mission, flown from 1973 May 28 to 1974 Feb 7, contained two grazing incidence telescopes with 2″spatial resolution, one from *AS&E* (3.5-60 Å), and one from *Marshall Space Flight Center (MSFC)* (3-53 Å). The *AS&E* X-ray telescope recorded 32,000 photographs (on film), that documented extensively the discoveries of coronal holes (in SXR) and X-ray-bright points, and allowed qualitative studies of the morphology and evolution of coronal structures. In the 80's, soft X-ray imaging was performed by the two spacecraft, *Solar Maximum Mission (SMM)* (1980 Feb 14 - 1989 Dec 2) and *Hinotori* (1981 Feb 26 - 1982 Oct 8). Besides the broadband SXR imagers (*HXIS/SMM* with 6 energy bands over 0.4-3.5 Å and 8″resolution; *SOX,* the soft X-ray crystal spectrometer on *Hinotori,* 1.8-1.9 Å), they contained also X-ray polychromators which resolved individual spectral lines. The *X-Ray Polychromators (XRP)* on *SMM* consisted of two instruments: the *Bent Crystal Spectrometer (BCS,* 1.7-3.2 Å), and the *Flat Crystal Spectrometer (FCS,* 0.4-3.5 Å), with a resolution of 14″. Currently operating spacecraft with imaging capabilities in SXR are *Yohkoh* (launched in August 1991), and *Coronas* (launched in March 1994). At the time of writing, *Yohkoh* has already recorded several 100,000 SXR images in the 3-45 Å range, with spatial resolutions (pixel size) of 2.45″and 4.9″.

1.2 UV/EUV Imaging

Ultraviolet (UV; > 90 nm) and Extreme Ultraviolet (EUV; < 90 nm) cover roughly the wavelength range of 150 to $\approx$7 nm. The solar UV/EUV spectrum is dominated by emission from resonance lines of H I (Lyman series), He I, He II,

Table 1. Imaging instruments used in joint SXR/EUV and radio observations

Instrument	Wavelength	Energy	Peak Temperature	Field of View	Spatial Resol.
Skylab/S-054	2-54 Å	0.2-6 keV	> 1.0 MK	Full Sun	2″
OSO-8	1-6 Å	2-15 keV	> 3.0 MK	Full Sun	60″
AS&E	8-65 Å	0.2-1.6 keV	> 1.0 MK	Full Sun	2-3″
P78-1	3-25 Å	0.5-4.1 keV	2 − 10 MK	5′	20″
SMM/XRP					
- S XV	5.0 Å	2.5 keV	15 MK	4′	14″
- Si XIII	6.7 Å	1.86 keV	10 MK	4′	14″
- Si XII	6.7 Å	1.85 keV	7 MK	4′	14″
- Fe XVIII	14.2 Å	0.87 keV	6.5 MK	4′	14″
- Mg XI	9.17 Å	1.35 keV	6.5 MK	4′	14″
- Fe XVII	15.0 Å	0.83 keV	4 MK	4′	14″
- Ne IX	13.45 Å	0.92 keV	3.5 MK	4′	14″
- O VIII	18.97 Å	0.65 keV	3 MK	4′	14″
Yohkoh/SXT	3-45 Å	0.3-4 keV	> 1.0 MK	Full Sun	2.45″
Skylab/S-055					
- C II	1335.7Å	9.2 eV	0.043 MK	5′	5″
- C III	977.0Å	12.7 eV	0.079 MK	5′	5″
- O IV	554.4Å	22.4 eV	0.21 MK	5′	5″
- O VI	1031.9Å	12.0 eV	0.35 MK	5′	5″
- Ne VII	465.2Å	26.7 eV	0.63 MK	5′	5″
- Mg X	625.3Å	19.8 eV	1.4 MK	5′	5″
- Si XII	521.1Å	23.8 eV	2.2 MK	5′	5″
- Fe XVI	335 Å	37.0 eV	2.5 MK	5′	5″
- S XIV	417 Å	29.7 eV	3.0 MK	5′	5″

Instrument	Wavelength	Frequency	Magnetic field (s=3)	Field of View	Spatial Resol.
Nançay	177 cm	0.169 GHz		Full Sun	3.7′
	74 cm	0.408 GHz		Full Sun	1.7′
Culgoora	188 cm	0.16 GHz		Full Sun	2′
Stanford 5-elem.	2.8 cm	10.7 GHz	1270 G	4.2′EW	16″
NRAO 3-elem.	3.7 cm	8.1 GHz	960 G	0.2′EW	3.4″
	11.1 cm	2.7 GHz	320 G	0.5′EW	10″
WSRT	6.15 cm	4.9 GHz	580 G	9′	9″
RATAN-600	1.7-30 cm	1-17 GHz	120-2000G	40′EW	15-270″
VLA (Conf.A-D)	90 cm	0.33 GHz	40 G	3-70′	6-200″
	20 cm	1.5 GHz	180 G	0.6-15′	1.4-44″
	6 cm	4.8 GHz	570 G	0.2-5′	0.4-14″
	3.6 cm	8.5 GHz	1000 G	0.1-3′	0.2-8″
	2 cm	15 GHz	1800 G	0.1-1.5′	0.1-4″
Nobeyama	1.8 cm	17 GHz	2000 G	Full Sun	10″

intermediate ionization levels of C, N, O, Si and S, and highly-ionized levels of Si, Ne, Mg and Fe (see Table 1). The temperature diagnostics of UV/EUV images covers the entire regime between 10^5 K and 10^7 K, and thus probes the solar transition region and active region loops in a wide range of temperatures. An excellent account of solar EUV observations can be found in the textbook of Bray et al. (1991).

EUV spectrographs with imaging capabilities were often flown on rocket flights together with SXR imagers. The first EUV spectroheliogram from orbit was obtained in 1965 by *OSO-2*. In 1966, *NRL* achieved a spatial resolution of 10″for EUV rocket images. *OSO-6* returned spectroheliograms with 35″resolution in 1969. Recent rocket flights with EUV equipment include *SERTS (GSFC)*, *HRTS (NRL)*, *SPDE (LPARL)*, and *MSSTA* (Stanford).

Skylab (1973) was the first spacecraft mission with EUV equipment, consisting of an EUV spectrometer/spectroheliometer (*HCO*, 28-135 nm) and an EUV spectroheliograph (*NRL*, 17-63 nm), achieving a spatial resolution of 5″. Other solar-dedicated spacecraft with UV/EUV imagers are *SMM*, *Coronas*, and *SOHO* (to be launched in 1995).

Table 2. Simultaneous SXR(or EUV) and radio observations

Observations	SXR/EUV	Radio	Reference
1973 May 31	Skylab,OSO-7	Nançay	Drago (1974)
			Chiuderi-Drago et al.(1977)
1973 Jun 9	Skylab	NRAO	Kundu et al. (1980)
1973 Aug 21	Skylab	Culgoora	Dulk and Sheridan (1974)
1973 Aug 28-Sep 13	Skylab	Stanford	Pallavicini et al.(1979,1981)
1977 Jul 25,78 May 12	OSO-8	VLA	Kundu et al. (1981)
1979 Nov 16	AS&E	VLA	Webb et al. (1983)
1980 Mar 29,30	SMM/XRP	VLA	Schmahl et al. (1982)
1980 May 25,26	SMM/XRP	WSRT	Strong et al. (1984)
1980 Jun 10,11	SMM/XRP	VLA,WSRT	Chiuderi-Drago et al.(1982)
1980 Jun 13-16	SMM/XRP	WSRT	Shibasaki et al. (1983)
1981 Feb 13	AS&E	VLA	Webb et al. (1983,1987)
1984 Jun 4, Jul 8	SMM/XRP	VLA	Lang et al. (1987a)
1985 Jun 7	SMM/XRP	VLA	Lang et al. (1987b)
1987 Nov 28	SMM/XRP	VLA	Nitta et al. (1991)
1987 Dec 4	SMM/XRP	VLA	Schmelz et al. (1992)
1987 Dec 18	SMM/XRP	VLA	Brosius et al. (1992)
1991 May 7	SERTS	VLA	Brosius et al. (1993)
1992 Jan 9,10, May 1	Yohkoh	RATAN-600	Lang et al. (1993)
1992 Apr 2-10	Yohkoh	VLA	Aschwanden et al. (1995)
1992 Apr 24	Yohkoh	VLA	Gopalswamy et al. (1994)
1992 Aug 18	Yohkoh	Nobeyama	Shibasaki et al. (1994)

1.3 Radio Imaging

An account of reported observations with simultaneous SXR/EUV and radio coverage is provided in Table 2, along with relevant literature references. Instrumental characteristics of the radio observations are listed in Table 1.

Imaging at radio wavelengths is done by interferometry, and thus, requires always a reconstruction of a synthesized map from Fourier components (uv-plane) that are measured from the correlated flux (visibilities) between pairs of antennas (baselines). Early radio interferometers were often configured in form of linear arrays of radio dishes, allowing primarily one-dimensional scans of the Sun, such as the *Stanford 5-element interferometer*, the *NRAO 3-element interferometer* in Greenbank, the *Westerbork Synthesis Radio Telescope (WSRT)*, or the early version of the *Nançay Radioheliograph*. Other radiotelescopes are one-dimensional because of the linear extension of their primary collecting area, such as *RATAN-600* in the Caucasus. Two-dimensional imaging of the quiet Sun can be achieved with 1-dim arrays by *Earth-rotation synthesis*, provided that transient solar phenomena or ionospheric fluctuations can be filtered out.

Two-dimensional configurations of radio interferometers allow rapid synthesis imaging with virtually unlimited time resolution. However, high-quality maps with deep dynamic range still require long (full-day) exposures, to optimize the uv-coverage. Two-dimensional solar radio interferometers were operated at *Culgoora* (Australia) and at the *Clark Lake Radio Observatory (CLRO)*. Currently operating radio synthesis telescopes include the *Very Large Array (VLA)*, the *Owens Valley Radio Observatory (OVRO)*, the *Nançay Radioheliograph*, the *Berkeley-Illinois-Maryland-Array (BIMA)*, the *Siberian Solar Radio Telescope (SSRT)*, the *Nobeyama Radio Observatory (NRO)* in Japan, and the *Giant Metric Radio Telescope (GMRT)* in India.

Radio imaging requires a sophisticated technique that has its own limitations. It is important to understand the basic limitations to judge the quality of a radio map. The spatial resolution of an image is basically given by the longest baseline, while the size of the largest visible structure is limited by the shortest baseline, which always has a smaller angular extent than the field of view given by the beamwidth of an individual antenna. Since the radio emission of solar active regions sometimes have a larger angular extent than the limit imposed by the shortest baseline, large-scale structures are then overresolved and invisible in a synthesized map, leading to a missing flux problem (in the total flux). In radio maps of active regions with high angular resolution, this effect can lead to an underestimate of the brightness temperature, which can constitute discrepancies in the comparison with temperatures derived from SXR. Improved maps of large-scale structures can be obtained with *Mosaic techniques*.

The quality of a radio map is often characterized by the dynamic range, i.e. the ratio of the peak flux to the r.m.s. noise, which strongly depends on the uv-coverage. A high dynamic range is easier to obtain for simple source structures, while complex extended source structures add to confusion and reduce the dynamic range. Another problem of aperture synthesis technique is the time variability of sources during the sampling time, which has carefully to be flagged out. Standard techniques of image restoration are the *(Högbom) CLEAN algorithm*, which performs an iterative deconvolution with the instrumental point spread function, and the *Maximum Entropy Method (MEM)*, which uses positive flux and smoothness constraints. An alternative method that improves the uv-coverage is *frequency-synthesis*, which is applicable if the source structure is invariant at different frequencies.

2 Stereoscopy

Stereoscopy, a method of creating the illusion of depth perception by recording two images of the same three-dimensional (3D) object from slightly different aspect angles, can be applied to solar observations in two ways: either using two different vantage points (requiring two spacecraft), or using solar rotation. The second approach - combining the information of two images of the Sun recorded with the same instrument at different times - is quite feasible for studies of quasi-stationary features of the quiet Sun. Stereoscopic viewing of pairs of *Skylab* images, recorded ≈12 hours apart, has been demonstrated by Batchelor (1994). Numerical methods to retrieve quantitative information on the 3D position of coronal structures have been attempted with soft X-ray images from *Skylab* and *Yohkoh*, as well as with radio images from the *VLA*.

2.1 Stereoscopy with linear path elements

The topology in coronal SXR images seems mostly to consist of loop elements, which can be parametrized by linear path elements (along a quasi-circular trajectory) in the 3D geometric space. Berton & Sakurai (1985) described a method to determine the 3D shape of coronal loops (from *XUV/Skylab 3* images), which was recently also applied to *Yohkoh* images (Slater et al. 1994). In the method of Berton & Sakurai (1985), a loop is parametrized by a planar trajectory (similar to the method described by Loughhead et al. 1984), whose 3D coordinates are found by a least-square procedure, after the loop coordinates from two subsequent days have been transformed into a co-rotating reference system. Berton & Sakurai (1985) determined with this method the inclination of the loop plane, asymmetries of the loop shape, and found that the asymmetry was consistent with potential-field calculations. However, the authors point out that special attention must be paid to choosing structures stable enough during the time interval considered for restoration (about 1 day).

2.2 Stereoscopy with gaussian deconvolution

In contrast to SXR images that show myriads of arcades and loops, radio images of the solar corona at 20 cm (1.5 GHz) are characterized by amorphous structures with gaussian and elliptical shapes. These gaussian components have an average diameter of $70'' \pm 20''$ with a sharp cutoff at $\lesssim 20''$ in the size distribution (Aschwanden & Bastian 1994b). Bastian (1994) has interpreted the lack of fine

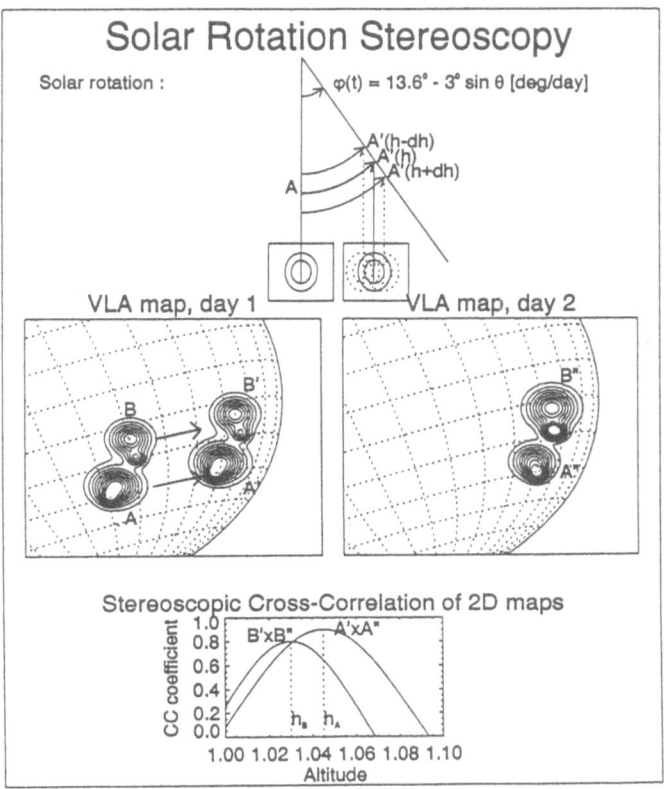

Fig. 1. Scheme of solar rotation stereoscopy method: A radio map is deconvolved into gaussian components (A,B). The map segment (A) from day 1 is transformed to day 2 (A') by applying the differential rotation, and then cross-correlated with the corresponding map segment from day 2 (A"). The coordinate transformation is performed with different altitudes h, and the maximum of the cross-correlation coefficient of $[A \times A"]$ indicates the correct altitude (h_A) for source A. [from Aschwanden & Bastian 1994b]

structures ($\lesssim 15''$ at a wavelength of 20 cm) in terms of scattering of radio waves on coronal inhomogeneities. The isotropy of this scattering mechanism, which blurs every compact source into a seeing disk with spherical symmetry, has one positive side-effect: it makes a 20 cm radio map most suitable for a deconvolution into spherically symmetric gaussian components. In fact, Aschwanden & Bastian (1994a) have demonstrated that the content of a VLA map can be conveniently decomposed into a relatively small number ($\lesssim 200$) of such gaussian components, which represent suitable elements for stereoscopic correlations.

An algorithm for stereoscopic correlation between pairs of solar radio maps (Fig. 1) has been developed in Aschwanden & Bastian (1994a; 1994b), allowing determination of the exact 3D position of quasi-stationary radio sources (with identical heliographic positions, mapped at different times). The algorithm consists of the following two steps: (1) iterative decomposition of radio maps into gaussian components, and (2) cross-correlation of map segments (that encompass individual gaussian components) by performing coordinate transformations (that account for the differential rotation) with a variable source altitude, until a maximum of the cross-correlation is obtained. This way, the altitude of each gaussian component is evaluated, and map transformations in time can be performed by recomposing the gaussian components with the correct 3D projections.

From the systematic application of stereocopic correlations, a number of new physical results has been derived (Aschwanden & Bastian 1994b): e.g., (1) the distribution of source altitudes, (2) systematic center-to-limb variations in source altitude and brightness temperature, and (3) the rotation rate of coronal sources at 20 cm has been found to be consistent with the photospheric differential rotation rate.

Limitations and problems inherent to stereoscopic correlation methods are: (1) time variability, (2) ambiguity in tracking of sources that have closely-spaced longitudes but identical latitude, (3) aspect angle-dependent opacity effects in longitude, and (4) source motion relative to the assumed differential rotation rate.

3 Tomography

Tomography is a method to infer a 3D-density function from a number of 2D-images obtained from different aspect angles. Classical tomography is performed by scanning an object with equidistant angular spacings, which allows reconstruction of the 3D-density function with the Radon-Transform (in Fourier space) or with algebraic backprojection (in geometric space). These principles are widely used by the medical community for imaging the internal structure of the body with techniques such as computer-aided tomography (CAT) scanners and nuclear magnetic resonance (NMR) imagers. Tomography was also applied to white-light emission of the solar corona (Altschuler 1979).

Tomography represents a mathematical deconvolution problem, where a 3D-density distribution is deconvolved from many line-of-sight integrals of a measured quantity. This measured quantity is usually an intensity or brightness, that is proportional to the emissivity or absorption integrated over the column depth along the line-of-sight. In medical tomography, for instance, the attenuation of an X-ray source by the absorption of water (in the scanned organ) is used to infer the column depth of the absorber. For solar tomography, in contrast, the emissivity of free-free bremsstrahlung from an optically thin plasma has been proposed to infer the volumetric density distribution of the radiating plasma. Thus, tomography of the solar corona via free-free emission seems to be most promising in soft X-ray wavelengths, where the emission is always optically thin and is detectable almost everywhere (except in coronal holes).

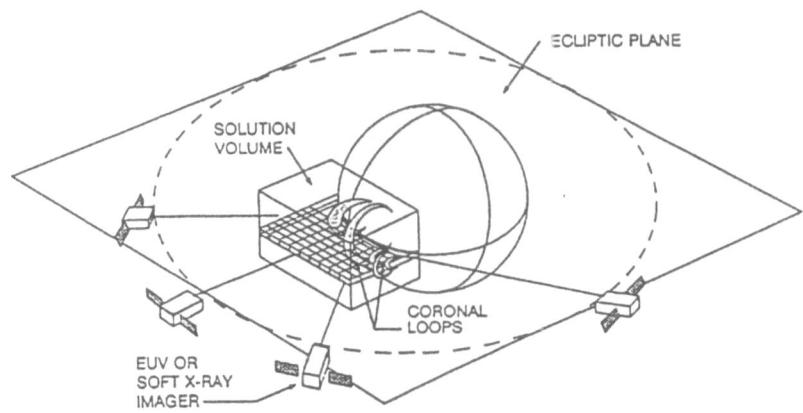

Fig. 2. The concept of space-borne tomography is to obtain images of coronal structures from a number of different angular vantage points. The tomographic reconstruction can be performed independently in each plane parallel to the ecliptic, where all the spacecraft are placed. [from Davila 1994]

3.1 Multiple spacecraft tomography

Solar tomography with multiple spacecraft has been recently proposed and simulated by Davila (1994). The main difference of space-borne tomography from medical tomography is the extremely sparse angular coverage that has to be dealt with. Davila (1994) simulated a cross-section of a solar active region containing 2 resolved and 4 unresolved loop structures, and reconstructed this distribution with an algebraic back-projection technique, addressing the questions of numerical convergence, optimum angular spacing, and optimum number of observing spacecraft. He found a satisfactory reconstruction of the original data with 4-6 spacecraft, spread over an angular range of 140^0 (Fig. 2). Davila simulated also 3D-views of nested loops, and demonstrated that the fidelity of tomographic reconstruction can considerably be improved with *Cleaning* (e.g. with Richardson-Lucy technique), that deconvolves the point-spread function of the algebraic back-projection algorithm.

The same method of tomographic reconstruction can also be applied to reconstruct sharp images from "rotation-smeared" raw data obtained from a spinning spacecraft. This technique has been demonstrated for a solar He 304 Å image by Davila & Thompson (1992). Future solar imagers may benefit from this method since spin-axis-stabilized spacecraft are less costly than a three-axis-stabilized version.

Concepts of future spacecraft missions with tomographic capabilities (Grigoryev 1993) were the subject of a recent workhop on *Special Perspectives Investigations (SPINS)*, held at the *Space Environment Laboratory (SEL)*, Boulder/Colorado (Pizzo 1994).

3.2 Solar rotation tomography

For quasi-stationary features, solar tomography can also be performed by employing the solar rotation to vary the aspect angle. Standard *Computed Tomography (CT)* (filtered Fourier back-projection) has been recently applied to SXR data from *Yohkoh* by Hurlburt et al. (1994). In order to obtain equispaced angular coverage and an unobstructed reconstruction volume, Hurlburt et al. (1994) used daily SXT maps of the solar corona above a tangential plane at the solar South pole, over the period of a half solar rotation. The resulting tomogram shows static structures of the polar corona, where the boundaries of a coronal hole represents the most prominent topological feature. The absence of "streaking" artifacts, which would result from sudden topological changes or from rapid intensity fluctuations, gives some confidence that the selection of maps contains only static structures. Future improvements are planned to implement differential rotation corrections, and to develop limited angle reconstruction techniques to reduce effects of time evolution, in particular for reconstruction of active regions at mid-latitudes.

3.3 Frequency tomography

A different approach of tomographic reconstruction, using the frequency dependence (instead of the angular dependence) of the measured brightness, has been described by Bogod et al. (1994). The authors consider free-free emission of the quiet corona at microwave wavelengths, measured at 30 different wavelengths with RATAN-600, and calculate the inversion of the frequency-dependent opacity integral

$$T_B(p) = \int_0^\infty T_e(t) \, \exp(-t/p) \, dt/p, \quad t = \tau(1cm), \quad p = 1/\lambda_{cm}^2$$

where $T_B(p)$ is the observed brightness temperature measured at different wavelengths. The inversion of $T_e(t)$ can be done analytically by using a Taylor series expansion (or Laplace transform), and the temperature dependence $T_e(t)$ as function of the opacity t is obtained. This method could be useful to infer the coronal temperature $T_e(h)$ and density profile $n_e(h)$, if complementary information from SXR and EUV images are incorporated.

4 Physical Aspects of Coronal Structures

We discuss some physical aspects of coronal structures that have been largely inferred from simultaneous imaging in SXR/EUV and radio. Our focus is to characterize the **inhomogeneity** of the corona and the observational limitations to infer the basic physical properties. Such information serves to define suitable building blocks for 3D models of coronal structures (e.g. active regions) that can be investigated with stereoscopic and tomographic methods.

4.1 Topology

The topology of the solar corona is strongly structured by the magnetic field. Depending on whether the local magnetic field lines are open or closed, the plasma is structured by gravitational stratification (in coronal holes), or confined in magnetic loops (in active regions). The approach of modeling stellar atmospheres with homogeneous layers and a barometric scale height, is problematic in regions with high magnetic fields, where the plasma is confined in closed structures and is governed by the dynamics of local heating and cooling processes that are almost isolated from the rest of the corona. A coarse criterion for such inhomogeneous structures is when the plasma parameter becomes less than unity, i.e. $\beta = 8\pi n_e kT/B^2 < 1$. Thus, each magnetic loop in an active region should be considered as a "mini-atmosphere" rather than as a part of a homogeneous corona. This was already realized from the first *Skylab* pictures (for a review see Vaiana & Rosner 1978). It turns out that the construction of appropriate inhomogeneous models is still the most crucial keypoint for deriving correct physical parameters of the solar corona. Combined observations in soft X-rays, EUV, and radio have clearly demonstrated that inhomogeneous models (both in density and temperature) are required to satisfy all observational constraints.

Coronal structures observed in SXR were categorized into 6 classes from the *Skylab* data (Vaiana, Krieger, & Timothy 1973): 1) active regions, 2) active region interconnections, 3) large loop structures associated with unipolar magnetic regions, 4) coronal holes, 5) bright points, 6) filament structures. This terminology is still widely in use today to describe topological elements of the corona. More sensitive images (e.g. from *NIXT* and *Yohkoh*) revealed also fainter SXR components, such as the diffusive quiet Sun outside active regions, faint arcades of loops, polar crowns, helmet streamers, cusp-shaped structures, etc. As time lapse movies from *Yohkoh* observations show, most of these structures are not in static equilibrium, but in continual, slow evolution (e.g. Uchida et al. 1992, 1993; Tsuneta et al. 1992, 1993; McAllister et al. 1992)

4.2 Magnetic Field

The three-dimensional (3D) coronal magnetic field can be calculated by extrapolating the photospheric (vector or longitudinal) field, using either the potential field (e.g. Sakurai 1982) or force-free (e.g. constant α) approximations. These theoretical calculations do not always provide a realistic representation of the coronal magnetic field, and thus, observational comparisons with SXR, EUV and radio data are very crucial.

Active Regions represent large-scale bipolar magnetic features, often emerging along global neutral lines. Since the hot coronal plasma is confined in magnetic loops (for $\beta < 1$), SXR and EUV loops provide useful tracers of the magnetic field. Such a quantitative comparison with potential field calculations was performed by Sakurai et al. (1992) for a SXR loop observed before and after a flare. The comparisons clearly demonstrated that the S-shaped (sheared)

loop before the flare corresponded to a non-potential configuration, while the
(relaxed) post-flare loop matched a dipole potential field. However, comparing
SXR loops (e.g. from *Yohkoh* images) with ground-based vector magnetograph
data, show that there is not a simple relation between the SXR structures and
the distribution of vertical electric currents in an active region (Metcalf et al.
1994).

Microwave observations (of gyroresonance emission) are the only direct measurement of the coronal magnetic field strength. Magnetic field models in active
regions have been computed and compared with the predicted gyroresonance
emission by Pallavicini et al. (1981), Schmahl et al. (1982), and Brosius et al.
(1992). Microwave observations can provide fiducial points for testing magnetic
field extrapolations. Maximum magnetic field in sunspots reach typically values of 2000-3000 G, decreasing with a gradient of $dB/dh \approx 0.3$ G/km in height
(e.g. Brosius et al. 1992; Lee, Hurford & Gary 1993). Multi-frequency mapping of
gyroresonance sources in combination with stereoscopic altitude measurements
render directly the altitude dependence of the magnetic field strength $B(h)$.
Such attempts are currently in progress (Aschwanden, Bastian & Nitta 1995;
Lim, Aschwanden & Gary 1995).

4.3 Electron Density

Early measurements of the electron density of the quiet corona were derived
from coronograph observations of the scattered white light above the limb, assuming a spherically symmetric, gravitationally stratified atmosphere. However,
the density structure of the corona is highly inhomogeneous, as revealed by SXR,
EUV and radio observations. In active regions, electron densities were found to
be enhanced up to a factor of 10-100 compared with the quiet corona. The electron density of active region loops can be inferred from the emission measure in
soft X-rays, i.e. $EM = \int n_e^2(s)ds$. The electron density distribution $n_e(s)$ along
the line-of-sight is often approximated with a constant value over the column
depth s, which can be estimated from the loop diameters as seen perpendicular
to the line-of-sight. This way, Webb et al. (1987) measured the electron density of active region loops resolved in SXR (with diameters of 3.5-12 Mm) to
$n_e = 0.3 - 2.3 \cdot 10^{10}$ cm^{-3}.

In coronal holes, density depressions of $\approx 20 - 30\%$ of the quiet Sun density
were found, with densities of $n_e \approx 3 \cdot 10^7 - 3 \cdot 10^8$ cm^{-3} at the base of the
corona. The density constrast between the quiet corona and coronal holes was
first inferred from combined SXR/EUV and radio observations (Drago 1974).
Density models of coronal holes can be found in Kopp & Orrall (p.179-224 in
Zirker 1977). Comparisons of densities in coronal holes measured with radio
versus SXR/EUV were believed to suffer from systematic errors such as (1)
instrumental scattering of SXR/EUV detectors, (2) radio calibration problems,
(3) the assumption of ionization equilibrium for EUV emission (see discussion
in Lantos 1980; Chambe 1978).

The solar corona is very inhomogeneous down to the smallest measurable
spatial scales (of $\approx 1''$). It is likely that active region loops consist of unresolved

fine structure. Depending on their volume filling factor, the electron density of the finest structures can be arbitrarily higher than the maximum values measured sofar. Density fluctuations (turbulence) on sub-arcsecond scales in the lower corona are believed to contribute significantly to wave scattering not only at metric but also at decimetric wavelengths (Bastian 1994). This effect is believed to be the main reason for the non-detection of fine structure in radio at scales of $\lesssim 15''$ (at 20 cm). Much smaller structures should be seen with the VLA, since it has an angular resolution of $1.2''$ at 20 cm, and $0.12''$ at 2 cm in A configuration.

4.4 Electron Temperature

Although a large fraction of the coronal volume is characterized with a plasma temperature of 1-2 MK, temperature inhomogeneities over 3 orders of magnitude occur. The quiet corona outside active regions has an average temperature of ≈ 1.5 MK, but drops to 0.6-0.8 MK in coronal holes (seen in radio, but invisible in SXR). Temperatures of active region loops vary from 10^4 K (seen in Hα and UV) up to $2 - 5 \cdot 10^6$ K (seen in SXR and EUV; e.g. Hara et al. 1992). Temperature inhomogeneities in active regions have been most clearly demonstrated from EUV observations onboard *Skylab*. Resolved loops were detected in the cool ($< 1.5 \cdot 10^4$ K) Lyα and CII lines up to the hotter Mg X ($T > 1.4 \cdot 10^6$ K) and Si XII ($T > 2.0 \cdot 10^6$ K) lines. The presence of cool plasma ($T \lesssim 1$ MK) in active regions, which was previously mainly detected in EUV, was recently also postulated from comparisons of SXR temperatures and coincident radio brightness temperatures (CoMStOC observations: Webb et al. 1987; Nitta et al. 1991, Brosius et al. 1992, Schmelz et al. 1992). An additional argument in favor of the coexistence of a multi-temperature plasma in active regions was brought forward from the interpretation of the statistical center-to-limb darkening of the radio brightness temperature at 20 cm in active regions (Aschwanden & Bastian 1994b). Multi-temperature diagnostics of active regions loops in the range of 1.0-3.5 MK can also be studied in optical lines (FeXIV[green], FeX[red], and CaXV[yellow]; Ichimoto et al. 1994).

Coronal holes are regions of lower temperature ($\approx 0.6 - 0.8$ MK; Lantos & Avignon, 1975; Dulk & Sheridan, 1974; Habbal, Esser & Arndt, 1993). Because this temperature is still well above the transition region temperature, coronal holes are a purely coronal phenomenon, and virtually undetectable in transition region lines. Chiuderi-Drago et al. (1977) inferred from combined EUV and radio (169, 408 MHz) observations that even for coronal holes an inhomogeneous model is required: Besides the prevailing temperature of T≈ 0.8 MK, a small admixture (of $\approx 10\%$) of a hot 2 MK plasma was needed to reconcile the EUV and radio parameters. Also Dulk et al. (1977) conclude from combined EUV and radio (Culgoora) observations that standard temperature models fail (for both the quiet Sun and coronal holes), and postulated models with thermal diffusion, mass outflow, or departures from ionization equilibrium. The presence of a hot temperature component (T=2 MK) in coronal holes (besides the prevailing cool component of 0.8 MK) seems also to be confirmed by recent *Yohkoh* observations (Tsuneta, Kosugi, Martens, Strong, private communication).

Systematic temperature inhomogeneities occur also above sunspots. The lower temperature in a sunspot in the photosphere is reflected as a temperature depression in coronal heights. The temperature above the penumbra (2 MK) drops to 1.2 MK above the umbra in an altitude of $\approx$ 6 Mm, as inferred from 6 cm microwave observations: (Strong et al. 1984; Siarkowski et al. 1989). Because SXR detectors are generally blind to temperatures cooler than $\lesssim$ 2 MK, SXR loops seem to "float" in the corona; their (cool) footpoints above the umbra are invisible in SXR, and appear to be truncated above the penumbra (e.g. Golub 1990, comparing Ha and SXR data; Gopalswamy et al. 1994, comparing *VLA* and *Yohkoh* data).

Despite the multi-temperature diagnostic capabilities of EUV spectrographs, the radial temperature structure across an active region loop is still controversial. The major difficulty lies in the problem of whether hot and cold loop structures seen in EUV are coincident or adjacent. Foukal (1975) inferred that loops have a cool core and a hot envelope, where the hot plasma is generally denser and appears more diffuse. However, no convincing spatial correlation between hot and cool loop structures has been established sofar, but instead, hot and cool EUV loops were shown not to be co-axial (Cheng 1980; Dere 1982; Habbal, Ronan & Withbroe 1985; Karovska et al. 1994).

Stereoscopic and tomographic reconstruction of coronal plasmas are most promising for optically thin free-free emission. However, since the opacity is strongly dependent on the temperature, a realistic model requires a sensible description of the multi-temperature plasma, which needs to be quantified in terms of the differential emission measure (see e.g. Pallavicini et al. 1981; Raymond & Foukal 1982; Arndt et al. 1994).

4.5 Spatial Scales

The largest uniform structures of the corona are coronal holes, which cover up to 20% of the solar surface, while the total area of active regions seldom exceeds 1-2% of the solar surface (during the maximum of the solar cycle). Active regions itself consist of arcades and bundles of nested loops, with typical footpoint separations of $10 - 100$ Mm, and loop diameters of $1 - 10$ Mm. The measured size of loop diameters depends strongly on the emission mechanism or wavelength used: Diameters down to 1″are measured in Ha and SXR (Fig. 3), but radio measurements at 1.5 GHz do not show sources smaller than 15″. Small-scale structures have been seen in soft X-rays down to the resolution of the instruments, which is about 1″for NIXT images (Golub 1993), or 0.7″for MSSTA (Walker 1993).

From theoretical arguments, a lower limit to the transverse scale size of loops seen in SXR would be the skin-depth of current penetration due to magnetic diffusivity over the lifetime of a loop, which is estimated to 1 km (or 0.001″) (Golub, 1993).

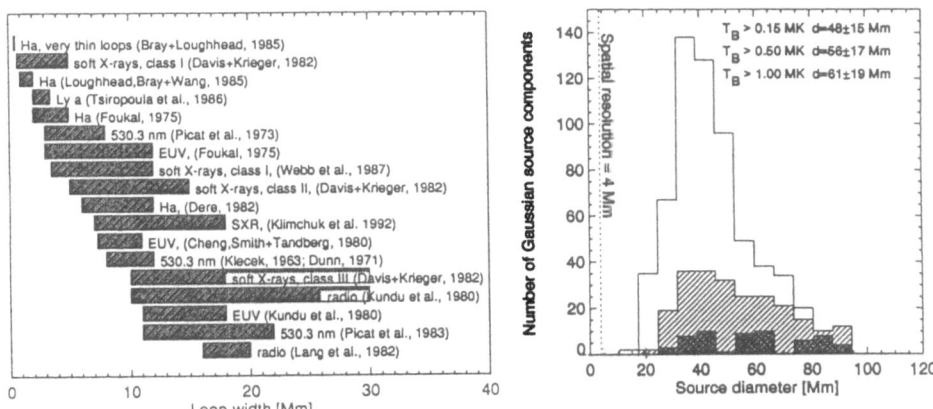

Fig. 3. Reported measurements of diameters of active region loops in Hα, EUV, SXR and radio cover a large range of 0.5-30 Mm (left hand side). Note the wavelength dependence of the measured ranges. The distribution of radio source sizes measured in a VLA map at 1.5 GHz covers a range of 20-100 Mm (right hand side). Note the distinct absence of source sizes below 15 Mm down to the instrumental resolution of 4 Mm [from Aschwanden & Bastian 1994b].

A fundamental limitation to the resolution of radio mapping at decimetric frequencies is probably the scattering of electromagnetic waves on a turbulent spectrum of fluctuations in the electron density in the lower corona (Bastian 1994). Radio wave scattering blurs closely-spaced loops in active regions into a diffuse, amorphous brightness distribution, whose outer contours coincide with the extent of SXR emission. Based on the statistics of measured loop diameters and widths at different wavelengths (see compilation in Fig.3) it appears that the smallest resolved features at 1.5 GHz encompass emission from entire loop arcades rather than from a single loop. Because these unresolved arcades (at 1.5 GHz) often have a curved shape and were found above neutral lines, they may have been mistaken for individual loops in earlier reports of VLA observations (Lang, Willson & Rayrole 1982; Lang, Willson, & Gaizauskas 1983; Kundu & Lang 1985).

4.6 Altitudes

Altitude measurements of coronal sources are important to deconvolve coincident structures along the line-of-sight and to obtain information on the density profile along the line-of-sight. Early altitude measurements relied (1) on direct measurements near the limb, (2) on the relative displacement to photospheric features (e.g. in the case of sunspot-associated radio sources), or (3) on the relative rotation speed with respect to the photospheric rotation rate (Donati-Falchi et al. 1978; Christiansen et al. 1957). From the displacement relative to the sunspot at different days, the altitude of sunspot-associated (compact) microwave sources (at 6 cm) was determined to h=6 Mm (Shibasaki et al. 1983). The altitude of soft X-ray sources (OSO-8, 1-4 keV) was measured from the occultation curve during limb-crossings: 50% of the heights were found below 20 Mm, and 90% below 57 Mm (Mosher 1979). Bornmann & Matheson (1990) determined the heights of SXR sources to h=30-100 Mm.

The most accurate altitude measurements can be achieved from stereoscopic correlation of source centroids (Aschwanden & Bastian 1994b). Uncertainties resulting from deviations from the standard differential rotation rate can be compensated with redundant observations from multiple days. Statistics of 66 active region sources observed with the VLA at 20 cm yielded an average height of $h = 25 \pm 15$ Mm (90% were found below 40 Mm). Systematic stereoscopic correlations are currently performed with multi-day observations of active regions with the VLA and OVRO at multiple frequencies (Aschwanden, Bastian & Nitta, 1995). First stereoscopic measurements of microwave sources at 10-14 GHz yield source heights of 5-7 Mm, after careful evaluation of the rotation rate of the associated photospheric sunspot (Lim, Aschwanden & Gary 1995).

4.7 Rotation Rate

The standard value of the photospheric (synodic) differential rotation rate is $\Omega = 13.45^0 - 3^0 sin^2 B$ [deg/day] (Allen 1973). The exact value depends on the method or tracers used. Rotation rate differences among different tracers were found to depend on age of sunspots (Zappala & Zuccharello 1991), and may depend on the depth in the convection zone where they originate, by the mechanism of buoyant, emerging magnetic flux tubes (Stenflo 1989).

The measurement of the rotation rate of coronal sources is more difficult because the projected motion depends also on the coronal altitude. First attempts of coronal rotation rate measurements (Liu & Kundu 1976) contain, therefore, a considerable uncertainty because of the unknown altitude. Reliable values can only be determined from simultaneous measurements of position and altitude, e.g. from redundant (≥ 3) stereoscopic measurements. Such systematic stereoscopic measurements have been performed recently for 66 active regions observed with the VLA at 1.5 GHz. In the statistical average, these sources (at an altitude of $h = 25 \pm 15$ Mm) were found to corotate with the photospheric differential rotation rate (Aschwanden & Bastian 1994b). The relative motion with respect to the photospheric differential rotation rate (Allen 1973) showed a distribution of 0.001±0.193 deg/day in longitude, and 0.002±0.116 deg/day in latitude. The small residuals (≤ 0.001 deg/day in longitude) confirms that free-free emission at 1.5 GHz originates from magnetically confined plasma that is rigidly anchored in the photosphere. The standard deviation (of 0.2 deg/day in longitude), however, indicates that individual active regions have a considerable relative motion with respect to the average differential rotation rate.

Coronal holes have a rigid rotation (27 day synoptic rotation period), measured over lifetimes from 1-10 solar rotations (Timothy, Krieger & Vaiana, 1975). The solid rotation may reflect that they are anchored with magnetic structures at the bottom of the convection zone (Krieger, p.71 in Zirker 1977). They connect also to the interplanetary magnetic sector boundaries (which have a 28.5 day rotation period), and are not tied to the photospheric differential rotation. This requires a continuous closing/opening of field lines at the boundaries. Individual active regions seem to drift through a coronal hole (Wang & Sheeley, 1993). Wang & Sheeley (1993) explain the rotation of coronal holes by the interaction of axisymmetric polar fields and nonaxisymmetric flux associated with the boundaries.

4.8 Emission mechanisms

SXR emission in the solar corona is produced by free-free emission of the hot ($\gtrsim 1 - 2$ MK) plasma, with comparable contributions from the continuum and lines. The corona is everywhere optically thin in soft X-rays, thus, the free-free emissivity is proportional to the emission measure $EM = \int n_e^2(s)ds$ integrated along the line-of-sight. The conversion of the observed SXR or EUV flux into the emission measure requires assumptions on the ionization and excitation equilibria, and thus, on the elemental abundance. These assumptions are not needed at radio wavelengths, where the free-free opacity merely depends on the electron density and temperature. This makes radio observations a very useful complement to SXR and EUV measurements.

Radio emission at metric/decimetric wavelength of the **quiet Sun** is due to thermal bremsstrahlung from hot (1-3 MK) plasma. The observed brightness temperature of the quiet Sun depends on the coronal altitude where the opacity becomes unity: $T_B \approx 1$ MK at $\nu \lesssim 100$ MHz, $T_B \approx 50,000$ K at $\nu = 1$ GHz, $T_B \lesssim 15,000$ K at $\nu \gtrsim 10$ GHz. The decreasing brightness temperature with higher frequencies indicates that the coronal contribution becomes lesser, while the contribution from the cooler transition region becomes stronger. For reviews on the radio emission of the quiet Sun see Kundu (1965), Lantos (1980), Fürst (1980), Schmahl (1980), Marsh & Hurford (1982), Kundu (1985), Dulk (1985), Sheridan & McLean (1985), Bastian (1987).

Coronal Holes have been observed in SXR/EUV and radio (Dulk & Sheridan 1974; Dulk et al. 1977; Sheridan & Dulk 1980). Coronal holes are 2% brighter than quiet sun at wavelengths of 8.6 mm and 2 cm, but darker at 10 cm and 2.8 cm (see compilation of Bohlin p.27-70 in Zirker 1977). At metric frequencies, the radio brightness temperature of coronal holes was found to be depressed at 160 MHz, but not at 80 MHz (Sheridan & Dulk 1980). But Wang, Schmahl & Kundu (1987) found coronal holes to be dark at 73.8, 50, and 30.9 MHz. The discrepant results from CLRO and Culgoora may partially be caused by image restoration problems of the Culgoora data for the quiet Sun. Papagiannis & Baker (1982) found the largest contrast between quiet sun and coronal holes ($25 - 30\%$) at decimeter wavelengths.

Microwave emission from **Active Regions** consists of two components: (1) free-free emission from extended, low-brightness temperature, plage-associated sources, usually centered above the neutral lines, and (2) gyroresonance emission (at ≈ 3 harmonic), from compact, high-brightness temperature, highly polarized, sunspot-associated sources (Pallavicini et al. 1979, 1981; Kundu et al. 1980; Chiuderi-Drago et al. 1982; Schmahl et al. 1982). Bipolar active regions generally display an asymmetry in temperature and density, and a stronger magnetic field above the leading sunspot, which needs to be considered in models of gyroresonance emission (Alissandrakis 1986; Gary & Hurford 1987; Brosius et al. 1992). The extended plage-associated component (free-free emission) can also be mapped at higher microwave frequencies (VLA 8.5 and 15 GHz, White et al. 1992; Nobeyama 17 GHz, Shibasaki, private communication). The detection of this low-brightness component ($T_B < 50,000$ K) requires a high dynamic range in the images.

Although the electron temperature of active region loops is 2-5 MK according to SXR measurements, the observed radio brightness temperature is generally lower ($T_B \approx 0.5 - 2$ MK) at 1.5 GHz. Cospatial measurements in SXR and radio (Webb et al. 1987; Nitta et al. 1991; Brosius et al. 1992; Schmelz et al. 1992; see also radio/SXR temperature ratios in Fig.4 bottom left) indicate that there are contributions of free-free emission from cool ($\lesssim 1$ MK) plasma loops, which become marginally optically thick at 1.5 GHz. The inclusion of these cool loops, which can be diagnosed in EUV, became an important ingredient of realistic (inhomogeneous) active region models, without which the observed radio brightness temperature cannot be understood.

4.9 Center-to-limb opacity effects

A consequence of the optically thin free-free emission from the quiet Sun is that limb brightening is expected at microwave frequencies (because the corona is optically thin), which is absent at metric frequencies (where the corona is almost optically thick). Compilations of center-to-limb variations of the quiet Sun versus frequency can be found in Fürst (1980) and Bastian (1987).

A center-to-limb variation of the radio flux from active regions was found in early observations at 20 cm with poor resolution (3). This effect was attributed to a projection effect of horizontally stratified sources, which varies with $\cos(\alpha)$ of the aspect angle α (Christiansen et al. 1957).

A center-to-limb variation of the radio brightness temperature was recently discovered at 20 cm from the statistics of 66 active region sources observed with the VLA (Aschwanden & Bastian 1994b), and is independently confirmed with other data (Fig.4 bottom left). The effect can be statistically characterized by: $T_B(\alpha)/T_B(0) \approx 0.4 + 0.6 \, cos^2\alpha$ Since the observed active region sources are clearly resolved, this effect cannot be attributed to a geometrical projection effect. Optically thin free-free emission would produce a limb brightening for horizontally stratified sources. Optically thick free-free emission of an isothermal source yields a constant brightness temperature for different aspect angles and cannot explain the observed limb darkening either. The only plausible alternative is an interpretation in terms of inhomogeneous models, both in electron density and temperature. Cool loops have a smaller scale height than hot loops and absorb free-free emission more efficiently at the limb than at disk center because of the vertical orientation of the densest loop parts. Quantitative active region

models with nested multi-temperature loops (Fig. 4) are able to reproduce the
observed limb darkening.

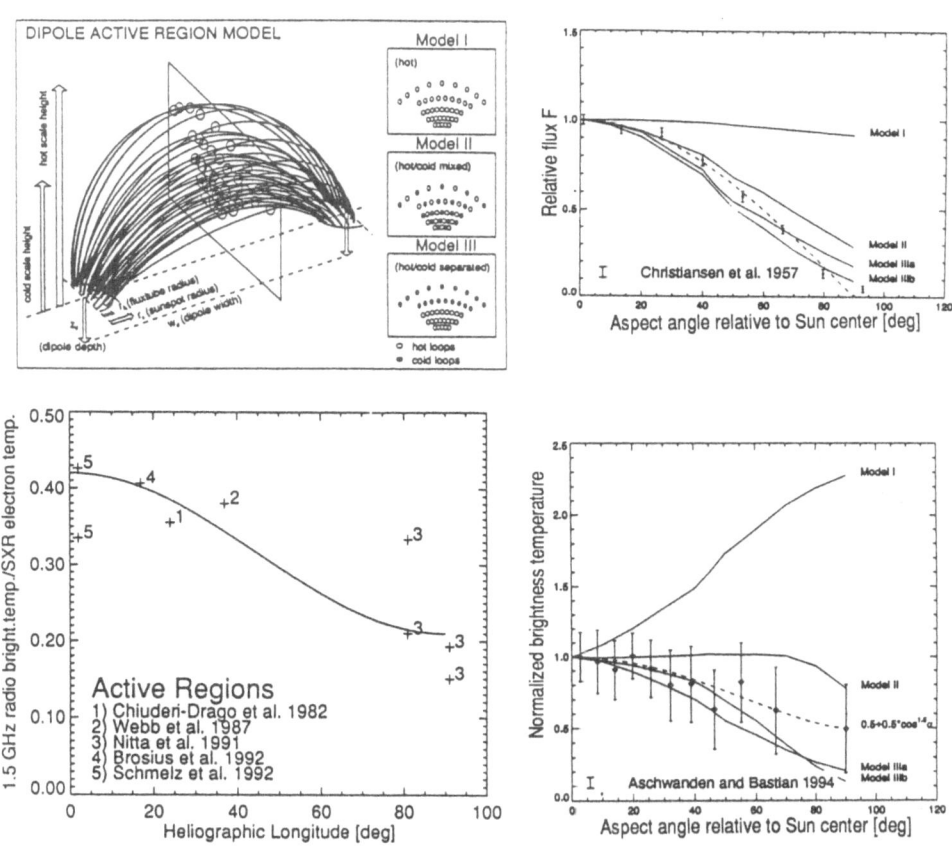

Fig. 4. A realistic model of an inhomogeneous active region model contains a number
of nested (e.g. dipole) loops with different temperatures and density scale heights (top
left). Such a model (Model II or III) can reproduce the center-to-limb darkening of the
total radio flux (top right), and also the center-to-limb darkening of the peak brightness
temperature (bottom right), while isothermal models (Model I) fail. The aspect-angle
dependent variation of the radio peak brightness temperature and total radio flux was
numerically computed for free-free emission with typical active region parameters.

5 Concluding Remarks

The complexity of the physical properties of the highly inhomogeneous solar corona demonstrates that multi-wavelength observations (in SXR, EUV and radio) are extremely important because they provide complementary information which cannot be obtained at any single wavelength: (1) SXR provides the emission measure and electron temperature of hot plasmas, (2) UV/EUV gives also the differential emission measure of cool plasmas, (3) radio emission is sensitive to absorption by cool plasma (via free-free emission), and allows measurement of the coronal magnetic field (via gyroresonance emission). Additional constraints on aspect-angle dependent opacity effects can be inferred from the center-to-limb darkening observed in radio. Realistic models of active regions require an appropriate description of temperature and density inhomogeneities in a three-dimensional geometry which is largely constrained by the coronal magnetic field. Realistic 3D models of coronal structures are a necessary prerequisite to exploit stereoscopic and tomographic methods, because the interpretation of the reconstructed topology crucially depends on the assumed opacity models. In other words, stereoscopic or tomographic information is only as good as the knowledge of the implied opacity. Further unknowns are the volume filling factors of small-scale structures and the turbulence spectrum of electron density fluctuations, which significantly affect opacity and scattering at radio wavelengths.

Acknowledgements: The author thanks Stephen White, Tim Bastian, Ed Schmahl, Gordon Holman, Joe Davila, and Werner Neupert for helpful discussions and comments. This work was partially supported by the grants NAGW-3456 and NAG-52352.

References

Allen,C.W. 1973, Astrophysical Quantities, London: Athlone
Alissandrakis,C.E. 1986, Sol. Phys., 104, 207
Altschuler,M.D. 1979, in *Image Reconstruction from Projections*, G.T.Herman (ed.), Berlin:Springer, p.105
Arndt,M.B., Habbal,S.R. and Karovska,M. 1994, Sol. Phys., 150, 165
Aschwanden,M.J. and Bastian,T.S. 1994a, ApJ, 426, 425
Aschwanden,M.J. and Bastian,T.S. 1994b, ApJ, 426, 434
Aschwanden,M.J., Bastian,T.S., and Nitta,N. 1995, in preparation
Bastian,T.S. 1987, PhD Thesis, University of Colorado
Bastian,T.S. 1994, ApJ, 426, 774
Batchelor,D.A. 1994, Sol. Phys., in press
Berton,R., and Sakurai,T. 1985, Sol. Phys., 96, 93
Bogod,V.M., Grebinsky,A.S., Opeikina,L.V. 1995, CESRA workshop, Caputh/ Potsdam, Germany, May 16-20, 1994
Bornmann P.L. and Matheson L.D. 1990, A&A, 231, 525
Bray,R.J., Cram,L.E., Durrant,C.J., and Loughhead,R.E. 1991, *Plasma Loops in the Solar Corona*, Cambridge: Cambridge University Press
Bray,R.J. and Loughhead,R.E. 1985, A&A, 142, 199
Brosius,J.W., Davila,J.M., Thompson,W.T., Thomas,R.J., Holman,G.D., Gopalswamy, N., White,S.M., Kundu,M.R., Jones,H.P. 1993, ApJ, 411, 410
Brosius,J.W., Willson,R.F., Holman,G.D., and Schmelz,J.T. 1992, ApJ, 386, 347
Chambe,G. 1978, A&A, 70, 255
Cheng,C.C. 1980, ApJ, 238, 743
Cheng,C.C., Smith,J.B., Tandberg-Hanssen,E. 1980, Sol. Phys., 67, 259

Chiuderi-Drago,F., Avignon,Y., and Thomas,R.J. 1977, Sol. Phys., 51, 143
Chiuderi-Drago,F., Bandiera,R., Falciani,R., Antonucci,E., Lang, K., Willson, R.F.,
    Shibasaki,K., and Slottje,C. 1982, Sol. Phys., 80, 71
Christiansen,W.N., Warburton,J.A., and Davies,R.D. 1957, Austr.J.Phys. 10, 491
Davila,J.M. 1994, ApJ, 423, 871
Davila,J.M. and Thompson,W.T. 1992, ApJ, 389, L91
Davis,J.M. and Krieger,A.S. 1982, Sol. Phys., 80, 295
Dere,K. 1982, Sol. Phys., 75, 189
Donati-Falchi,A., Felli,M., Pampaloni,P. and Tofani,G. 1978, Sol. Phys., 56, 335
Dulk,G.A. 1985, Ann.Rev.Astron.Astrophys.23, 169
Dulk,G.A. and Sheridan,K.V. 1974, Sol. Phys., 36, 191
Dulk,G.A., Sheridan,K.V., Smerd,S.F., and Withbroe,G.L. 1977, Sol. Phys., 52, 349
Dunn,R.B. 1971, in *Physics of the Solar Corona*, ed. C.J. Macris, pp.114-129, Dor-
    drecht:Reidel
Drago,F. 1974, in "Skylab Solar Workshop", ed. G.Righini, Osservazioni e Memorie
    Osservatorio di Arcetri (Firenze), n.104, 120
Foukal,P. 1975, Sol. Phys., 43, 327
Fürst 1980, in "Radio Physics of the Sun", M.R.Kundu and T.E.Gergely (eds.), p.25
Gary,D.E. and Hurford,G.J. 1987, ApJ, 317, 522
Golub,L. 1993, in *Physics of Solar and Stellar Coronae*, eds. J.F.Linsky and Serio,S.,
    Dordrecht: Kluwer, p.71
Golub,L., Herant,M., Kalata,K., Lovas,I., Nystrom,G., Pardo,F., Spiller,E., and Wilczyn-
    ski,J. 1990, Nature 344, 842
Gopalswamy,N., Schmahl,E.J., Kundu,M.R., Lemen,J.R., Strong,K.T., Canfield,R.C.,
    and De La Beaujardiere 1994, Proc. Kofu Symp., eds. S.Enome and T.Hirayama,
    NRO Report 360, p.347
Grigoryev,V.M. 1993, Sol. Phys., 148, 389
Habbal,S.R., Esser,R., and Arndt,M.B. 1993, ApJ, 413, 435
Habbal,S.R., Ronan,R., and Withbroe,G.L. 1985, Sol. Phys., 98, 323
Hara,H., Tsuneta,S., Lemen,J.R., Acton,L.W., and McTiernan,J.M. 1992, PASJ, 44,
    L135
Holman,G.D. 1995, in *The Many Faces of the Sun: The Scientific Results of the Solar
    Maximum Mission*, eds. Strong,K.T., Saba,J., and Haisch,B., Berlin:Springer
Hurlburt,N.E., Martens,P.C.H., Slater,G.L. and Jaffey,S.M. 1994, in "*Solar Active Re-
    gion Evolution: Comparing Models with Observations*", ed. K.S. Balasubra-
    manian, Amer.Soc.Pac., *Volume reconstruction of magnetic fields using solar
    imagery*
Ichimoto,K., Kumagai,K., Sakurai,T., Hara,H., Takeda,A., and Yohkoh SXT Team
    1994, in Proc. Kofu Symp., (eds. S.Enome and T.Hirayama), NRO Report, No.
    360, p.113
Karovska,M., Blundell,S.F. and Habbal,S.R. 1994, ApJ, 428, 854
Klecek,J. 1963, Bull.Astr.Inst.Czech. 14, 167
Klimchuk,J.A., Lemen,J.R., Feldman,U., Tsuneta,S., and Uchida,Y. 1992, PASJ, 44,
    L181
Kundu,M.R. 1965, "Solar Radio Astronomy", New York: Interscience Publishers, 660p
Kundu,M.R. 1985, Sol. Phys., 100, 491
Kundu,M.R. and Lang,K.R. 1985, Science 228, 9
Kundu,M.R., Schmahl,E.J., and Gerassimenko,M. 1980, A&A, 82, 265
Kundu,M.R., Schmahl,E.J., and Rao,A.P. 1981, A&A, 94, 72
Lang,K.R., Willson,R.F., Kile,J.N., Lemen,J., Strong,K.T., Bogod,V.M., Gelfreikh,G.B.,
    Ryabov,B.I., Hafizov,S.R., Abramanov-Maximov,V.E., and Tsvetkov,S.V. 1993,
    ApJ, 419, 398
Lang,K.R., Willson,R.F., and Gaizauskas,V. 1983, ApJ, 267, 455
Lang,K.R., Willson,R.F., and Rayrole,J. 1982, ApJ, 258, 384.
Lang,K.R., Willson,R.F., Smith,K.L., and Strong,K.T. 1987a, ApJ, 322, 1035
Lang,K.R., Willson,R.F., Smith,K.L., and Strong,K.T. 1987b, ApJ, 322, 1044
Lantos,P. 1980, in *Radio Physics of the Sun*, IAU-Symp., Kundu,M.R. and Gergely,T.E.
    (eds.), p.41
Lantos,P. and Avignon,Y. 1975, A&A, 41, 137
Lee,J.W., Hurford,G.J., and Gary,D.E. 1993, Sol. Phys., 144, 45
Lim,J., Aschwanden,M.J., and Gary,D.E. 1995, in preparation
Liu,S.Y. and Kundu,M.R. 1976, Sol. Phys., 46, 15
Loughhead,R.E., Chen,C.L., and Wang,J.L. 1984, Sol. Phys., 92, 53
Loughhead,R.E, Bray,R.J., and Wang,J.L. 1985, ApJ, 294, 697
Marsh,K.A. and Hurford,G. 1982, Ann.Rev.Astron.Astrophys.20, 497

McAllister,A., Uchida,Y., Tsuneta,S., Strong,K.T., Acton,L.W., Hiei,E., Bruner,M.,E.,
     Watanabe,T., and Shibata,K. 1992, PASJ, 44, L205
Metcalf,T.R., Canfield,R.C., Hudson,H.S., Mickey,D.L., Wuelser,J.P., Martens, P.C.H.,
     and Tsuneta,S. 1994, ApJ, 428, 860
Mosher,J.M. 1979, Sol. Phys., 64, 109
Nitta,N., White,S.M., Kundu,M.R., Gopalswamy,N., Holman,G.D., Brosius, J.W.
     Schmelz, J.T., Saba, J.L.R. and Strong,K.T. 1991, ApJ, 374, 374
Pallavicini,R., Sakurai,T., and Vaiana,G.S. 1981, A&A, 98, 316
Pallavicini,R., Vaiana,G.S., Tofani,G., and Felli,M. 1979, ApJ, 229, 375
Papagiannis,M.D. and Baker,K.B. 1982, Sol. Phys., 79, 365
Picat,J.P, Fort,B., Dantel,M., Leroy,J.L. 1973, A&A, 24, 259
Pizzo,V.J. (ed.) 1994, "The findings of the SPINS science workshop", SEL workshop
     Report January 1994, "SPINS : Special Perspectives Investigations", NOAA/
     SEL, Boulder, Colorado, November 3-5, 1993
Raymond,J.C. and Foukal,P. 1982, ApJ, 253, 323
Sakurai,T. 1982, Sol. Phys., 76, 301
Sakurai,T., Shibata,K., Ichimoto,K., Tsuneta,S., Acton,L.W. 1992, PASJ, 44, L123
Schmahl,E.J. 1980, in "Radio Physics of the Sun", IAU Symp. 86, eds. Kundu,M.R.
     and Gergely,T., p.71
Schmahl,E.J., Kundu,M.R., Strong,K.T., Bentley,R.D., Smith,J.B.Jr.,and Krall,K.R.
     1982, Sol. Phys., 80, 233
Schmelz,J.T., Holman,G.D, Brosius,J.W., and Gonzalez,R.D. 1992, ApJ, 399, 733
Siarkowski,M., Sylwester,J., Jakimiec,J., and Bentley,R.D. 1989, Sol. Phys., 119, 65
Slater,G.L., Hudson,H.S., and Freeland,S.L. 1994, SPD/AGU spring meeting, May 23-
     27, Baltimore/MD
Sheridan,K.V. and Dulk,G.A. 1980, in "Solar and Interplanetary Dynamics", M.Dryer
     and E.Tandberg-Hanssen (eds.), IAU-Coll. 91, p.37
Sheridan,K.V. and McLean,D.J. 1985, in "Solar Radiophysics", Cambridge: Cambridge
     University Press, p.443
Shibasaki,K. et al. 1994, PASJ 46, L17
Shibasaki,K., Chiuderi-Drago,F., Melozzi,M., Slottje,C., and Antonucci,E. 1983, Sol.
     Phys., 89, 307
Stenflo,J.O. 1989, A&A, 210, 403
Strong,K.T., Alissandrakis,C.E., and Kundu,M.R. 1984, ApJ, 277, 865
Timothy,A.F., Krieger,A.S., and Vaiana,G.S. 1975, Sol. Phys., 42, 135
Tsiropoula,G., Alissandrakis,C., Bonnet,R.M., Gouttebroze,P. 1986, A&A, 167, 35
Tsuneta,S. and Lemen,J.R. 1993, Physics of Solar and Stellar Coronae, eds. J.F.Linsky
     and S.Serio, Dordrecht: Kluwer, p.113
Tsuneta,S., Takahashi,T., Acton,L.W., Bruner,M.E., Harvey,K.L., and Ogawara,Y.
     1992, PASJ, 44, L211
Uchida,Y., McAllister,A., Strong,K.T., Ogawara,Y., Shimizu,T., Matsumoto,R., and
     Hudson,H. 1992, PASJ, 44, L155
Uchida,Y. 1993, Physics of Solar and Stellar Coronae, eds. J.F.Linsky and S.Serio,
     Dordrecht: Kluwer, p.97
Vaiana,G.S. and Rosner,R. 1978, Ann.Rev.Astron.Astrophys.16, 393
Vaiana,G.S., Krieger,A.S., and Timothy,A.F. 1973, Sol. Phys., 32, 81
Walker,A.B.C.,Jr., Barbee,T.W.,Jr., Hoover,R.B., and Lindblom,J.F. 1988, Science
     241, 1781
Walker,A.B.C.Jr., Hoover,R.B. and Barbee,T.W.Jr. 1993, in Physics of Solar and Stel-
     lar Coronae, J.F.Linsky and S.Serio (eds.), Dordrecht: Kluwer, p.83
Wang,Y.M. and Sheeley,N.R.Jr. 1993, ApJ, 414, 916
Wang,Z., Schmahl,E.J., and Kundu,M.R. 1987, Sol. Phys., 111, 419
Webb,D.F., Davis,J.M., Kundu,M.R., and Velusamy,T. 1983, Sol. Phys., 85, 267
Webb,D.F., Holman,G.D., Davis,J.M., Kundu,M.R., and Shevgaonkar,R.K. 1987 ApJ,
     315, 716
White,S.M., Kundu,M.R., and Gopalswamy,N. 1992, ApJ.Suppl.Ser., 78, 599
Zappala,R.A. and Zuccarello,F. 1991, A&A, 242, 480
Zirker,J.B. 1977, in Coronal Holes and High Speed Wind Streamers, A monograph
     from Skylab Solar Workshop I, ed. J.B.Zirker

Initial Results from the Nobeyama Radioheliograph

Shinzo Enome

Nobeyama Radio Observatory, NAOJ, Minamisaku,
Nagano 384-13, Japan
e-mail address: senome@nro.nao.ac.jp

Abstract. The Nobeyama Radioheliograph started routine observations in late June, 1992, after two years of construction and system integration. In two years of observations, which contain a last part of the maximum phase of the solar cycle 22, it was possible to obtain almost every aspect of solar activity at 17 GHz. They include two X-class flares, tens of M-class flares, hundreds of flares from the very initial phase through the late decay phase, or from pre-flare enhancements through post-flare loops, slowly varying components from active regions, bright points, prominence eruptions, and so on, with a spatial resolution of 10 arcsec and a time resolution of 1 s and 50 ms for specific events.

Although the number of samples is not large enough, it appears as a general tendency that gradual phenomena in time domain are large in spatial domain. Gradual rise and fall type bursts are such example, and gradual-hard events are another example. This trend seems to hold even during a course of an event that the impulsive-phase source is smaller than the decay phase source. This will be more explicitly stated that explosive phenomena take place in a very limited space and extended structures change slowly.

We will describe burst structures in space and time, due to the gyrosynchrotron mechanism, and thermal emission due to the free-free mechanism and gyroresonance mechanism. Magnetobremsstrahlung or gyromagnetic radiation is unique to radio waves that maps magnetic structures in active region corona. We will present examples for each case and show how this instrument is effective to observe and resolve them. Future plan and prospect are briefly described.

1 Introduction

The Nobeyama Radioheliograph is the first dedicated instrument at a short centimeter wavelength to resolve transient solar radio emission down to 10 arcsec in space and 50 ms in time. This unique characteristic is revealing new aspects of solar activity in such regimes as dominated by strong magnetic fields and high temperature regions and low temperature regions. These works have not been done or pursued with the existing large interferometers for cosmic objects, since the time available for solar observations is limited to about 10 % or so of the total machine time.

In this paper, we will describe a summary on the basis of published papers and on the works in progress. This will enhance the understanding on limitations in performance and utilities of the Nobeyama Radioheliograph, which are not widely well known among solar physics communities.

After two years of the operation of the Nobeyama Radioheliograph I would like to mention and point out some impression, remarks, etc. here. It is now well known by scientists associated with Yohkoh and with the Nobeyama Radioheliograph as well as by specialists with dedicated solar instruments that solar activity in this cycle decreased very sharply at a few epochs, among which are April 1992, October 1992, and November 1993. For example, Yohkoh observed a good number of X-class events during the first year after the start of observations in October 1991. The Nobeyama Radioheliograph started its observation in June 1992, and both instruments shared a number of events in common for about one year, but not so many flares after the beginning of the second year of common observations.

We were very fortunate in this context that we could construct the Nobeyama Radioheliograph exactly along the master schedule. This was accomplished by the efforts of many people involved in the Nobeyama Radioheliograph project. It should be mentioned here that the redundancy of the array configuration was another accelerating factor to have all the antenna system fine-tuned in a very short time of two months from the first fringe on 1992 February 27 to the first image on 1992 April 23.

The duty cycle of Yohkoh, or a low-altitude satellite, is one hour in one and half hour period, whereas that of the Nobeyama Radioheliograph is eight hours in a day, or in twenty-four hours. These differences produce versatile effects to various activities on the Sun. For example, the coverage will be best for long duration events such as LDE and GRF, but the coverage is purely statistical for impulsive flares. The situation is worse for such events, which give rise in a secondary or weaker active region, since Yohkoh has an automatic observing region selection function (ARS)[Ts1] , which choose the most intense active region, where the partial frame is placed. In such a case, we cannot expect to have partial frame images for the very initial phase of the relevant event.

With these backgrounds in mind it will be easy to understand that an appreciable number of papers and studies have been presented so far at this stage and we can draw some definite conclusions for gradual phenomena, but the cases are not sufficient for other categories of events.

2 Major Specifications and Characteristics of the Nobeyama Radioheliograph

The Nobeyama Radioheliograph is a 17-GHz radio interferometer dedicated for solar observations, which was completed in two years at Nobeyama, Nagano about 160 km west of Tokyo in March 1992. It consists of eighty-four 80-cm-diameter antennas arranged in a Tee-shaped array extending 490 m in east- west and 220 m in north-south directions. Since late June 1992, radio full-disk images

Table 1. Characteristics and Performance of the Nobeyama Radioheliograph

Observing frequency	17 GHz ($\lambda = 1.7635$ cm)
Bandwidth	33.6 ± 0.9 MHz
	asymmetry among antennas:< 0.6 dB
Field of view	40 arcmin
Spatial resolution	10 arcsec
Temporal resolution	1 s for entire observation period
	50 ms for selected events
Dynamic range of images	≥ 20 dB for snapshot
	≥ 30 dB for rotational synthesis
Observing period	± 4 H around local noon at about 0245 UT
Sensitivity in solar observations	4.4×10^{-3} sfu* or 1300 K
	for 1-s snapshot
	7.3×10^{-5} sfu or 22 K
	for 1-h rotational synthesis
Polarization	Both circular polarizations
	(time sharing in every 25 ms)
Isolation of polarization switch	≥ 20 dB
Overall phase stability	≤ 0.3 dB (rms)
Overall gain stability	≤ 0.2 dB (rms)

* 1 sfu $= 10^{-22}$Wm^{-2}Hz^{-1}

of the Sun have been obtained for 8 hours every day. The spatial resolution is 10 arcsec and the temporal resolution is 1 s and also 50 ms for selected events. Every 10 s, correlator data are synthesized into images in real time and displayed on a monitor screen. The array configuration is optimized for the observation of the whole Sun with high spatial and temporal resolutions as well as a high dynamic range of the images. An image quality of better than 20 dB is realized by incorporation of technical advantages in hardware and software, such as 1) Low-loss and phase-stable optical-fiber cables for local reference signal and IF signals, 2) newly developed phase-stable local oscillators, 3) custom CMOS gate- array LSI's of 1-bit quadra-phase correlators for 4 x 4 combinations, and 4) new image processing techniques to suppress large sidelobe effects due to the solar disk and extended sources [Na1], [Ni2], [Ha2]. Major specifications and characteristics are summarized here (cf. Table 1).

3 Transient Phenomena

3.1 Bursts

Impulsive Components. Impulsive bursts or components of bursts have been morphologically defined on the basis of time profiles in early days, when only radiometers were the available instruments of radio observations. Even though, when they are observed with the Nobeyama Radioheliograph, they are characterized by a highly polarized emission as high as 30 % to 60 % for the sample events described below, and a compact structure of the emitting regions. These characteristics well outline the non-thermal nature of the radio emission, namely gyrosynchrotron emission from spiraling electrons in sunspot magnetic fields. It was often questioned, as mentioned in the requirement and design of the Nobeyama Radioheliograph [En1], whether or not the same nonthermal electron population is responsible for both radio and hard X-ray emission in an impulsive phase. Where is the location of the acceleration site of nonthermal electrons for impulsive radio bursts? Observations by the Nobeyama Radioheliograph do not yet reveal samples that can answer these questions. Based on the papers published so far and on some preliminary analyses, individual cases are reviewed hereafter. Impulsive flares often show a foot point source, or sources, in the rising and maximum phases, and then a loop top source appears in the decay phase [En3], [Ta1], [Sh3], [Sh4]. The concept of the foot point source and the loop top source assumes a magnetic loop or loops, thermal plasmas, and an ensemble of nonthermal electrons. Combination of a loop and nonthermal electrons will result a double source with opposite polarity. Observations, however, do not present such a typical example, but a double source with the same polarity [En3],[Ta1], a single source structure in the 1992 October 8 events [Sh3] and of Source 1 in the 1992 September 6 event [Ya1]. Possible explanations will be 1) a tiny and unresolved loop of 5 arcsec or less involved in the impulsive phase, and/or 2) a propagation effect. A tiny loop is speculated for the 1992 August 12 event [Ta1] and the 1992 August 17-18 event [En3]. A propagation effect is considered [Tk1] for the flare on 1992 October 27 to explain the geometry of the HXT double source and the left-handed circularly polarized single source.

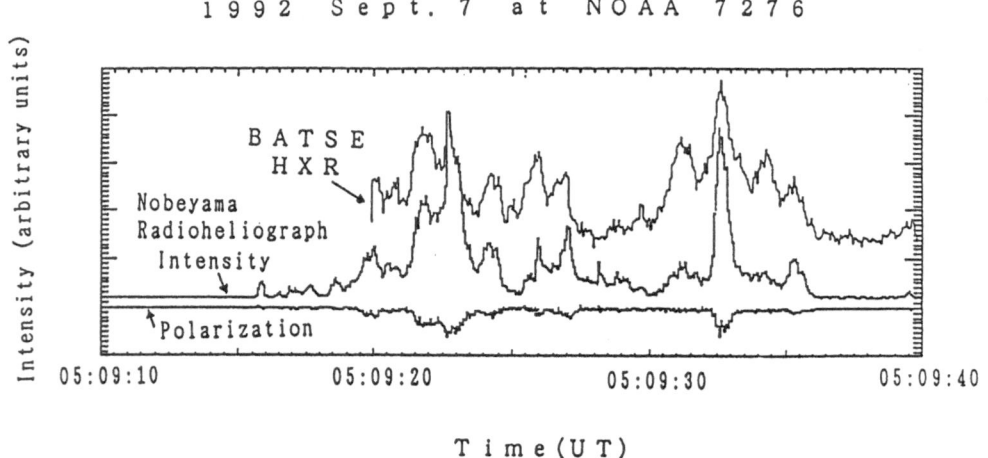

Fig. 1. Time profiles of a sub-second event on 1992 September 7.

Sub-Second Components. As described in Section 2 on the specifications and characteristics of the Nobeyama Radioheliograph, the Nobeyama Radioheliograph is featured with unprecedented high space-time resolutions during routine observations. Very simple or naive reasoning of 50 ms time resolution is that the spatial resolution of 10 arcsec divided by 50 ms gives 1.4×10^5 km s^{-1}, which corresponds to the possible velocity of the relevant electrons to emit 17 GHz radiation during the impulsive phase of flares through the gyro-synchrotron mechanism in a strong magnetic field. This means that if a bulk of electrons at deka keV energy is streaming as an ensemble through a magnetic loop, we shall be able to trace its motion with this performance. A short-term survey was conducted by [Ta2] using correlation time profiles. This program is conducted for a period of three months to select several sub-second spike events. Some of them are found to be associated with BATSE events of CGRO [Sc1]. One of such examples was observed at an early phase of an M1.0 flare at 0232 UT on 1992 August 12 in AR 7248. Typical rise and fall times are 50 - 100 ms with 1 - 2 s intervals. The size of the spikes is estimated to be 2 arcsec ($T_b \simeq 0.7$ MK and $P \simeq 10$ - 20 %) with the total span of the east-west linear array. An example of good time coincidence is illustrated in Figure 1 for 17 GHz and BATSE X-ray emission, which gives us a good reason to attribute incoherent mechanisms of gyro-synchrotron and bremsstrahlung for each radiation, respectively..

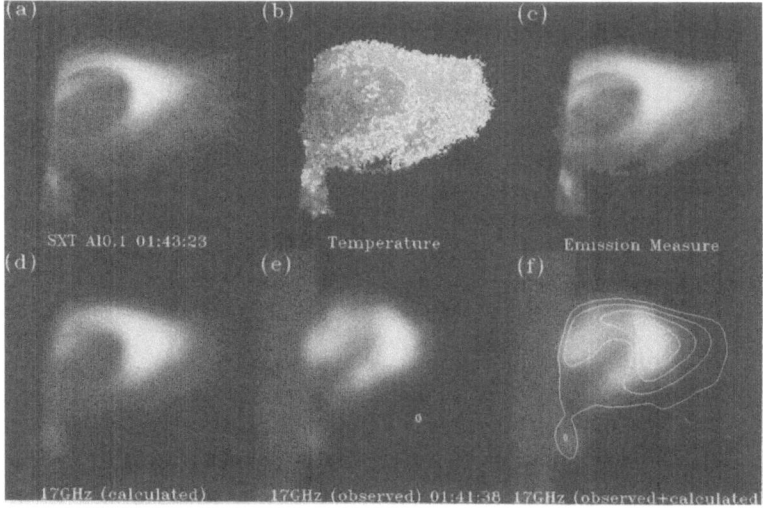

Fig. 2. (a) An SXT image of the LDE on 1993 March 16. (b) and (c) Temperature and emission measure maps derived from the SXT images. (d) Calculated intensity at 17 GHz. (e) Observed radio image of the LDE at 17 GHz. (f) Comparison between the observed (gray scale) and calculated (contour) radio images.

Long-Duration Events. A good sample of LDE events is obtained with the Nobeyama Radioheliograph. A part of it is analyzed by [Ha3]. It is shown that the foot-points are strongly polarized at 17 GHz and their brightness is well correlated with the soft X-ray brightness observed with the Yohkoh SXT. It is concluded that the 17-GHz emission is due to the gyro-resonance absorption at the third and fourth harmonics of the gyro-frequency, when high but not 100 % polarization is taken into account with a brightness temperature of one million degree. A temperature and emission measure analysis is made for the LDE on 1993 March 16 by employing the filter ratio method. With these physical quantities derived as a function of position, a calculation is made of the 17-GHz brightness distribution. A comparison of the observed and the calculated brightness distribution shows a brighter lower half for the former, which is interpreted by higher sensitivity for low-temperature plasma at the radio frequency, than for soft X-rays of SXT, Yohkoh, which implies the existence of a low-temperature plasma in the lower half of the LDE loop as shown in Figure 2.

X-Class Events. Two X-class flares were observed with the Nobeyama Radioheliograph on 1992 June 28 (X1.8) and 1992 November 2 (X9.0). Preliminary analyses [En2], [Na2] show that both flares occurred beyond the west limb of the Sun, therefore, the main bodies of the Hα flares were not seen. In radio waves, however, we can see weak activity about 20 and 10 minutes before the onset of the impulsive phase for both events. The degree of polarization at this pre-flare activities is higher than that at the main phase, and, if we look at the data carefully, the degree at the earlier pre-flare activity is higher than that at the later pre-flare activity. The height of the source is low at the earlier activity and high at the later activity, which is the same level as that at the main phase. Combination of polarization and position characteristics in the pre-flare phase activity and in the main phase emission suggest that the early pre-flare activity is due to heating at a gyro-resonance level, that the later pre-flare activity is due to lower-energy electrons at low harmonics, and that the main phase emission is due to higher-energy electrons at higher harmonics. A morphological analysis of the 1992 June 28 event is still in progress [Na2].

Gradual Components/Evaporation Source. On 1993 May 2 a gradual rise and fall type burst was observed with the Nobeyama Radioheliograph at the west limb. A preliminary analysis [Sh5] reveals that there occurs during the course of four hours the initial brightening at a loop top, gradual heating of one leg to its foot point, heating of the whole loop at the maximum phase, a gradual retreat to a one leg brightening of the original leg, and finally a brightening at the loop top with a rising motion. This sequence is illustrated in Figure 2. This is naively interpreted as an evaporation of chromospheric gas. For the determination of the temperature of the evaporating source simultaneous Yohkoh SXT observations are important, but unfortunately the SXT was not observing this event.

A similar event was also observed on 1992 June 28 [Ni1] at the east limb associated with an M-class flare. An evaporation source started from one footpoint of an arcade-like structure, where a nonthermal radio source was located. The frontend speed of the evaporating or elongating source is about 170 km s^{-1}.

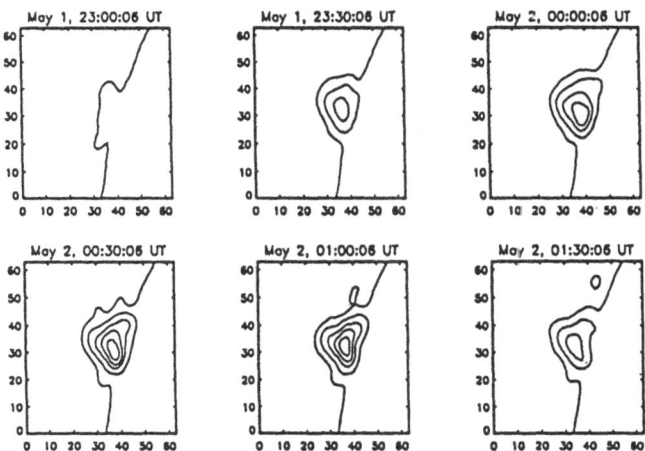

Fig. 3. One-leg evaporation source observed with the Nobeyama Radioheliograph at 17 GHz. Time interval is 30 minutes and total duration is 2 hours and 30 minutes.

The observed geometry suggests a magnetic loop tying the nonthermal and thermal sources, and the thermal source is generated from the nonthermal source. A quantitative analysis is presented on the energetics to show that a continuous supply of nonthermal electrons for 1000 s can maintain the hot plasma at the loop top [Ni1]. Another example of a gradual rise and fall type event was observed with the Nobeyama Radioheliograph, the Owens Valley Radio Observatory (OVRO) Solar Array (1 - 15 GHz), and the Soft X-ray Telescope of Yohkoh [Ga1]. It was also associated with some spikes or impulsive components in the initial phase at 1 - 7 GHz and 5 - 12 GHz, and in the maximum phase at 5 - 17 GHz. SXT images show a multiple loop structure of 20 - 80 arcsec, and the 17 GHz images show a global agreement with the SXT images, whereas the OVRO images show a double structure at the impulsive phases at 6 - 9 GHz with one subsource at one foot point of the loop. This is explained that in an explosive energy release phase a multiple loop system is involved, and a dense loop will be formed giving rise to the gradual phase as many other examples observed at OVRO. A more complex gradual event on 1993 March 4 was analyzed to involve four components [Sh3]. One foot point of the left-handed circular polarization source is connected to two foot points of right-handed circular polarization, with an unpolarized fourth source at the loop top.

Gradual-Hard Components. It is well established that there is a class of flares associated with "gradual microwave/hard X-ray bursts" as depicted by time profiles at microwaves and hard X-rays [Cl1]. A few examples have been

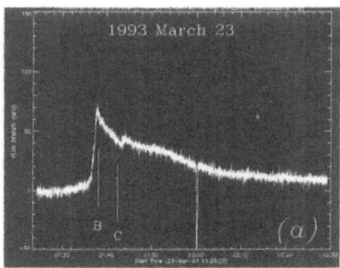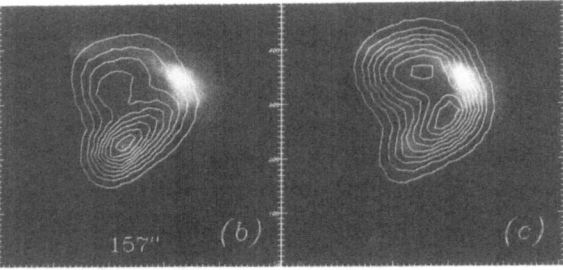

Fig. 4. Panel a – time profile, panel b – 17 GHz image (contour plot) and SXT image (gray scale) at the maximum phase, and panel c – at the declining phase.

analysed with images observed with the Nobeyama Radioheliograph and SXT and HXT of Yohkoh [Na3]. Images of the 1993 March 23 event present for the first time clear observational evidence that the radio source displays a loop-like extended structure of about 1 arcmin with weaker emission at the loop top, whereas soft X-ray images of Yohkoh at about 15 minutes earlier and 30 minutes later show a loop top source.

The non-uniform structure of the radio source is explained in terms of a non-uniform magnetic field structure as demonstrated by [Ch1]. This work is an extension of a uniform field model calculation [Du1] to apply a numerical integration of the radiative transfer equation for radio radiation from an ensemble of electrons trapped in magnetic fields generated from a solenoid coil model [Sa1]. The radio intensity is proportional to the fifth and seventh power of the field strength for Dulk's uniform field model and Choi's non-uniform model for a spatially homogeneous energetic electron distribution.

Initial Phase Enhancement. An impulsive flare of 1992 September 6 was studied with the Bragg Crystal Spectrometer (BCS) onboard Yohkoh, and compared with Hard X-ray Telescope data and with the Nobeyama Radioheliograph images [Ka1]. During the initial phase of the flare the Ca XIX spectra of BCS show a 0.8 keV temperature plasma with $100\,\mathrm{km\,s^{-1}}$ turbulence. During this initial phase hard X-ray emission was not detected by HXT, but 17 GHz emission was detected with the Nobeyama Radioheliograph with 50 % circular polarization, which was first reported by [Tn1]. In the later rising phase the polarization degree decreases very rapidly to 20 % and recovers to 30 % in the decay phase. This indicates that the origin of strongly polarized radio emission at the initial phase is due to gyro-resonance absorption by the thermal plasma detected with BCS, whereas radio emission at the rising phase is coincident with hard X-ray emission and is due to gyro-synchrotron emission from nonthermal electrons. Radio emission in the decay phase will be a combination of thermal and nonthermal emission.

Transient Brightening. A new class of transient brightening was discovered [Sh4], which did not correlate with soft X-ray emission, but occurred within the umbra of the strong and complex sunspot of AR 7654 on 1994 January 21. This is interpreted as transient heating at the 2000 Gauss level, which emits gyro-resonance absorption radiation. The source of this transient heating is considered as reconnection of magnetic fields. The emission is visible only with 5-arcsec one-dimensional array but not with the 10-arcsec two-dimensional pencil beam. This suggests that the size of the source is 5 arcsec or less. The flux density is of the order of a few sfu, which corresponds to a few to several MK brightness temperature. If we consider the fact that the polarization degree is for many cases 100 %, the electron temperature of the heated region will be higher than the corresponding brightness temperature. If this estimation is incorporated with the typical duration of a few minutes for this type of events, heating of the 2000 Gauss atmosphere up to a few tens MK and cooling in such a time period is a current physical picture derived from this new phenomenon. In this phenomenon the foot points are located within the umbra, whereas in other cases [Go1] the footpoints are not rooted in the umbra but in the penumbra, and the radio brightening observed by the VLA is associated with spirals in the SXT images.

3.2 Active Regions

Gyroresonance Sources. A purely polarized slowly varying component at 17 GHz was detected with the Nobeyama Radioheliograph [Sh1]. This radio emission was associated with the active region 7260, which was accompanied by a large sunspot. The field strength of this sunspot was measured at the Mees Solar Observatory, University of Hawaii. The contour level of 2000 Gauss is in good agreement in shape and area with the polarized radio emission at 17 GHz. This implies that the radio emission is due to the gyro-resonance absorption at the third harmonic. It should be pointed out that soft X-ray emission was not detected with the Soft X-ray Telescope (SXT) of Yohkoh corresponding to the polarized radio source, which is consistent with the fact that appreciable emission was detected due to the free-free mechanism. It is also evident that spiraling loops as a part of a whirl concentrating along the axis of the sunspot is seen in SXT images. This implies that a strong electric current is existing along the axis, which is consistent with the vector magnetogram data.

Polarization Reversal. It has been well known that when a propagation condition is satisfied as a function of the magnetic field direction and intensity, and electron density, the sense of polarization does change, which is called mode coupling, that does not preserve the magnetic field and polarization sense at the origin of radiation [Co1]. Observational evidence has not been presented yet. The Nobeyama Radioheliograph was the first to demonstrate the evidence of mode coupling at 17 GHz for the two active regions AR 7260 and 7654 [Sh6]. The AR

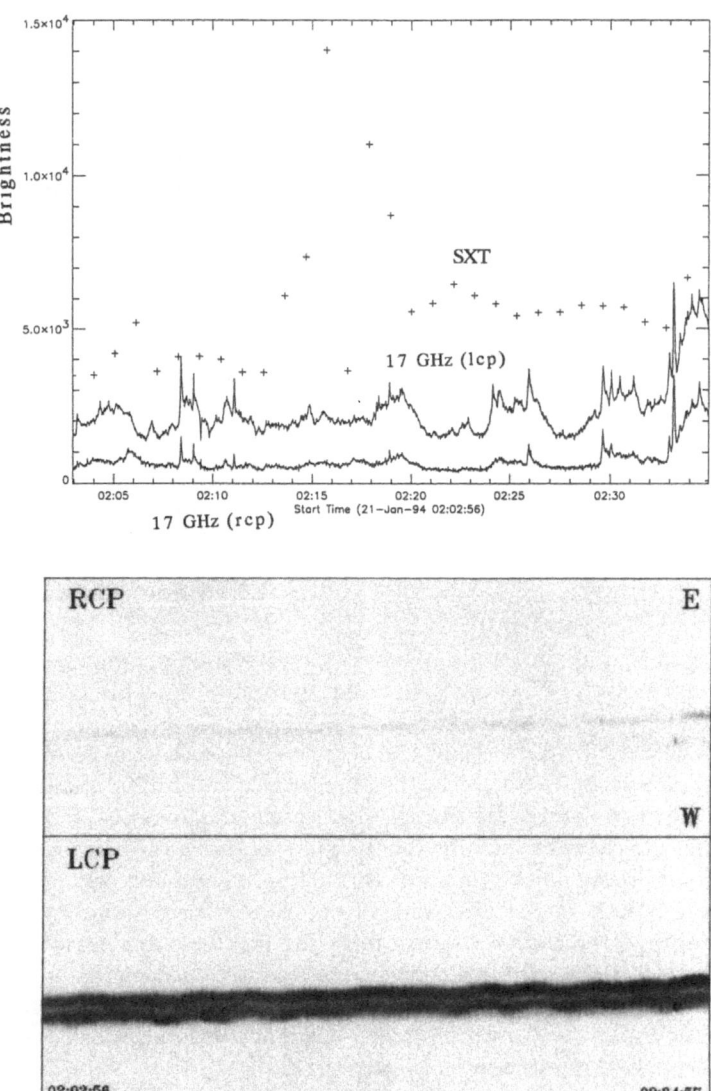

Fig. 5. a: Time profiles of the transient event on 1994 January 21 in soft X-ray emission, radio correlation outputs in LCP and RCP are plotted in arbitrary scale as a function of time for 30 minutes on 1994 January 21. **b:** Output of 16th harmonics for full span of the east-west one-dimensional array in vertical direction as a function of time for the above time period. The vertical scale for the frame is 150 arcsec. The transient brightenings are seen in this figure superposed on the stationary emission, which corresponds to the slowly-varying component of the AR 7654.

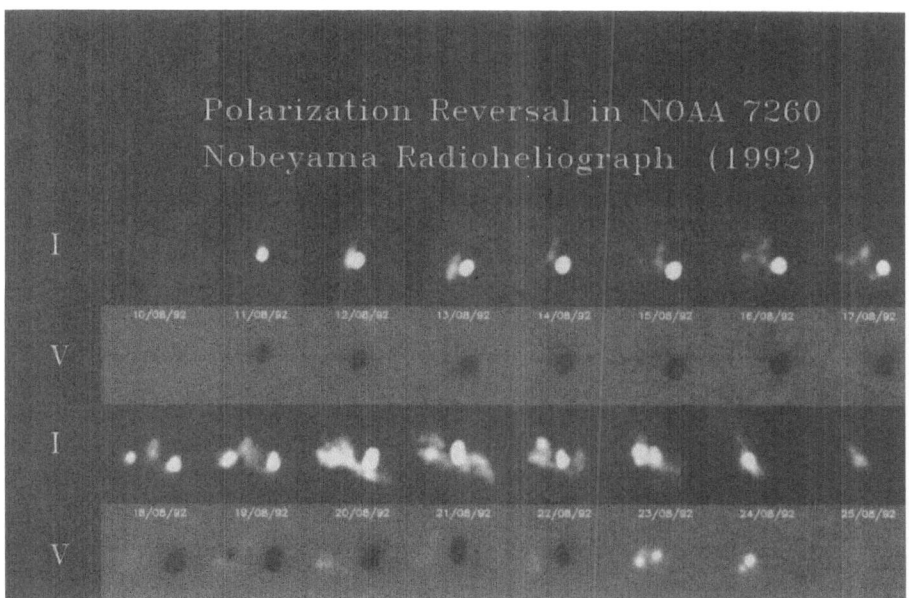

Fig. 6. Evolution of active region 7260 across the solar disk mapped at 17 GHz in intensity and polarization (R-L)

7260 passed the central meridian on 1992 August 18 and the polarization reversal occurred between August 22 and 23, when the spot was located at 60 degree west. The preceding spot had negative magnetic polarity, which corresponds to the left-handed circular polarization at 17 GHz before August 22 and to the right-handed circular polarization after August 23 as illustrated in Figure 6. In the case of AR 7654 the reversal occurred at 50 degree west. Both active regions did not show much change of the magnetic field and had a very strong degree of radio polarization. A quantitative study is in progress for the field configuration, field intensity, and ambient electron density, etc..

3.3 Flare-productive Active Regions

A preliminary study is presented on the relations between the evolution of the active region AR 7515 and its flare productivity [Ku1] in terms of active region loop configurations, interaction of loops, burst time profiles, source structures in radio and X-rays, etc..

Another example of an evolution study is briefly described concerning AR 7321 by [Ni3]. The radio features of this active region are delayed compared with the soft X-ray features, which is interpreted that heating of plasma preceded the magnetic field appearance above the active region. During the passage of the active region across the disk from 1992 October 25 through 1992 November 2 more than 20 bursts were recorded at Nobeyama and at Toyokawa, which showed a high turnover frequency, and a number of absorptions was observed. These facts suggest the existence of a strong magnetic field of 1000 Gauss and of a continuous emergence of strong fields.

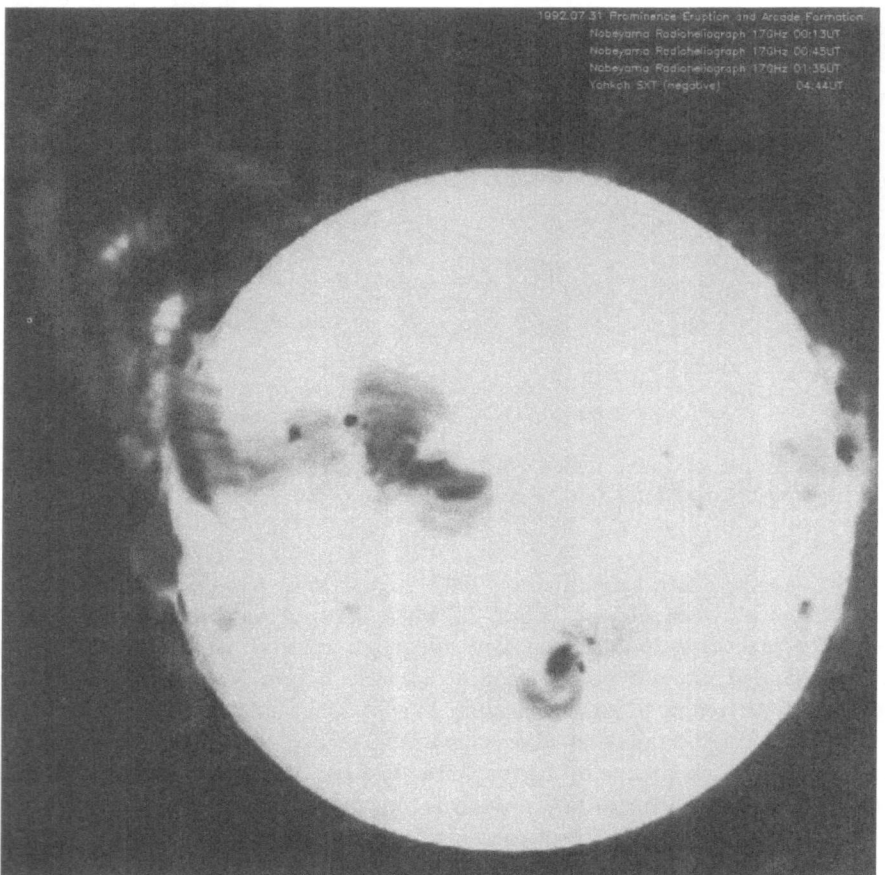

Fig. 7. A composite image of the eruptive prominence on 1992 July 30–31 in radio waves at 17 GHz outside the disk at 0013, 0045 UT and in soft X-rays on the disk at 0135UT.

3.4 Prominence Eruptions

CSHKP Model/Disparition Brusque. A prominence eruption followed by a coronal brightening was simultaneously observed in radio (17 GHz), soft X-rays, and Hα on 1992 July 30–31. The observations were performed by newly developed high-performance instruments such as the Nobeyama Radioheliograph, the SXT onboard Yohkoh, the Flare Monitoring Telescope of the Hida Observatory, and some other Hα telescopes [Ha1]. This event gives us an unprecedented wealth of observational pieces of information for this type of phenomena, namely disparition brusque, when occurred on the disk. The prominence erupted at a velocity of 100 km s^{-1} seen at radio and at Hα. During its eruption the total radio flux of the prominence did not change very much, which suggests a decrease of the surface filling factor of the prominence. The associated coronal arcade was clearly observed in soft X-rays by SXT/Yohkoh and seen in radio waves at 17 GHz. Temperature and emission measure time profiles were obtained by combining soft X-ray and radio brightness. The mean temperature was 3.8 MK in an early phase and decreased to 2.6 MK in seven hours. The emission measure at the maximum was 1.6×10^{48} cm^{-3} after 3 hours from the start of the arcade brightening. A magnetic field configuration is inferred in terms of the Carmichael/Sturrock/Hirayama/Kopp-Pneumann (CSHKP) model.

4 Radiation Mechanisms

Micro-processes, Energy Distributions, and Radiation Mechanisms. It is well known that when a charged particle is accelerated, emission of radiation takes place and decelerated, absorption of radiation gives rise. This is true for an ensemble of electrons which are involved in the process. Acceleration or deceleration is executed by the electric force between charged particles or by the magnetic force when a charged particle is moving in the field. The former and latter process are relevant to the collision between particles and to the Lorentz force to a gyrating electron. When an ensemble of electrons are considered, 1) collisions in a thermal energy distribution gas will produce free-free emission, 2) magnetic gyration of electrons in a thermal energy distribution gas will produce gyro-resonance emission and absorption, 3) magnetic gyration of electrons in a power-law energy distribution gas will produce gyro-synchrotron emission and absorption. All these processes are incoherent, and superposition of fundamental processes is possible. In order to treat emission and absorption processes simultaneously we will introduce the transfer equation.

At radio frequencies a thermal plasma is opaque for the temperature range from 10^4 K to 10^7 K, so we can detect free-free emission from a prominence, absorption from a dark filament, free-free emission from an active region and from a flare plasma at 10^7 K. These radiations are characterized by a low degree of circular polarization, since the collision is the primary source of the radiative process.

If we have a very strong magnetic field at the radiation site, thermal emission and absorption will be strongly affected by the magnetic field and radiations show a high degree of circular polarization. The sense of polarization is coupled with the magnetic polarity, so that positive and negative magnetic polarity corresponds to left-handed and right-handed circular polarization, respectively. This rule corresponds to the magneto-ionic wave modes and is related to the sense of electron gyration. In an active region the third harmonic gyro-radiation is effective and this corresponds to a magnetic field of 2000 Gauss at 17 GHz. In the post-flare loops the electron temperature will be much higher than that in an active region, and radiations at the fourth harmonic of the gyro-frequency will be effective. Pre-flare, pre-impulsive, and post-flare loop heatings are suggested to be associated with this emission mechanism.

When an ensemble of electrons are gyrating above a sunspot, gyro-synchrotron emission will be detected, the degree of polarization is not so high as in gyro-resonance radiation from a thermal plasma, since the harmonic number will be much higher than in the thermal case. So far we do not have any single phenomenon, which is not explained by one of these three radiation mechanisms. This means that at 17 GHz we have not yet observed a coherent phenomenon. The sub-second source was a suspected candidate, but it turned out that it is incoherent emission since it is associated with X-ray emission detected by BATSE/CGRO as mentioned earlier [Ta2], [Sc1].

5 High-Dynamic Imaging

A theoretical analysis is made of the correlator outputs of the Nobeyama Radioheliograph to demonstrate the evaluation of its imaging performance [Ko1]. After removing antenna-based errors by the self-calibration method based on the redundancy of the antenna configuration, correlator-based errors are analysed in terms of four factors 1) unequal bandpass characteristics, 2) antenna pointing errors, 3) delay errors, and 4) noise at the correlator outputs. The expected dynamic range of synthesized snap-shot images is found to be 30 dB based on these correlator-based errors, while the actual images have a 25 dB dynamic range. This difference is mostly due to the image-restoration procedure currently adopted [Ha2]. Efforts have been made to improve this difference in theoretical and actual performances. It is farther important to obtain high-dynamic imaging for the chromospheric features, which are diffuse in structure and of low contrast in brightness in positive and in negative directions to the mean level of the quiet Sun.

A promising example of the high dynamic imaging is obtained by improving the AIPS imaging software system and applying the maximum entropy method (MEM) to a set of calibrated correlator-outputs of the Nobeyama Radioheliograph [Ba1]. The following image is one of such tests. This map shows fine structures on the chromosphere such as a dark filament/prominence system, chromospheric networks, patchy polar-cap brightenings, as well as emission associated with active regions including gyro-resonance and free-free sources. The noise level of this image is estimated to be as low as 200 K in contrast to the quiet Sun brightness temperature of 10^4 K, which corresponds to 27 dB. The

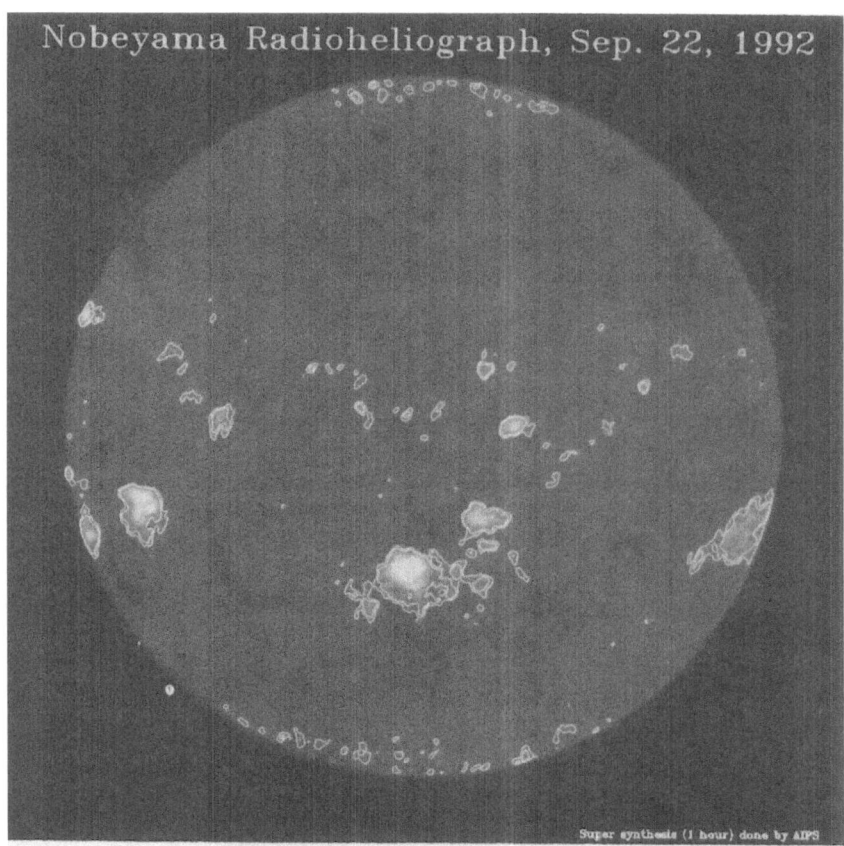

Fig. 8. An example of a high-dynamic imaging map of 1992 September 22 observed with the Nobeyama Radioheliograph processed by Tim Bastian (NRAO) with the AIPS and MEM. The estimated image quality is 27 dB. Several features are clearly visible in this map such as patchy polar-cap brightenings, chromospheric networks, a dark filament/prominence system, active regions, a gyro- resonance slowly-varying component, a giant spicule, etc.

AIPS image processing software package is being installed at the Nobeyama Radio Observatory for the version 15JULY94.

6 Collaborations

Open-use policy has been presented at the Kofu meeting for the Nobeyama Radioheliograph data [Sh2]. Several collaborative studies along this policy are in progress between the host group and guest groups [Ku1], [Ku2], [Ku3], [Wh1]. Anyone interested in monitoring the Nobeyama Radioheliograph data can access

them through anonymous FTP, if such facility is available. Results of such collab-
orations will shortly appear on scientific journals. First detection of radio bright
points at 17 GHz was explored [Ku2]. A second example of "radio-rich" events
is discovered [Ku3]. Soft X-ray active-region transient brightenings (ARTBs) are
studied at 17 GHz to have common thermal emissions without nonthermal radio
waves nor hard X-rays [Wh1], which suggests a fundamental difference between
flares and ARTBs that the former is considered to be a process of nonthermal
particle acceleration and of thermal relaxation, whereas the latter is a thermal
process. Several other proposals have been accepted as collaborative works, and
these studies are in progress.

7 Future Plan

Dual Frequency Observation. A dual-frequency observation program is in
progress as a two-year program by employing a frequency-selective surface (FSS)
as a subreflector and a 34-GHz receiver for each antenna. In this geometry of
optics the Coude focus is used at 17 GHz and the prime focus is for the 34
GHz. A new set of 84 34-GHz receivers is planned to be fabricated this year, and
fabrication of 84 units of FSS is planned with installation of each FSS and each
receiver to each antenna for the second year. Current, and possibly final goal
and requirement of specification is that installation of FSS will not deteriorate
the current performance of the 17 GHz system.

8 Conclusions

We have summarized in this article initial results of the Nobeyama Radioheliograph. It is shown that extended structures such as long duration events,
gradual-hard events, gradual rise-and-fall events, etc. are well resolved spatially
as well as temporally. It is, however, not well established yet, what are the spatial developments observationally for very impulsive phenomena especially at
the initial phase, since for those events the spatial structure is generally very
small and sometimes beyond the limitation of spatial resolution and beyond the
dynamic range of image quality. Introduction of AIPS software and application
of MEM will enhance the image quality of our maps up to the instrumental limit
or 5 dB for isolated sources and more for diffuse or extended structures. This
image processing will be extensively applied to the correlation data observed so
far to deduce much improved images.

The installation of a dual-frequency observation system will enable us to
increase the spatial resolution by a factor of two and to obtain spectral information.

Collaborative researches with other radio instruments, which have characteristics complementary to the Nobeyama Radioheliograph are very important to
increase the frequency coverage, spatial resolution coverage, etc., which give us
sufficient pieces of information for plasma diagnostics.

Acknowledgments The Nobeyama Radioheliograph antenna site is available on the basis of the agreement between the Department of Agriculture, Shinshu University and the Nobeyama Radio Observatory, NAOJ.

References

[Ba1] Bastian, T., et al., to be published

[Cl1] Cliver, E. W.: Secondary Peaks in Solar Microwave Outbursts, Solar Phys. **84** (1983) 347-359

[Ch1] Choi, Y. S.:PhD Dissertation:A Model Calculation of Solar Microwave Burst Structure and a Comparison with Observations, University of Tokyo, 1994

[Co1] Cohen, M. H.: Magnetoionic mode coupling at high frequencies, Astrophys. J. **131**(1960) 664-680

[Du1] Dulk, G. A.: Radio Emission from the Sun and Stars in Ann. Rev. Astron. Astrophys.(1985) **23** 169-224

[En1] Enome, S.: Nobeyama Radioheliograph in *Lecture Notes in Physics*, **387** Flare Physics in Solar Activity Maximum 22, (1991) 330-337

[En2] Enome, S. and Radioheliograph Group: An X9 Flare of Nov. 2, 1992 in *X-Ray Solar Physics from Yohkoh*, eds. Uchida, Y., Shibata, K., Watanabe, T., and Hudson, H. S., Universal Academy Press (Tokyo), 109-110

[En3] Enome, S., Nakajima, H., Shibasaki, K., Nishio, M., Takano, T., Hanaoka, Y., Torii, C., Shiomi, Y., and 12 coauthors: Alignment of Radio, Soft X-ray, Hard X-ray Images of Sources in Impulsive and Gradual Phases of the Flare of 1992 August 17-18. Publ. Astron. Soc. Japan **46**(1994) L27-L31

[Ga1] Gary, D., Enome, S., and Bruner, M.: OVRO and NRO observations of the solar flare on 1993 June 3 in *Proceedings of Kofu Symposium: New Look at the Sun with Emphasis on Advanced Observations of Coronal Dynamics and Flares*, eds. Enome S. and Hirayama, T., NRO Report No. 360 (1994) 165-168

[Go1] Gopalswamy, N., Schmahl, E. J., Kundu, M. R., Lemen, J. R., Strong, K. T., Canfield, R. C., and Beaujardiere, J. de La: Study of Active Region Magnetic Field Structures using VLA Radio, Yohkoh X-ray and Mees Optical Observations in *Proceedings of Kofu Symposium: New Look at the Sun with Emphasis on Advanced Observations of Coronal Dynamics and Flares*, eds. Enome S. and Hirayama, T., NRO Report No. 360 (1994) 347-351

[Ha1] Hanaoka, Y., Kurokawa, H., Enome, S., Nakajima, H., Shibasaki, K., Nishio, M., Takano, T., Torii, C., and 12 coauthors: Simultaneous Observations of a Prominence Eruption Followed by a Coronal Arcade Formation in Radio, Soft X-rays, and Hα Publ. Astron. Soc. Japan **46**(1994) 205-216

[Ha2] Hanaoka, Y., Shibasaki, K., Nishio, M., Enome, S., Nakajima, H., Takano, T., Torii, C., Sekiguchi, H., and 9 coauthors: Processing of the Nobeyama Radioheliograph Data in *Proceedings of Kofu Symposium: New Look at the Sun with Emphasis on Advanced Observations of Coronal Dynamics and Flares*, eds. Enome S. and Hirayama, T., NRO Report No. 360 (1994) 35-43

[Ha3] Hanaoka, Y.: Long Duration Events Observed with the Nobeyama Radioheliograph in *Proceedings of Kofu Symposium: New Look at the Sun with Emphasis on Advanced Observations of Coronal Dynamics and Flares*, eds. Enome S. and Hirayama, T., NRO Report No. 360 (1994) pp. 181-184

[Ka1] Kato, T., and Fujiwara, T., BCS Group: BCS spectra from flares on 6th September 1992 in *Proceedings of Kofu Symposium: New Look at the Sun with Emphasis on Advanced Observations of Coronal Dynamics and Flares*, eds. Enome S. and Hirayama, T., NRO Report No. 360 (1994) 191- 194

[Ko1] Koshiishi, H., Enome, E., Nakajima, H., Shibasaki, K., Nishio, M., Takano, T., Hanaoka, Y., Torii, C., and 6 coauthors: Evaluation of the Imaging Performance of the Nobeyama Radioheliograph. Publ. Astron. Soc. Japan 46(1994) L33-L36

[Ku1] Kundu, M. R., Shibasaki, K., Enome, S., Nitta, N., Burner, M., Sakao, T., and Kosugi, T.: Evolution of an Active Region and Flare Productivity in *Proceedings of Kofu Symposium: New Look at the Sun with Emphasis on Advanced Observations of Coronal Dynamics and Flares*, eds. Enome S. and Hirayama, T., NRO Report No. 360 (1994) 353- 356

[Ku2] Kundu, M. R., Shibasaki, K., Enome, S., and Nitta, N: Detection of 17 GHz Radio Emission from X-ray Bright Points. Astrophys. J. Letter (1994) to be published

[Ku3] Kundu, M. R., White, S. M., Nitta, N., Shibasaki, K., Bruner, M., and Enome, S., Sakao, T., and Kosugi, T.: "Radio-Rich" Solar Microwave Bursts. Astrophys. J. Letter (1994) submitted

[Na1] Nakajima, H., Nishio, M., Enome, S., Shibasaki, K., Takano, T., Hanaoka, Y., Torii, C., Sekiguchi, H., and 9 coauthors: The Nobeyama Radioheliograph. Proceedings of IEEE 82(1994), 705-713

[Na2] Nakajima, H., Enome, S., Shibasaki, K., Nishio, M., Takano, T., Hanaoka, Y., Torii, C., Shiomi, Y., and 6 coauthors: Time Development of the 1992 June 28 X-class Flare in *X-Ray Solar Physics from Yohkoh*, eds. Uchida, Y., Shibata, K., Watanabe, T., and Hudson, H. S., Universal Academy Press (Tokyo), 63-64

[Na3] Nakajima, H., Enome, S., Shibasaki, K., Nishio, M., Takano, T., Hanaoka, Y., Torii, C., Shiomi, Y., and 5 coauthors: Morphological development of gradual nonthermal microwave flares in *Proceedings of Kofu Symposium: New Look at the Sun with Emphasis on Advanced Observations of Coronal Dynamics and Flares*, eds. Enome S. and Hirayama, T., NRO Report No. 360 (1994) 185-189

[Ni1] Nishio, M., Nakajima, H., Enome, S., Shibasaki, K., Takano, T. Hanaoka, Y., Torii, C., Sekiguchi, H., and 9 coauthors: Radio Imaging Observations of the Evolution of Thermal and Nonthermal Source during a Gradual Solar Burst. Publ. Astron. Soc. Japan 46(1994) L11-L15

[Ni2] Nishio, M., Nakajima, H., Enome, S., Shibasaki, K., Takano, T., Hanaoka, Y., Tori, C., Shiomi, Y., and 23 coauthors: The Nobeyama Radioheliograph -Hardware System- in *Proceedings of Kofu Symposium: New Look at the Sun with Emphasis on Advanced Observations of Coronal Dynamics and Flares*, eds. Enome S. and Hirayama, T., NRO Report No. 360 (1994) 19-33,

[Ni3] Nishio, M., Takakura, T., Ikeda, H., Nakajima, H., Enome, S., Shibasaki, K., Takano, T., Hanaoka, Y., and 2 coauthors: Evolution and Radio Activity of a Flare Productive Active Region NOAA 7321 in *Proceedings of Kofu Symposium: New Look at the Sun with Emphasis on Advanced Observations of Coronal Dynamics and Flares*, eds. Enome S. and Hirayama, T., NRO Report No. 360 (1994) 151-155

[Sc1] Schwartz, R.: (1994) private communication. We acknowledge the BATSE team headed by Dr. Gerald Fishman, and C. Meegan, R. Wilson, and W. Paciesas as well as the Compton Observatory Science Support Center for BATSE high time resolution data available.

[Sa1] Sakurai, T. and Uchida, Y.: Magnetic Field and Current Sheets in the Corona above Active Regions. Solar Phys.52(1977), 397-416

[Sh1] Shibasaki, K., Enome, S., Nakajima, H., Nishio, M., Takano, T., Hanaoka, Y., Torii, C., Sekiguchi, H. and 8 coauthors: A Purely Polarized S-Component at 17 GHz. Publ. Astron. Soc. Japan 46(1994) L17-L20

[Sh2] Shibasaki, K., Enome, S., Nakajima, H., Nishio, M., Takano, T., Hanaoka, Y., Torii, C., Sekiguchi, H., and 5 coauthors, 1994: The Nobeyama Radioheliograph Data use in *Proceedings of Kofu Symposium: New Look at the Sun with Emphasis on Advanced Observations of Coronal Dynamics and Flares*, eds. Enome S. and Hirayama, T., NRO Report No. 360 (1994) 45-51

[Sh3] Shibasaki, K., Enome, S., Nakajima, H., Nishio, M., Takano, T., Hanaoka, Y., Torii, C., Sekiguchi, H., and 5 coauthors: Structural Changes of radio sources during early phase of small bursts in *Proceedings of Kofu: Symposium: New Look at the Sun with Emphasis on Advanced Observations of Coronal Dynamics and Flares*, eds. Enome S. and Hirayama, T., NRO Report No. 360 (1994) 205-208

[Sh4] Shibasaki, K., Takano, T., Enome, S., Nakajima, H., Nishio, M., Hanaoka, Y., Torii, C., Sekiguchi, H., and 5 coauthors: Space Sci. Rev. **68** (1994) 217-224

[Sh5] Shibasaki, K., et al.: (1994) to be submitted to the Proceedings of COSPAR

[Sh6] Shibasaki, K., et al.: (1994) private communication

[Ta1] Takano, T., Enome, S., Nakajima, H., Shibasaki, K., Nishio, M., Hanaoka, Y., Torii, C., Sekiguchi, H., and 9 coauthors: Behavior of Accelerated Electrons in a Small Impulsive Solar Flare on 1992 August 12. Publ. Astron. Soc. Japan **46**(1994) L21-L25

[Ta2] Takano, T. et al.: (1994) private communication

[Tk1] Takakura, T. and the HXT group: Nishio, M. and the Radioheliograph group: Evolution of Flare Source Inferred from Hard X-ray and Radio Observations: Solar Burst on 27 October 1992 in *Proceedings of Kofu: Symposium: New Look at the Sun with Emphasis on Advanced Observations of Coronal Dynamics and Flares*, eds. Enome S. and Hirayama, T., NRO Report No. 360 (1994) 157-160

[Tn1] Tanaka, K: Impact of X-ray Observations from the Hinotori Satellite on Solar Flare Research. Publ. Astron. Soc. Japan **39**, (1987) 1-45

[Ts1] Tsuneta, S., Acton, L., Bruner, M., Lemen, J., Brown, W., Caravalho, R., Catura, R., and 3 coauthors: The Soft X-ray Telescope for the Solar-A Mission. Solar Phys. **136**(1991) 37-67

[Wh1] White, S. M., Kundu, M. R., Shimizu, T., Shibasaki, K., and Enome, S.: The Radio Properties of Solar Soft X-ray Active Region Transient Brightenings. Astrophysical J.(1994) to be submitted

[Ya1] Yaji, K., Kosugi, T., Sakao, T., Masuda, S., Inda-Koide, M., and Hanaoka, Y.: A comparison of hard X-ray, soft X-ray, and microwave sources in solar flares in *Proceedings of Kofu Symposium: New Look at the Sun with Emphasis on Advanced Observations of Coronal Dynamics and Flares*, eds. Enome S. and Hirayama, T., NRO Report No. 360 (1994) 143-146

Long-Duration Non-Thermal Energy Release in Flares and Outside Flares

Karl-Ludwig Klein

Observatoire de Paris, Section de Meudon, DASOP & CNRS-URA 1756,
F-92195 Meudon Principal Cedex, France

Abstract. X-ray and radio signatures of long-lived particle populations in the corona are presented, based on imaging observations and on model calculations of X-ray and microwave emission: gradual hard X-ray and associated radio emission from centimetric to hectometric waves, Moving Type IV bursts, Type II bursts, storm continua and noise storms. The energy budget of suprathermal particles is estimated, possible acceleration sites are investigated, and similarities and differences between impulsive flares, gradual events and noise storms are discussed.

1 Introduction

Energy release processes in the solar corona range from permanent heating to short-lasting (~minutes) impulsive flares. Impulsive flares are the best studied phenomenon of transient energy release, occurring on time scales from a few seconds down to some tens of milli-seconds. The significance of these time scales was recently discussed by Benz (1994) and Krüger et al. (1994).

Transient energy release in the corona is not restricted to impulsive flare durations: various radiation signatures show the presence of suprathermal particles in the corona over much longer durations (Section 2). This paper reviews such signatures during flares, with the example of gradual hard X-ray/radio bursts including metric Type IV bursts (Section 3), and outside flares as revealed by noise storms (Section 4). The characteristic radiative signatures favour attempts to ascribe long-lasting energy release to different processes than impulsive flares. The most widely accepted idea has been that re-acceleration of part of the energetic particles produced during the impulsive phase gives rise to prolonged hard X-ray and radio emission in some events. Using observations over a wide spectral range and model calculations of electromagnetic radiation, it is examined to which extent this distinction is borne out by the data. Earlier reviews relevant to the subject were given by Trottet (1986) and Melrose and Dulk (1987).

2 Long-Duration Radio and Hard X-ray Emission

The best established diagnostics of energetic particle populations in the corona are their γ-ray, hard X-ray and radio emission. Hard X-ray burst durations ($h\nu >25$ keV, Fig. 1, top right) show a continuous distribution up to at least

1 hour. Recent γ-ray observations reveal persistent emission up to several hours at hν >50 MeV (Leikov et al. 1993, Kanbach et al. 1993). Long before hard X-ray and microwave imaging became available, metric Type II bursts due to large-scale coronal shock waves and centimetric-to-metric Type IV emission had revealed the presence of non thermal electrons in the corona above ~0.5 R$_\odot$ over time periods much longer than typical impulsive flares. Figure 1 (left and bottom) displays different components of a Type IV burst (Wild 1970, Pick 1986):

1. Flare continuum is a broadband radio emission from centimetric to metric waves (12:39–12:47 UT). It will be discussed in Section 3.

2. After the flare continuum fades, post-gradual emission at decimetric/metric radio waves may continue for several tens of minutes or even hours ("storm continuum" or "stationary Type IV burst"). Storm continuum shares several characteristic features with noise storms: the long duration, the spectral range and the high degree of circular polarization. Noise storms are the topic of Section 4.

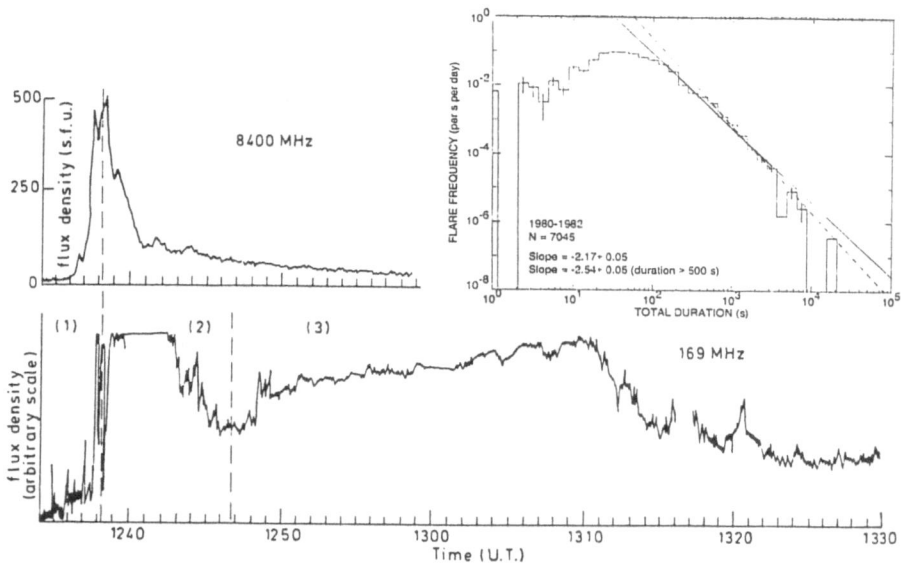

Fig. 1. Distribution of durations of excess hard X-ray emission above 25 keV detected by HXRBS/SMM (top right; Crosby et al. 1993). Microwave (top left) and metre wave (bottom) lightcurve of a flare (Klein et al. 1983) displaying impulsive (1), gradual (2) and post-gradual (3) emission phases.

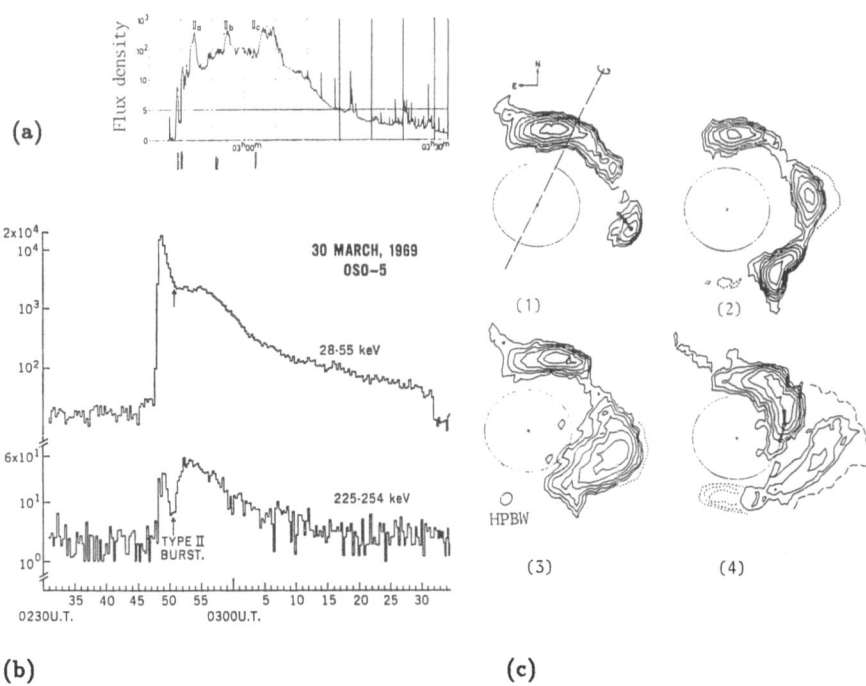

Fig. 2. Time history of the metric radio flux density (a; 8C MHz, Smerd 1970) and hard X-ray count rate (b; adapted from Frost and Dennis 1971) of a gradual burst. c. Source configuration at four instants (numbered in order of increasing time and indicated by vertical bars below the flux density curve in a) during the metric continuum emission. (Smerd 1970).

3 "Gradual" Hard X-ray/Radio Bursts

The relation between hard X-ray and metric radio emission is illustrated in Fig. 2. The hard X-ray count rate (Fig. 2.b) displays an initial impulsive peak, after which (~3:51 UT) a new rise occurs with significantly harder (flatter) photon spectrum, lasting more than 20 min above 200 keV and longer at lower energies. This long-lasting emission is generally referred to as "gradual" hard X-ray emission. The 80 MHz emission (Fig. 2.a) consists of a slowly evolving flare continuum, whose time profile resembles the gradual hard X-rays, and several more intense peaks. Because of the limited dynamic range of film records only the latter show up in the dynamic spectrum and were referred to as Type II (Smerd 1970). The flare continuum is visible on fixed frequency records, from microwaves (Badillo and Salcedo 1969) to 80 MHz.

Due to its prominent role in the dynamic spectrum, the "Type II" emission was subsequently considered as evidence that the coronal shock wave is the accelerator of the energetic electrons which emit the hard X-rays and the radio continuum (Frost and Dennis 1971). However, Figs. 2.a, b show the presumed Type II emission has no counterpart in the high-energy photon count rate, contrary to the flare continuum. Since the hard X-rays are due to electrons whose lifetime is much shorter than the duration of the event, Klein et al. (1983) concluded that the electrons emitting the gradual hard X-ray burst and the flare continuum are accelerated over a prolonged time period in the low corona, and that the coronal shock wave does not play a crucial role in this process. Kahler (1984) came to the same conclusion, arguing that the hard X-rays are emitted in confined sources in the low corona, far below the extended coronal shock wave.

The long-lasting hard X-ray events are characterized by smooth evolution: above 100 keV fluctuations shorter than about 10 s are extremely rare (Vilmer et al. 1994). The photon spectra are in general hard (flat) and often harden throughout the event (Bai and Dennis 1985, Cliver et al. 1986), with radiation produced by the most energetic electrons (high-energy X-rays, microwaves) peaking later than that emitted by low-energy electrons. The production of suprathermal electron populations seems to be a general feature of long-lasting energy release in the corona, since even the longest and most slowly evolving soft X-ray brightenings are accompanied by hard X-ray emission in their rising phase (Hudson et al. 1994).

Gradual X-ray bursts at $h\nu \sim 20$ keV come from extended sources in the corona at heights of several 10^4 km, similar to the soft X-ray sources (Pallavicini et al. 1977, reviews by Tanaka 1987, Dennis 1988). The sources are located near the top of magnetic arcades inferred from the extrapolation of the photospheric magnetic field. These regions have typical field strengths of 50 Gauss (Sakurai 1985) and ambient electron density $\simeq 10^9 - 10^{10}$ cm^{-3}. Nakajima (1983) observed simultaneous gradual and impulsive hard X-ray and microwave emission from two spatially separated sources. The gradual microwave emission had a lower spectral turnover frequency, indicating lower magnetic field strength than the impulsive one. The imaging observations therefore show that gradual hard X-ray and microwave emission comes from electrons interacting in the low corona with a more dilute plasma, at greater altitudes than during the impulsive phase. Type IV flare continuum emission comes from overlying sources (Fig. 2.c), showing that accelerated electrons have rapid access to different structures in the corona, or else that acceleration processes at widely separated sites operate in a coordinated way. The involved structures may be highly dynamic, as gradual hard X-ray/radio bursts are associated with coronal mass ejections (Pallavicini et al. 1977, Cliver et al. 1986, Chertok et al. 1992).

3.1   Non-thermal Electrons in Gradual Hard X-Ray/Radio Bursts

Time Evolution of Hard X-Ray and Microwave Spectra: Trapping and Time-Extended Acceleration Although the idea that the large-scale coronal shock wave accelerates electrons radiating hard X-rays and metric continuum in the low and middle corona became unattractive, it is still discussed as a candidate for accelerating particles upstream (cf. review by Kallenrode 1993).

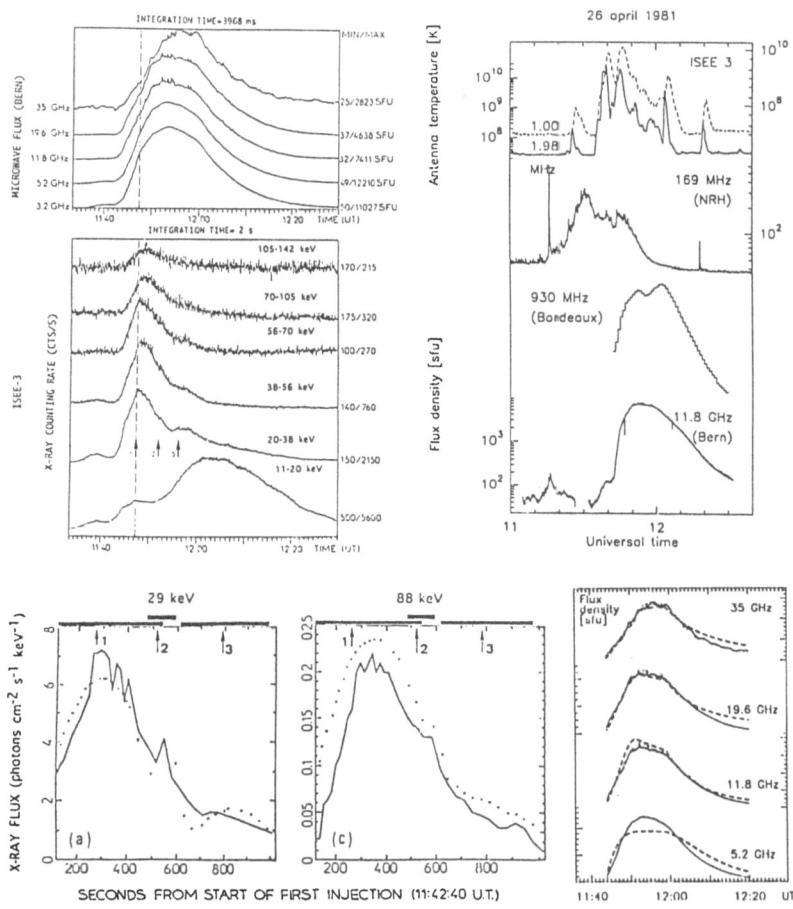

Fig. 3. Time history of hard X-ray and radio emission during the gradual event of 1981 April 26. *Top left*: Microwaves and hard X-rays (Bruggmann et al. 1994). *Top right*: Centimetre to hectometre waves (Klein and Trottet 1994). *Bottom*: Model computations (dashed) and observations (solid lines) in hard X-rays (two panels on the left) and microwaves (Bruggmann et al. 1994).

Furthermore, the idea of different processes of acceleration for different types of events has remained alive, and attempts to classify acceleration processes with the help of different types of radiative signatures persist (e.g. Bai and Sturrock 1989).

As an alternative, the hardening of the electron spectrum due to Coulomb interactions with thermal electrons in the corona has been invoked (Bai and Ramaty 1979, Vilmer et al. 1982, Vilmer 1987). Since gradual hard X-ray and the interaction of electrons with different media is the key to the interpretation of differences between these types of energetic particle signatures.

Electron Injection into the Middle and High Corona The flare continuum emission has different characteristics in different wavelength ranges, i.e. at different altitudes in the corona (Fig. 3, top right):

- At decimetric waves (930 MHz) the emission traces the successive electron injections identified in hard X-rays and microwaves.
- At metric wavelengths (169 MHz) the gradual event appears as a new increase of the emission from a previous impulsive event. After 11:50 UT the 169 MHz flux density decreases while the higher frequency emission persists. At that time the previously stable 169 MHz source starts to rise in the corona (Moving Type IV). The premature end of the metric emission is therefore not a counterexample to the case of Fig. 2, but is likely due to the emitting structures becoming disconnected from the particle source in the course of restructuring of the coronal magnetic field.
- The hectometric emission (2 and 1 MHz, emitted at 10-20 $R_\odot$ above the photosphere) is more rapidly variable. The rapid drift towards lower frequencies, apparent in the lightcurves and the dynamic spectrum (courtesy J.L. Steinberg), shows that the emission is Type III-like, due to electron beams propagating outward along open magnetic flux tubes.

Except at metric wavelengths the excess emission has similar duration over a wide spectral range, despite the different radiation processes, the different types of electron populations and the different magnetic field structures where the radiation is emitted. While trapping affects the X-ray and microwave spectra, the repetitive injection of electron beams into the high corona is another indication that acceleration persists during the whole event. The acceleration is structured in time and probably in space, as indicated by the fact that the azimuth angle displayed large fluctuations during some intense events (S. Hoang, pers. comm.). This is similar to impulsive phase acceleration (e.g. Benz 1994, Trottet 1986, 1994). The previously assumed acceleration of the beams by large-scale coronal shock waves (Cane et al. 1981, Kahler et al. 1989a; but cf. discussion by Kundu and Stone 1984) is not consistent with the similar injection time history to hard X-ray and microwave emitting electrons which seems to be a general property of gradual hard X-ray/radio bursts (Klein and Trottet 1994). The name "shock associated events" given to this type of hectometric emission therefore seems misleading.

This picture of time-extended acceleration also pertains to the long-lasting γ-ray events discovered by Kanbach et al. (1993) and Leikov et al. (1993). Trapping of impulsively accelerated particles can account for these long durations (Mandzhavidze and Ramaty 1992), but the co-evolution of the γ-ray and the microwave emission clearly favours time-extended acceleration of mildly relativistic electrons and protons (Akimov et al. 1994, Kocharov et al. 1994, Ramaty and Mandzhavidze 1994).

microwave emission is observed to come from a more dilute plasma than impulsive emission, the different particle dynamics are expected to yield different temporal evolution of the radiative signatures. Figure 3 shows one of the most conspicuous gradual events observed so far. The microwave peaks are delayed by several minutes with respect to the hard X-ray maxima. On the hypothesis that the emissions come from electrons injected during extended time periods into a coronal trap, Bruggmann et al. (1994) were able to reproduce the time evolution of the hard X-ray as well as the microwave spectra throughout the event (Fig. 3, bottom). Hulot et al. (1992) did the same for the hard X-ray and γ-ray evolution of another gradual event. It is important to notice that both time-extended acceleration and trapping are necessary to account for the observations. The collisional lifetime of electrons is too short to explain the duration of the events by trapping alone (Bai and Dennis 1985).

Table 1. Nonthermal electrons during impulsive and gradual events (Bruggmann 1994)

Date	$N_{tot}(25)$	$W_{tot}(25)$ [erg]	τ [s]	N_{tot}/τ [s^{-1}]	W_{tot}/τ [erg s^{-1}]	Remarks
7 Jun 1980	$> 2 \times 10^{36}$	$> 1.4 \times 10^{29}$	8.5	2.3×10^{35}	1.6×10^{28}	impulsive (1/7 peaks)
29 Jun 1980	1.8×10^{37}	1.1×10^{30}	73	2.0×10^{35}	1.5×10^{28}	impulsive
14 Aug 1979	1.1×10^{37}	7.6×10^{29}	15	7.3×10^{35}	5.1×10^{28}	impulsive
	1.9×10^{37}	1.5×10^{30}	140	1.4×10^{35}	1.1×10^{28}	gradual (2 peaks)
26 Apr 1981	1.2×10^{38}	7.4×10^{30}	1000	1.2×10^{35}	7.6×10^{27}	gradual
27 Apr 1981	$> 1.4 \times 10^{38}$	$> 1.2 \times 10^{31}$	212	6.7×10^{35}	5.7×10^{28}	gradual (1/3 peaks)
13 May 1981	1.0×10^{38}	5.7×10^{30}	360	2.8×10^{35}	1.6×10^{28}	gradual

The Energy Budget The energy contained in suprathermal electrons above 25 keV is compared in Table 1 for a set of impulsive and gradual events (Bruggmann et al. 1994). The energy was derived from a non thermal trap+precipitation model where the trapping efficiency increases from impulsive (upper half of Table 1) to gradual events. The energy content in non thermal electrons tends to be higher in gradual events. Division by the time τ during which the electrons are injected shows, however, that the power transferred to the non thermal electron population is comparable for impulsive and gradual events. The energy estimation depends on the low-energy cutoff which was set to 25 keV. While this makes the absolute value of the energy content uncertain, the comparison between different event types analyzed with the same model is expected to yield a significant result.

The above comparison reaches the same conclusion as Ramaty et al. (1993) who found no difference between impulsive and gradual flares in their modelling of γ-ray continuum (hν >300 keV) and line emission. This supports the idea that

3.2 Imaging Observations of Energy Release in the Low Corona

Scenarios for gradual phase energy release are generally devised in terms of reconnection of large-scale magnetic field structures previously opened by an erupting prominence (see review by Švestka and Cliver 1992). The gross features of this scenario are in agreement with recent imaging observations of the thermal emission during a prominence eruption (Hanaoka et al. 1994). Cliver et al. (1986) extended this scenario to particle acceleration which they located in the neutral sheet underlying the rising prominence.

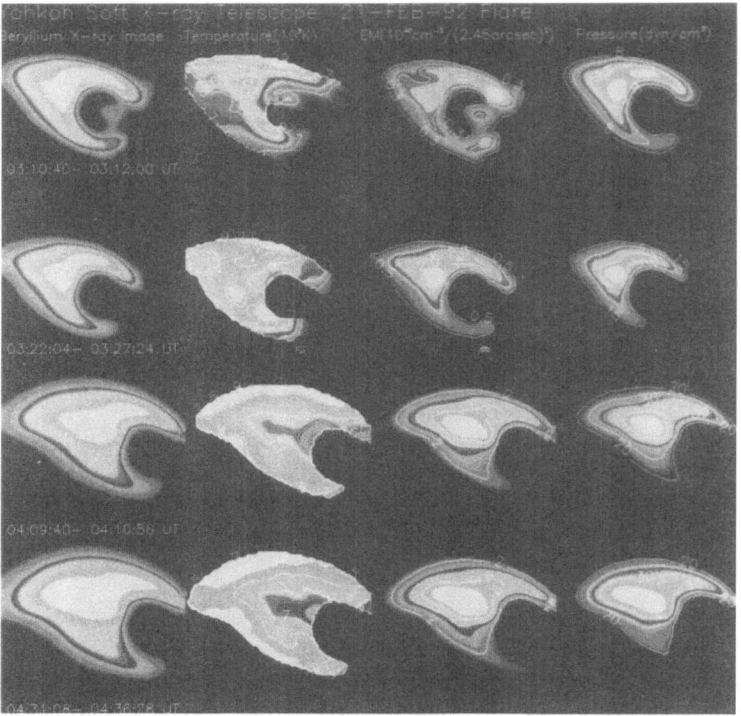

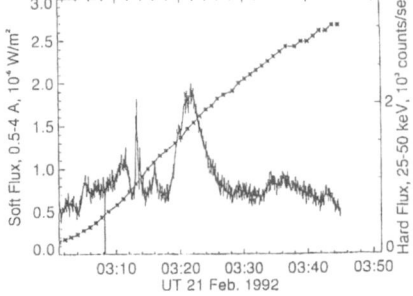

Fig. 4. *Top*: Maps of (from left to right) soft X-ray intensity, temperature, emission measure and pressure during and after a long-lasting flare (SXT/Yohkoh; Tsuneta et al. 1992). The highest values are coded in white. *Right*: Lightcurve at soft X-ray (GOES, monotonously rising curve) and hard X-ray (hν >25 keV, BATSE/CGRO) wavelengths (Hudson et al. 1994).

Figure 4 shows the source configuration in soft X-rays and the time evolution of soft and hard X-ray emission during a limb event lasting several hours (SXT/Yohkoh, Tsuneta et al. 1992, and BATSE/CGRO, Hudson et al. 1994). At the top of the figure maps of brightness, temperature, emission measure and pressure are displayed during the early rise of the soft X-ray flux and a peak in the hard X-ray count rate. The second row refers to a time interval later in the rise, comprising the maximum and decay phase of the strongest hard X-ray peak. The geometry of the soft X-ray *images* (left column of Fig. 4) basically remains the same throughout the event, while a variety of structural changes had been seen prior to the brightening, including the formation of large-scale magnetic structures and the acceleration of matter to high velocities (Tsuneta et al. 1992, Hudson 1994b). No particular event stands out at the onset of the brightening in hard and soft X-rays. During the hard X-ray emission only minor changes are seen in the soft X-ray images underneath the bright arch of Fig. 4 (Tsuneta et al. 1992). Yet, major dynamical processes must occur during this time, since the distributions of temperature and emission measure undergo marked changes (first and second row of Fig. 4), including the development of distinct hot regions within the overall structure and the displacement of the region with highest emission measure from one leg to the top. If confirmed by future work, the temperature inhomogeneities demonstrate that energy release proceeds in various confined regions within a large emitting volume, but not predominantly near the top or cusp where reconnection is supposed to occur in the models of post-flare loop formation. The typical post-flare loop behaviour ascribed to the reconnection of large-scale magnetic field lines is only displayed in the late phase of the soft X-ray event, around and after the maximum (two bottom lines in Fig. 4).

3.3   The Large-Scale Coronal Shock Wave

Type II bursts and the large-scale coronal shock wave they reveal have been removed from their central role as a particle accelerator during gradual hard X-ray/radio events. This seems to be true for electrons confined in the corona as well as for those escaping towards interplanetary space (Simnett 1986, Klein and Trottet 1994). The Type II emission nevertheless reveals coronal electron acceleration, as discussed by Mann (1995). The generation of these large-scale shock waves is not a specific signature of gradual events, since flares of any energy content or impulsiveness may give rise to Type II bursts (Pearson et al. 1989). Aurass et al. (1994) traced the generation of Type II related shock waves back to signatures of energy release during the impulsive acceleration of high-energy electrons.

3.4   Radio Evidence of Large-Scale Instabilities

Early evidence for the dynamics of coronal structures during gradual hard X-ray/radio bursts came from metre wave Moving Type IV bursts (cf. reviews by Goldman and Smith 1985, Stewart 1985, and references therein). They are emitted by suprathermal electrons in magnetic structures which rise through the corona. The radio emission lasts between some minutes and ~1 hour. Often the source detaches from a previous complex of flare continuum emission.

When compared with optical observations, Moving Type IV sources coincide either with features of the white-light coronal mass ejection (e.g. Stewart et al. 1982, Hildner et al. 1986, Gopalswamy and Kundu 1989) or with the prominence (Wild 1969, McLean 1973, MacQueen 1980, Stewart 1985, Klein and Mouradian 1990) which is often seen behind the outer white-light feature.

The shape of the radio sources resembles arch-like features (Wild 1969) or - more often - apparently isolated sources (Smerd and Dulk 1971). In some cases such isolated sources have been tracked out to several solar radii (e.g. Sheridan 1970, Riddle 1970, Dulk and Altschuler 1971, Klein and Mouradian 1990). The measured plane-of-sky velocities (200–1600 km/s, Robinson 1978) are similar to rapid white-light CMEs (Burkepile and St. Cyr 1993, their Fig. 6).

It is presently unclear how and where the Moving Type IV sources are formed and where/when the electrons which eventually emit the radio waves are accelerated. Trottet (1986, Fig. 9) reports a case where the Moving Type IV and the hard X-ray emission display similar fluctuations. The Type IV emission in Fig. 3 starts fading with the rise of the radio source. In both cases the moving radio source seems to be part of the evolving structures into which electrons are injected over a prolonged duration. In the course of the evolution of the large-scale magnetic field structures the moving source may disconnect from the acceleration sites and become an "isolated" source. Isolated sources are often described as closed magnetic structures which are generated by reconnection, in analogy with reconnection in the Earth's magnetotail. By these processes beams can be injected onto open field lines and escape to the high corona, whereas isotropic particle populations are found inside the closed magnetic field structure (2D) or a magnetic flux rope (3D) which is ejected tailward in the magnetosphere and upward in the solar corona (cf. Scholer et al. 1984). The flux rope configuration allows for a gradual transition from the case of connection with acceleration sites in the low or middle corona to complete disconnection (cf. Hughes and Sibeck 1987).

4   Radio Noise Storms

Noise storms persist during times ranging from an essentially unknown lower limit (some tens of minutes, say) to several days and are not systematically related with flares. The spectrum consists of broad-band continuum emission at decimetric to metric wavelengths, and superposed short (≤ 1 s), narrow-band (≤ 10 MHz) Type I bursts. Both bursts and continuum have brightness temperatures far above coronal electron temperatures and are in general nearly 100% polarized, with the exception of sources near the solar limb. At decametric wavelengths noise storms are often accompanied by storms of Type III bursts, where several bursts occur per minute over a duration which may attain several days. Impulsive low-energy (≤ 10 keV) electron events. (Lin 1985, Dulk 1990) in interplanetary space are generally associated with Type III storms at decametre and longer wavelengths, and the trajectory of the electrons can be traced back to active regions where noise storms occur (Kayser et al. 1987).

Noise storms were reviewed by Elgarøy (1977) and Kai et al. (1985), and were the topic of a workshop (Benz and Zlobec 1982).

4.1   The Energy Budget of Noise Storms

Noise storms are emitted through poorly understood collective processes where suprathermal electrons (some keV) produce Langmuir waves which are subsequently converted to electromagnetic radiation near the local electron plasma frequency (e.g. Melrose 1980, Benz and Wentzel 1981, Spicer et al. 1981, Wentzel 1986). The electromagnetic power emitted by a noise storm, of the order of 10^{17} erg/s (Elgarøy 1977, ch. 9.3), is therefore a lower limit of the power which the acceleration process must provide to the radiating electrons. Raulin and Klein (1994) estimated the energy contained in a population of suprathermal electrons trapped in coronal structures with a density range of about $3 \cdot 10^9$ to $3 \cdot 10^8 \mathrm{cm}^{-3}$ as indicated by the spectral range of the radio continuum emission. Only electrons with at least 10 keV energy survive Coulomb collisions when travelling between the two density levels. The collisional lifetime of such an electron is about 30 s. Supposing that the observed volume of the radio source of $10^{30} \mathrm{cm}^3$ (typical observed source surface times height inferred from the density range in the hydrostatic model atmosphere) is filled with 10 keV electrons, the total energy is $10^{31} \frac{n_f}{n_e}$ erg, where n_f and $n_e = 5 \cdot 10^8 \mathrm{cm}^{-3}$ are the number densities of the fast electrons and the thermal electrons, respectively. In theoretical work $n_f = 10^{-5} n_e$ is often assumed. The latter value leads to an energy of 10^{26} erg and a power of 10^{24} to 10^{25} erg/s in the electron population at $\simeq 10$ keV.

An independent estimate can be inferred from the associated decametric Type III storms (e.g. Benz and Wentzel 1981) and impulsive 2-10 keV electron events at 1 AU. Lin (1985; cf. also Jackson and Leblanc 1991) estimates that 10^{25} to 10^{26} erg are released during some minutes to these electrons, implying an energy release rate of 10^{23} erg/s.

With these crude estimations the energy budget of a noise storm amounts to about 10^{27} erg/hour or 10^{28} to 10^{29} erg/day. Noise storms hence have lower dissipation rate than a flare, but a total energy budget comparable with a small flare. The required rate of energy release can be supplied from the low atmosphere, e.g. through expanding magnetic flux (e.g. Spicer et al. 1981).

4.2   Noise Storms and Evolving Coronal Structures

Because of their low energy dissipation rate noise storms have long been attributed to energy release due to "slow" coronal evolution with uncertain link to the underlying active region. Based on the association of noise storms with day-to-day evolution of active regions (Kai and Sekiguchi 1973, Sakurai 1975), and on their joint imaging observations at radio and EUV/X-ray wavelengths, Stewart et al. (1986) argued that noise storms occur when magnetic flux that slowly expands into the atmosphere attains the appropriate coronal heights about one or two days after its emergence in the photosphere. On the other hand noise storms may follow flares within some tens of minutes, and noise storms which occurred simultaneously with energy release in the underlying active region, traced by soft X-ray brightenings, were reported by Webb and Kundu (1978), Lantos et al. (1981), Švestka et al. (1982) and Kundu and Gopalswamy (1990). While these events can be traced back to a transient change of the coronal structure, most often no such change could be identified prior to noise storms.

Kerdraon et al. (1983) were the first to show that the onset or enhancement of noise storms and storm continua are systematically associated with a restructuring of the corona within at most a few hours - a limit imposed by the observing technique of the SMM coronagraph. The restructuring showed up as features of enhanced white-light intensity which were much narrower than loop-type coronal mass ejections. Howard et al. (1985) identified such small-scale features as the most frequent (and least energetic) type of transient coronal brightenings in the images of the Solwind coronagraph, and called them "spike"-type coronal mass ejections. Raulin et al. (1991) and Raulin and Klein (1994) subsequently showed that noise storm onsets were systematically preceded, by a time lapse between zero and some tens of minutes, by energy release detected in soft X-rays and/or microwaves. Hence noise storms arise while an active region and the overlying corona are perturbed or restructured on time scales of some minutes, rather than during a presumed slow, day-to-day evolution of the corona.

Imaging and Spectroscopic Evidence for Dynamical Structures During Noise Storms Radio imaging has revealed source motions at fixed frequency during some short-lasting noise storms. Figure 5 shows three examples, related to a prominence eruption (a), to the slow disappearance of a filament with a subsequent coronal mass ejection (b), or to less violent changes in the underlying atmosphere (c). The common aspect of the observations is that these noise storm sources are not stationary, but undergo systematic shifts. The motions seem to be transverse rather than parallel to the field lines of extrapolated (c) or inferred magnetic configurations.

Besides direct imaging, spectral analysis may also give clues to dynamical processes in noise storm sources. Type I bursts often cluster in chains which undergo a spectral drift mostly towards lower frequencies with smaller drift rates than Type II bursts in the same frequency range (cf. Elgarøy 1977, Kai et al. 1985, and references therein). Similar drift rates are reported from correlation analyses at two separate frequencies (Elgarøy 1977, ch.6.1, Aurass et al. 1986). The onset of noise storm continua may also occur with a delay between neighbouring frequencies: Böhme (1993) found progressive delays at frequencies below 234 MHz. Furthermore, the 510 MHz emission was also frequently

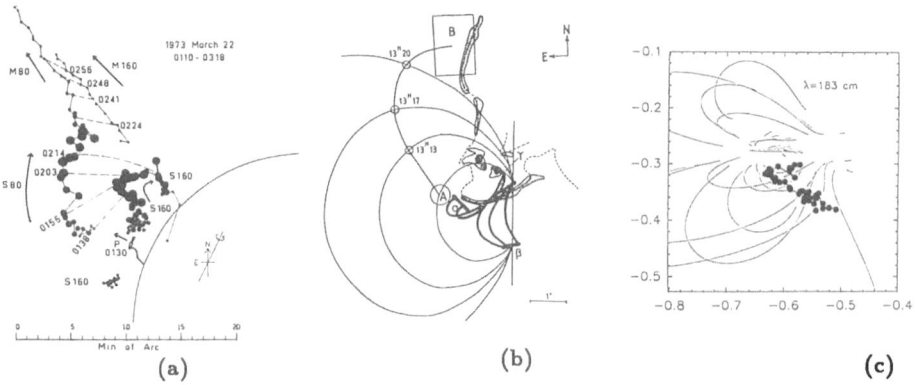

Fig. 5. Motions observed during noise storms at fixed frequencies. a.: Successive positions (denoted by large filled circles) of the source centroids at 80 (S80) and 160 (S160) MHz during a noise storm associated with a prominence eruption (McLean 1973). The ascending prominence is marked "P". After the noise storm faded, moving Type IV sources (M80, M160) start moving outward (small filled circles). b.: Positions at 169 MHz (open circles) during a noise storm overlying soft X-ray arches (fat solid lines) and filaments (hatched surfaces), together with a set of field lines whose geometry was inferred from the white-light observation of a subsequent coronal mass ejection (thin solid lines; Lantos et al. 1981, Hildner et al. 1986). c: Successive positions of a noise storm source centroid at 164 MHz (filled circles, axes in $R_\odot$). Field lines extrapolated from the longitudinal field in the photosphere are plotted as continuous lines (Raulin et al. 1991). The source motion is directed towards the centre of the magnetic field configuration (from south-west towards north-east and then west).

delayed with respect to 234 MHz. A schematic spectrum of a noise storm onset as seen by Böhme is plotted in Fig. 6.a. Low-frequency delays were also observed by McLean (1973), Lantos et al. (1981) and Raulin and Klein (1994) between ~160 MHz and 300–400 MHz, and during some storm continuum events by Robinson (1985). Figures 6.b, c compare the measured spectral drift rates of Type I chains (Fig. 6.b) and noise storm onsets. Type I chains and the onsets of noise storm continua drift at comparable rates.

It had become general use to ascribe such drifts, in analogy with Type II bursts, to an ascending exciter or accelerator (e.g. McLean 1973, Spicer et al. 1981). This interpretation seems hard to reconcile with (1) broadband fluctuations of the continuum (Gnezdilov and Fomichev 1987, Raulin and Klein 1994) which, provided they are signatures of the acceleration/injection process, suggest a quasi-simultaneous access of energetic electrons to different regions rather than the successive acceleration at growing altitudes, and (2) the source motions displayed in Fig. 5. Raulin and Klein proposed that if the radio sources are confined in expanding coronal structures where the density decreases, an appropriate site of emission at constant frequency will change its place, and time

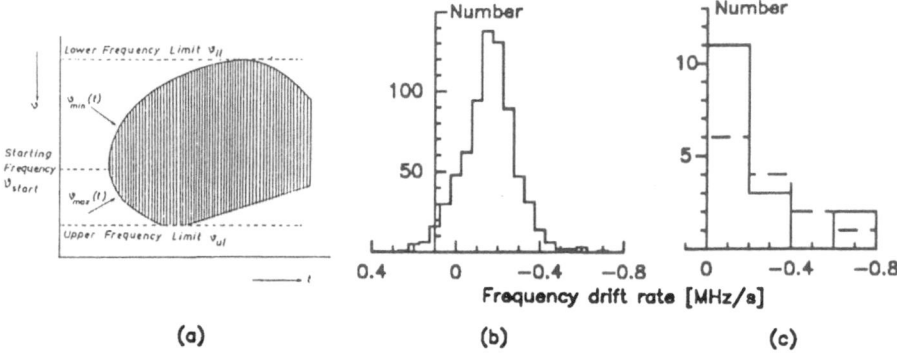

Fig. 6. a. Schematic drawing of the dynamic spectrogram of a noise storm onset (Böhme 1993, cf. also Robinson 1985, his Fig. 15.5). **b.** Distribution of frequency drift rates of Type I burst chains in the range 160–320 MHz (after deGroot et al 1976). Negative numbers give drifts towards lower frequencies. **c.** Distribution of frequency drift rates of noise storm continuum onsets (between 234 and 113 MHz: solid line, Böhme 1993; between 327 and 236 and 236 and 164 MHz: dashed line, Raulin and Klein 1994).

delays of low frequencies may arise because a sufficiently low plasma density has first to be established. Watari et al. (1994) directly observed an expanding and brightening soft X-ray loop during a drifting metric continuum. To give a qualitative impression, the onset delays of noise storm continua would imply an expansion velocity of 100–200 km/s if the source were a homogeneous expanding plasma slab (Raulin and Klein 1994). If matter is injected into the corona, positive frequency drifts may occur, and upward rising sources as at the onset of some storm continua are expected (Robinson 1985). This is consistent with the density evolution in the feet of a loop-type coronal mass ejection (Sime et al. 1984). Which kind of drift will actually be observed may depend on dynamical processes on smaller spatial scales than resolved by contemporary imaging instruments.

It is hence plausible to localize the onset of noise storms in a complex system of expanding coronal arches. Such structures may be destabilized by an ascending prominence (McLean 1973), a flare (Švestka et al. 1982) or by less conspicuous perturbations of the underlying atmosphere, e.g. as suggested by the brightening of a new source in the underlying active region (Raulin et al. 1991). Once the noise storm is established, no conspicuous structural changes are observed in the white-light corona (Kerdraon et al. 1983, Duncan 1983), although energy must be quasi-continuously supplied to maintain the radio emission. It is possible that the relevant processes occur on spatial scales smaller than the extent of the noise storm source. At the workshop Malik and Mercier presented evidence for dynamic changes on scales $\leq 1'$ during long-lasting noise storms. Simultaneous imaging observations in EUV/SXR, white light and at radio wavelengths during the SOHO mission will provide us with a comprehensive set of data that will allow us a more detailed discussion.

4.3   Impulsive Phenomena in the Noise Storm Emission. Comparison with ms-Spikes

Although noise storms usually last much longer than any enhanced level of electromagnetic radiation during flares, acceleration processes operating on second-time scales, similar to impulsive flares, are revealed by the decametric Type III bursts. Sources of Type I and Type III bursts are in general close to each other (Gergely and Kundu 1975, Duncan 1981). Furthermore, spectral observations show occasionally a one-to-one correspondence between individual Type I and Type III bursts (Aubier et al. 1978, Leblanc et al. 1983). These findings suggest a close relationship between Type I bursts and energy release processes in the noise storm source.

Since the early spectral observations with high time resolution it has become evident that the dynamic spectrum of a section of a noise storm is similar to flare-related decimetric spike bursts and metric Type III bursts (e.g. Tarnstrom and Philip 1972, Benz et al. 1982): Type III bursts in the low-frequency range of the spectrum are accompanied by short-lived, narrow-band emission at higher frequencies. A comparison of published work on the duration (Güdel and Benz 1990) and spectral width (Csillaghy and Benz 1993) of individual spikes with Type I bursts (various data sets collected in Elgarøy 1977) shows that while the bandwidths are comparable, Type I durations are longer than spike durations in the range 200–400 MHz. Benz et al. (1982) identify several observational differences between Type I bursts and Type III burst associated spikes in this wavelength range (later called "metric" spikes by Güdel and Zlobec 1991): the spikes are not completely polarized, are randomly distributed on the disk, and may display temporal fine structure on time scales below the duration of an individual spike. More recent work on decimetric spikes reveals differences with the metric spikes of Benz et al. (1982): while being less than 100 % polarized on average, decimetric spikes show a degree of polarization which is nearly 100 % at central meridian and decreases towards the limb (Güdel and Zlobec 1991). The spike sources are not uniformly distributed on the disk, but display a centre-to-limb decrease. Both features are similar to Type I bursts. On the other hand, the sense of circular polarization is opposite for decimetric spikes and their associated Type III bursts (Güdel and Zlobec 1991), contrary to the case of noise storms. In conclusion, Type I bursts show similarities with flare-associated metric and decimetric spikes, but the observed marked differences preclude interpretations in terms of the same processes of both energy release and generation of radio emission.

As an alternative, early models proposed that bursts of both types I and III are due to the same electron beams (Aubier et al. 1978). It is tempting to compare the spectrum with the model calculations of Vlahos and Raoult (1994): electron beams that are randomly injected into coronal fibers with random density distribution give radiative signatures of narrow bandwidth and short duration close to the injection site, but more slowly evolving Type III bursts at greater distances, i.e. lower frequencies. The transition between the narrow-band high-frequency bursts and the low-frequency Type III bursts is sharp. This is qualitatively consistent with the spectra observed during Type I/Type III storms. Again the different degrees of circular polarization show, however, that the involved radiation processes are not the same.

The common aspects of the Type I models is the interpretation in terms of scattered small-scale sites of energy release (Benz and Wentzel 1981, Spicer et al. 1981). In this case the characteristics of Type I bursts would imply that particle acceleration occurs during noise storms in regions with plasma frequencies in the range 100–400 MHz, with individual acceleration sites being scattered over a comparable range of densities. While the idea is plausible, it has not yet been proven that Type I bursts are signatures of the elementary processes in which the electrons emitting the noise storm continuum are energized.

5   Summary and Conclusions

Suprathermal particle populations in the corona require time-extended acceleration, while trapping cannot conserve them over durations significantly longer than that of acceleration. Different sites of acceleration have been identified which contribute to different types of particle signatures:

1. Gradual hard X-ray/radio bursts require the most energetic and numerous particles. They are accelerated in the active region, by not yet identified processes which, judging from model computations of the power transferred to the particles and of the particle spectra, are similar to impulsive phase acceleration. A preliminary analysis of the soft X-ray morphology suggests that while the global geometry is consistent with energy release in reconnecting large-scale structures previously opened by e.g. a rising prominence (but cf. discussion by Hudson 1994a), dynamical processes and restructuring on smaller spatial scales occur at the time of the suprathermal particle signatures.

2. Hectometric fast-drift bursts accompany gradual hard X-ray/radio signatures. The similar duration as hard X-ray and flare continuum emission is a strong argument that the generating electron beams are produced together with the most energetic electrons. The low emitting frequency of the bursts suggests, however, that the acceleration occurs in the middle and high corona. How exactly the close correspondence with the acceleration of the hard X-ray emitting electrons is established is an open question.

3. Metric Type II bursts require energetic electrons which by all observational evidence are accelerated in a large-scale coronal shock wave. This shock wave does not supply significant numbers of electrons for the hard X-ray and flare continuum emission or of electrons escaping to the high corona and producing the previously discussed hectometric bursts.

4. Some isolated-source-type Moving Type IV bursts may be the consequence of magnetic reconnection of large-scale coronal structures, forming plasmoids or flux ropes that contain energetic particle populations in analogy with processes in the Earth's magnetotail. The scenario has not been developed beyond the state of a cartoon model. The reconnection might also produce electron beams which will become visible in the high corona, like the hectometric fast-drift bursts.

5. Storm continua and noise storms require acceleration of electrons to at least 10 keV in the middle corona. The onset of these events in the middle corona occurs together with energy dissipation in the underlying active region, which may or may not show up as a flare in Hα. It is therefore not appropriate to attribute noise storms to slow coronal evolution as opposed to a violent perturbation associated with a flare. In the later course of the noise storm the acceleration in the middle corona occurs without conspicuous changes in either the coronal structure or the underlying active region. The acceleration comprises the repetitive and time-extended production of electron beams at lower frequencies than during the impulsive phase, in agreement with the presumed acceleration site in the middle corona.

The energetic particle signatures occur, sometimes simultaneously, in different environments, as defined by the ambient density and magnetic field. Different acceleration processes may act: for example, Type II bursts appear to be good evidence for acceleration in an extended coronal shock wave, while the more energetic particles during gradual events are supplied by different processes in the low or middle corona. Electron beams bear evidence of small-scale acceleration processes, not only in impulsive flares, but similarly during gradual hard X-ray/radio events and noise storms. A correspondence between acceleration processes and different radiative signatures, e.g. in terms of steps or phases of acceleration, as proposed by Bai and Dennis (1985), Melrose and Dulk (1987), Bai and Sturrock (1989) and many others, is in general not enforced by the observations. Rapidly evolving coronal structures on different energetic scales are related with all signatures of energy release discussed here: from the relatively slow expansion of coronal structures where noise storms are emitted, associated with weak soft X-ray brightenings, low-energy transient coronal disturbances and low instantaneous energy content of energetic electrons, to gradual hard X-ray/radio bursts due to energetic particle populations and associated with major events of restructuring including prominence eruptions and large coronal mass ejections. These dynamic processes can perhaps be extended to lower energies, with the continuous expansion of active regions observed in soft X-ray images (Uchida et al. 1992). The common aspects suggested here are largely speculative for the time being. There is ample space for joint observations with SoHO in order to get deeper insight into the interplay between the restructuring of the corona and energy release processes on time scales from seconds to several days.

Acknowledgements : The author has benefitted from discussions with Drs. G. Trottet and N. Vilmer, and with different colleagues in the frame of a collaboration sponsored by the CEE under contract SC1 CT91 0727. He is grateful to Dr. A.O. Benz for his critical reading of the manuscript and many helpful comments, and to Drs. S. Hoang, H. Hudson and M. Pick for giving him access to unpublished work. Thanks are also due to the members of the Potsdam Astrophysical Institute for their organization of the CESRA workshop.

References

Akimov, V.V., Leikov, N.G., Belov, A.V. et al. (1994): in High-Energy Phenomena– A New Era of Spacecraft Measurements, ed. by J.M. Ryan, W.T. Vestrand, AIP Conf. Proc. **294**, 106

Aubier, M.G., Leblanc, Y., Møller-Pedersen, B. (1978): Astron. Astrophys. **70**, 685

Aurass, H., Kurths, J., Mann, G., Hofmann, A. (1986): Solar Phys. **107**, 123

Aurass, H., Mann, G., Magun, A. (1994): Spa. Sci. Rev. **68**, 211

Badillo, V.L., Salcedo, J.E. (1969): Nature **224**, 503

Bai, T., Dennis, B.R. (1985): ApJ **292**, 699

Bai, T., Ramaty, R. (1979): ApJ **227**, 1072

Bai, T., Sturrock, P.A. (1989): Ann. Rev. Astron. Astrophys. **27**, 421

Benz, A.O. (1994); Spa. Sci. Rev. **68**, 135

Benz, A.O., Wentzel, D.G. (1981): Astron. Astrophys. **94**, 100

Benz, A.O., Zlobec, P. (1982): Proc. 4th CESRA Workshop on Solar Noise Storms. Trieste Observatory Publication

Benz, A.O., Zlobec, P., Jaeggi, M. (1982): Astron. Astrophys. **109**, 305

Böhme A., 1993, Solar Phys. **143**, 151

Bruggmann, G., Vilmer, N., Klein, K.-L., Kane, S.R. (1994): Solar Phys. **149**, 171

Burkepile, J.T., St. Cyr, O.C. (1993): NCAR Technical Note NCAR/TN-369+STR

Cane, H.V., Stone, R.G., Fainberg, J.L. et al. (1981): Geophys. Res. Letters **8**, 1285

Chertok, I.M., Gnezdilov, A.A., Zaborova, E.P. (1992): in Proc. Conf. Solar Wind Seven, ed. by E. Marsch, R. Schwenn (Pergamon Press, Oxford), p. 607

Cliver, E.W., Dennis, B.R., Kiplinger, A.L. et al. (1986): ApJ **305**, 920

Crosby, N., Aschwanden, M.J., Dennis, B.R. (1993): Solar Phys. **143**, 275

Csillaghy, A., Benz, A.O. (1993): Astron. Astrophys. **274**, 487

de Groot, T., Loonen, D., Slottje, C. (1976): Solar Phys. **48**, 321

Dennis, B.R. (1988): Solar Phys **118**, 49

Dulk, G.A. (1990) Solar Phys **130**, 139

Dulk, G.A., Altschuler, M.D. (1971): Solar Phys. **20**, 438

Duncan, R.A. (1981): Proc. Astron. Soc. Australia **4**, 230

Duncan, R.A. (1983): Solar Phys. **89**, 63

Elgarøy, Ø. (1977): Solar Noise Storms. Pergamon Press, Oxford

Frost, K.J., Dennis, B.R. (1971): ApJ **165**, 655

Gergely, T., Kundu, M.R. (1975): Solar Phys. **41**, 163

Gnezdilov, A.A., Fomichev, V.V. (1987): Soviet Astron. Lett. **13**, 297

Goldman, M.V., Smith, D.F. (1985): in Physics of the Sun Vol. II, ed. by P.A. Sturrock, D. Reidel, Dordrecht, p.325

Gopalswamy, N., Kundu, M.R. (1989): Solar Phys. **122**, 91

Güdel, M., Benz, A.O. (1990): Astron. Astrophys. **231**, 202

Güdel, M., Zlobec, P. (1991): Astron. Astrophys. **245**, 299

Hanaoka, Y., Kurokawa, H., Enome, S. et al. (1994): Publ. Astron. Soc. Japan 46, 205

Hildner E., Bassi J., Bougeret J.L. et al. (1986): in Energetic Phenomena on the Sun, ed. by M.R. Kundu, B.E. Woodgate, NASA CP-2439, p. 6-1

Howard, R.A., Sheeley, N.R., Koomen, M.J., Michels, D.J. (1985): J. Geophys. Res. 90, 8173

Hudson, H.S. (1994a): in High-Energy Phenomena–A New Era of Spacecraft Measurements, ed. by J.M. Ryan, W.T. Vestrand, AIP Conf. Proc. 294, 151

Hudson, H.S. (1994b): Proc. Kofu Symposium, ed. by S. Enome, T. Hirayama NRO Report 360, Nobeyama Radio Observatory, Japan, p. 1

Hudson, H.S., Acton, L.W., Sterling, A.S. et al. (1994): in The New Solar Physics from Yohkoh, ed. by. Y. Uchida, H.S. Hudson, T. Watanabe, K. Shibata (Univ. Acad. Press, Tokyo), in press

Hughes, W.J., Sibeck, D.G. (1987): Geophys. Res. Letters 14, 636

Hulot, E., Vilmer, N., Chupp, E.L. et al. (1992): Astron. Astrophys. 256, 273

Jackson, B.V., Leblanc, Y. (1991): Solar Phys. 136, 361

Kahler, S. (1984): Solar Phys. 90, 133

Kahler, S.W., Cliver, E.W., Cane, H.V. (1989a): Solar Phys. 120, 393

Kahler, S.W., Sheeley, N.R., Liggett, M. (1989b): ApJ. 344, 1026

Kai K., Sekiguchi H. (1973): Proc. Astron. Soc. Aust. 2, 217

Kai, K., Melrose, D.B., Suzuki, S. (1985): in Solar Radiophysics, ed. by D.J. McLean, N.R. Labrum (Cambridge Univeristy Press, Cambridge), p. 415

Kallenrode, M.-B. (1993): Adv. Space Res. 13 no. 9, 341

Kanbach, G., Bertsch, D.L., Fichtel, C.E. et al. (1993): Astron. Astrophys. Suppl. 97, 349

Kayser, S.E., Bougeret, J.-L., Fainberg, J., Stone, R.G. (1987): Solar Phys. 109, 107

Kerdraon, A., Pick, M., Trottet, G. et al. (1983): ApJ 265, L19

Klein, K.-L., Mouradian, Z. (1990): in Flares 22 Workshop "Dynamics of Solar Flares". ed. by B. Schmieder, E.R. Priest, Paris Observatory Publication, p.185

Klein, K.-L., Trottet, G. (1994): in High-Energy Phenomena–A New Era of Spacecraft Measurements, ed. by J.M. Ryan, W.T. Vestrand, AIP Conf. Proc. 294, 187

Klein, L., Anderson, K., Pick, M. et al. (1983): Solar Phys. 84, 295

Krüger, A., Kliem, B., Hildebrandt, J., Zaitsev, V.V. (1994): Ap. J. Suppl. 90, 683

Kundu M.R., Gopalswamy N. (1990): Solar Phys. 129, 133

Kundu, M.R., Stone, R.G.(1984): Adv. Space Res. 4 no. 7, 261

Lantos P., Kerdraon A., Rapley G.G., Bentley R.D. (1981): Astron. Astrophys. 101, 33

Leblanc, Y., Poquérusse, M., Aubier, M.G. (1983): Astron. Astrophys. 123, 307

Leikov, N.G., Akimov, V.V., Volzhenskaya, V.A. et al. (1993): Astron. Astrophys. Suppl. 97, 345

Lin, R.P. (1985): Solar Phys. 100, 537

MacQueen, R.M. (1980): Phil. Trans. Roy. Astr. Soc. London Ser. A 297, 605

Mandzhavidze, N., Ramaty, R. (1992): ApJ 396, L111

Mann, G. (1995): this volume

McLean D.J. (1973): Proc. Astron. Soc. Aust. 2, 222

Melrose, D.B. (1980): Solar Phys. 67, 357

Melrose, D.B., Dulk, G.A. (1987): Physica Scripta T18, 29

Nakajima, H. (1983): Solar Phys. 86, 427

Pallavicini, R., Serio, S., Vaiana, G.S. (1977): ApJ 216, 108

Pearson, D.H., Nelson, R., Kojoian, G., Seal, J. (1989): Ap J 336, 1050

Pick, M. (1986): Solar Phys. **104**, 19

Ramaty, R., Mandzhavidze, N., Kozlovsky, B., Skibo, J. (1993): Adv. Space Res. **13** no. 9, 275

Ramaty, R., Mandzhavidze, N. (1994): Proc. Kofu Symposium, ed. by S. Enome, T. Hirayama, NRO Report 360, Nobeyama Radio Observatory, Japan, p. 275

Raulin, J.P., Willson, R.F., Kerdraon, A. et al. (1991): Astron. Astrophys. **251**, 298

Raulin, J.P., Klein, K.-L. (1994): Astron. Astrophys. **281**, 536

Riddle, A.C. (1970): Solar Phys. **13**, 448

Robinson, R.D. (1978): Solar Phys. **60**, 383

Robinson, R.D. (1985): Radio Physics of the Sun, ed. by D.J. McLean, N.R. Labrum, Cambridge University Press, Cambridge, p. 385

Sakurai, K. (1975): Planet. Spa. Sci. **23**, 1344

Sakurai, T. (1985): Solar Phys. **95**, 311

Scholer, M., Gloeckler, G., Klecker, F. et al. (1984): J. Geophys. Res. **89**, 6717

Sheridan, K.V. (1970): Proc. Astron. Soc. Australia **1**, 376

Sime, D.G., MacQueen, R.M., Hundhausen, A.J. (1984): J. Geophys. Res. **90**, 563

Simnett, G.M. (1986): Solar Phys. **104**, 67

Smerd, S.F. (1970): Proc. Astron. Soc. Australia **1**, 305

Smerd, S.F., Dulk, G.A. (1971): in Solar Magnetic Fields, IAU Symp. no. 57, ed. by R. Howard, D. Reidel, Dordrecht, p. 616

Spicer, D.S., Benz, A.O., Huba, J.D. (1981): Astron. Astrophys. **105**, 221

Stewart, R.T. (1985): Radio Physics of the Sun, ed. by D.J. McLean, N.R. Labrum, Cambridge University Press, Cambridge, p. 385

Stewart, R.T., Dulk, G.A., Sheridan, K.V. et al. (1982): Astron. Astrophys. **116**, 217

Stewart, R.T., Brueckner, G.E., Dere, K.P. (1986): Solar Phys. **106**, 107

Švestka, Z., Dennis, B.R., Pick, M. et al. (1982): Solar Phys. **80**, 143

Švestka, Z., Cliver, E.W. (1992): in Eruptive Solar Flares, IAU Coll. 133, ed. by B.V. Jackson, M.E. Machado, Z.F. Švestka (Springer, Berlin), p. 1

Tanaka, K. (1987): Pub. Astron. Soc. Japan **39**, 1

Tarnstrom, G.L., Philip, K.W. (1972): Astron. Astrophys. **16**, 21

Trottet, G. (1986): Solar Phys. **104**, 145

Trottet, G. (1994): Spa. Sci. Rev. **68**, 149

Tsuneta, S., Hara, H., Shimizu, T. et al. (1992): Pub. Astron. Soc. Japan **44**, L63

Uchida Y., McAllister A., Strong, K.T. et al. (1992): Pub. Astron. Soc. Japan **44**, L155

Vilmer, N. (1987): Solar Phys. **111**, 207

Vilmer, N., Kane, S.R., Trottet, G. (1982): Astron. Astrophys. **108**, 306

Vilmer, N., Trottet, G., Barat, C. et al. (1994): Spa. Sci. Rev. **68**, 233

Vlahos, L., Raoult, A. (1994): Astron. Astrophys., in press

Watari, S., Isobe, T., Yohkoh SXT Team (1994): Proc. Kofu Symposium, ed. by S. Enome, T. Hirayama, NRO Report 360, Nobeyama Radio Observatory, Japan, p. 417

Webb, D.F., Kundu, M.R. (1978): Solar Phys. **57**, 154

Wentzel, D.G. (1986): Solar Phys.**103**, 141

Wild, J.P. (1969): Solar Phys. **9**, 260

Wild, J.P. (1970): Proc. Astron. Soc. Australia **1**, 365

Characteristics of Two Simple Microwave Bursts

M. R. Kundu[1], S. M. White[1], N. Nitta[2], K. Shibasaki[3], & S. Enome[3]

[1] Department of Astronomy, University of Maryland, College Park, MD 20 742
[2] Lockheed Palo Alto Research Lab., Palo Alto, CA 94304
[3] Nobeyama Radio Observatory, NAO, Minimisaku, Nagano 384-13, Japan

Abstract. We present simultaneous microwave and X-ray data for two microwave bursts with simple impulsive time profiles. The 17 GHz images show compact sources, and in the one case for which we have simultaneous soft and hard X-ray images, they also show compact sources coincident with the radio source. One of the bursts is barely detected in soft X-rays, yet has a moderate 17 GHz flux.

1 Introduction

In this paper we study the properties of microwave events with simple impulsive time profiles similar to those which we have found to be characteristic of millimeter–wavelength bursts. In recent years we have used the BIMA 3-element interferometer to observe solar flares at 3 mm wavelengths. We have shown that millimeter emission of flares in the impulsive phase is a good diagnostic of the most energetic electrons accelerated in flares. Millimeter emission arises from electrons with energies in excess of 1-MeV – it is unlikely that electrons with energies below 0.5 MeV contribute any significant millimeter emission; such energetic electrons are produced in flares of all sizes, including flares so small that their x-ray signature is barely detectable; and their behavior is not consistent with their energy distribution being a simple extension of the distribution of electrons with energies below 200 keV which produce the observed HXR below 100 keV ([5]; [3]).

A second class of events was detected by us ([6]; [7]). In this class, we found a remarkable similarity in the time profiles of emission associated with the impulsive onset of a flare. In a large fraction of flares, the impulsive phase emission at mm wavelength consists of a rapid rise ($\sim$ 5 seconds) linear in time to a sharp peak, followed by an exponential decay with a decay constant of order 15 seconds.

It seems that for these events the production of MeV-energy electrons follows a very similar pattern. The simplest interpretation of the morphology seems to be that the electrons are accelerated, or somehow injected into a coronal loop, on a timescale of several seconds (the linear rise), but acceleration then ceases abruptly and the number of electrons in the corona decreases exponentially. This linear rise-exponential decay time profile is often not shared by the microwave emission or the hard x-ray emission in the same events. Since we interpret the nonthermal impulsive-phase millimeter emission as a diagnostic of MeV-energy electrons, this means that they have a property not shared by the lower-energy electrons responsible for the hard x-rays (typically 25 - 100 keV) and microwaves (probably 100 - 500 keV). There is also some evidence that the spectral energy distribution of the millimeter-emitting electrons differs from that of the hard x-ray-emitting electrons.

The above results regarding the MeV-electron producing "ordinary" flares and "simple" flares were based upon non-imaging 3 mm interferometer data. It is important to investigate how low in the frequency range these special "simple" flares extend, and what is their spectral and spatial characteristics. So, we selected two "simple" flares observed by the Nobeyama Radio Heliograph operating at 17 GHz. These simple 17 GHz flares were observed as part of an on-going study of 17 GHz active regions and their flare productivity. The spatial and polarization information for these flares comes from NRH, and the spectral information comes from the Toyokawa solar monitoring data at 9.4, 2.0 and 1.0 GHz.

2 Observations

The two simple flares observed on May 25 and May 29 originated in AR 7515 which was followed from May 23 to June 2, 1993, as part of a study of active region evolution and flare productivity. These two flares as observed at 17 GHz were chosen because of the similarity of their time profile to the 3 mm time profiles of a class of 3 mm bursts observed with the BIMA telescope: their sharp and linear rise to maximum followed by an exponential decay, and their short duration. Since these bursts were observed with NRH and Toyokawa patrol telescopes, we were able to study their spatial and spectral characteristics.

3 Event of May 25

Figure 1 shows the time profile of this burst observed at 17 GHz on May 25, 1993. Also shown in the figure are the total power time profiles at 9.4 GHz as well as the Yohkoh HXT and GOES time profiles. Based on the closeness of the 9.4 and 17 GHz fluxes and the lack of detection at 3.8 GHz, we believe that the turnover frequency lies between 9 and 17 GHz. Note that the 17 GHz time profile is the correlated power at the longest base lines, rather than the total power profile which was not sensitive enough to detect this burst.

The 25 May burst has a simple source structure. The 17 GHz flux peaks at 01:14:38 UT and is polarized (-8%) during the impulsive phase. The decay phase is unpolarized (< 1%). The HXT time profile agrees reasonably well with the 17 GHz profile; however, its almost equally rapid decay as the rise is followed by a small post-burst increase which does not seem to have a 17 GHz counterpart.

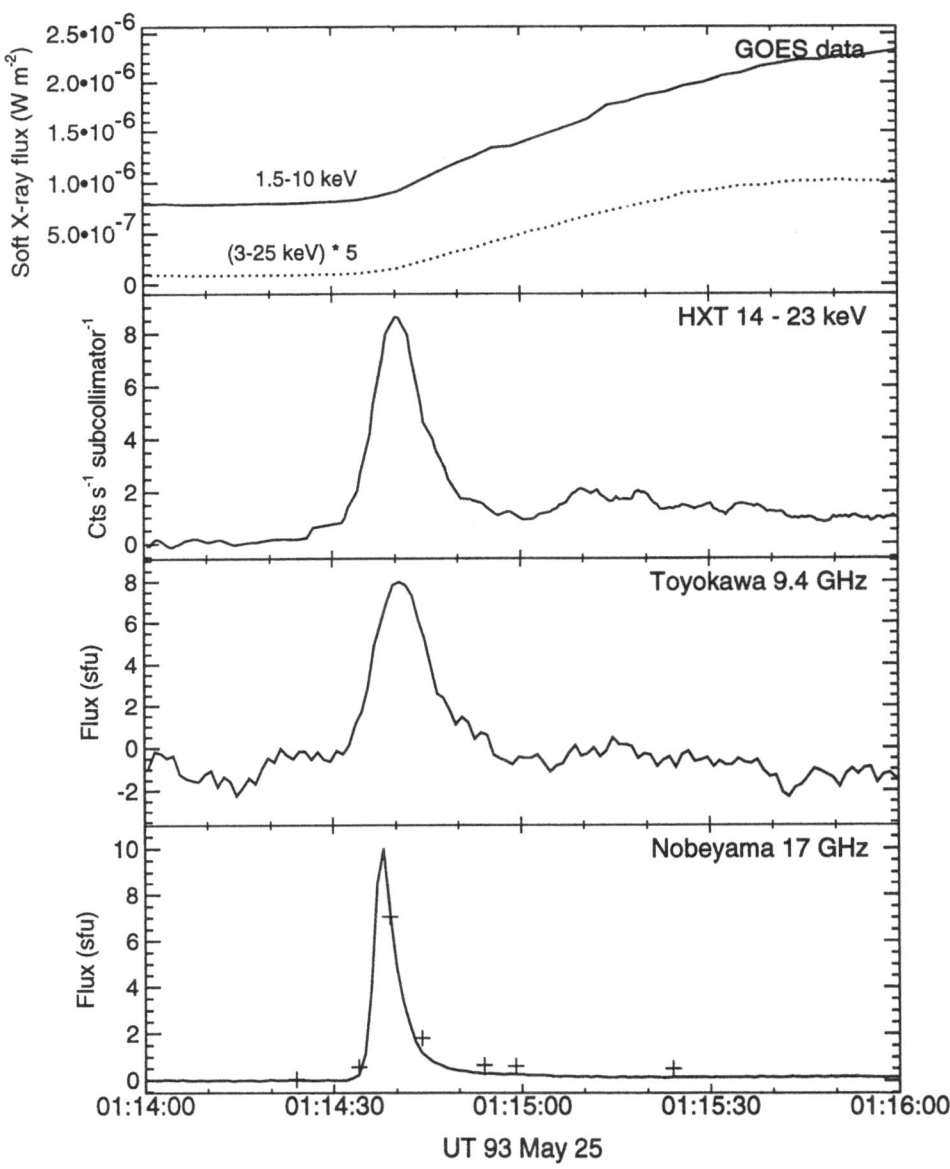

Fig. 1. The time profile of the 1993 May 25 01:14 UT flare in soft X–rays, hard X–rays, at 9.4 GHz and 17 GHz. The crosses in the bottom panel show the times of 17 GHz images used in the analysis; the solid line in the bottom panel is the time profile of the correlated power on the longest baselines of the Nobeyama array, corresponding to the flux in small spatial scales.

1993 May 25

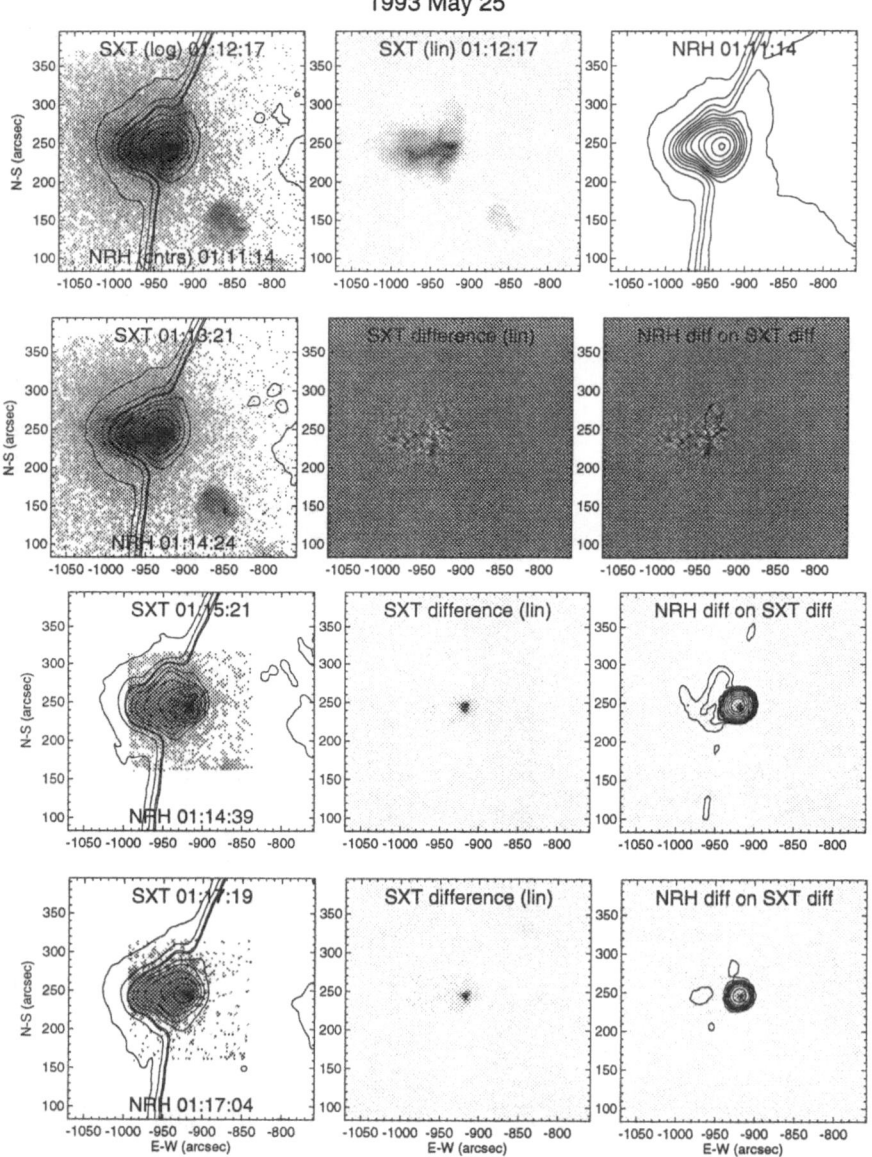

Fig. 2. Images of the 1993 May 25 01:14 UT flare in soft X–rays and at 17 GHz at selected times. The top row of panels shows the preflare 17 GHz images (17 GHz contours
on SXT image, log display, left panel; SXT image with linear intensity display, middle
panel; 17 GHz contours, right panel). In each subsequent row the left panel shows the
17 GHz contours overlaid on the SXT image at the corresponding time; the middle
panel shows the SXT image with the preflare image subtracted; and the right panels
show the same preflare–subtracted SXT image with contours of a preflare–subtracted
17 GHz image overlaid.

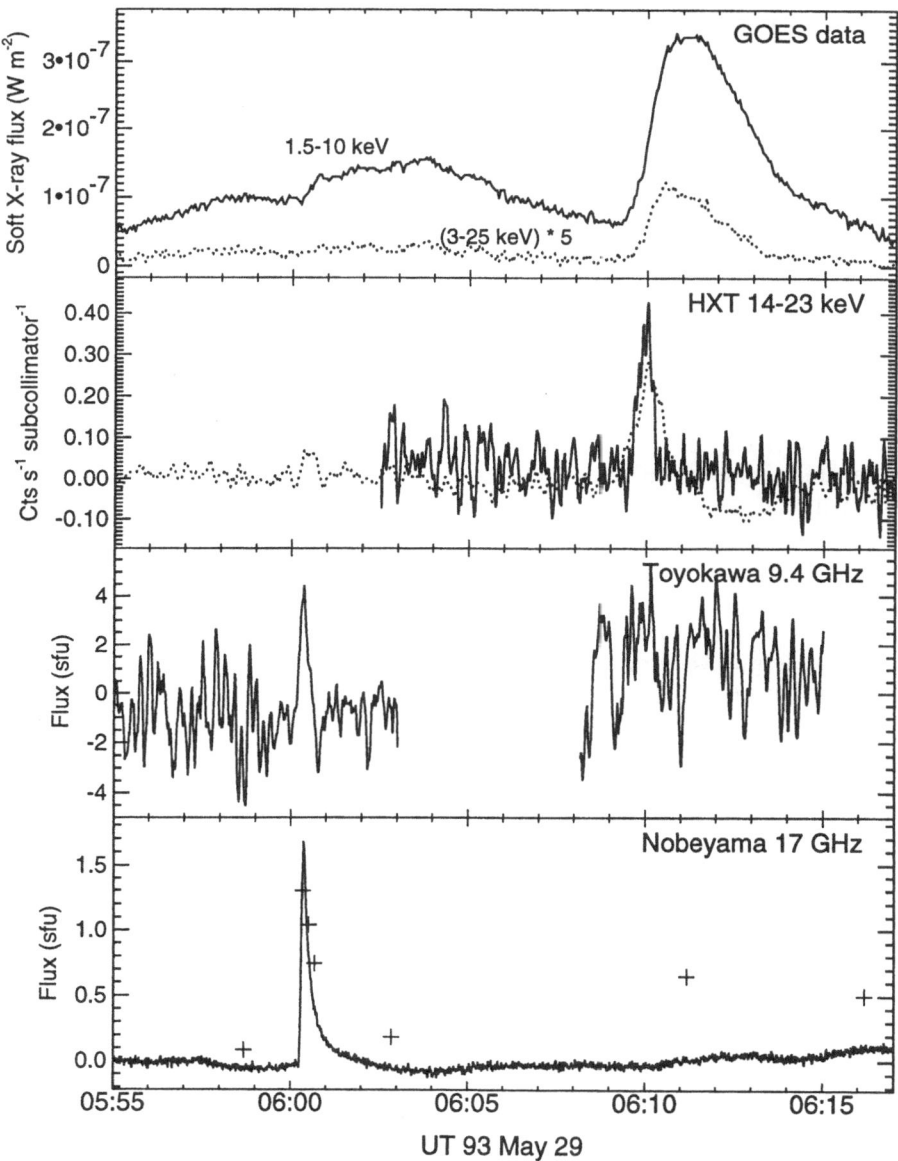

Fig. 3. The time profile of the 1993 May 29 06:00 UT flare in soft X–rays, hard X–rays, at 9.4 GHz and 17 GHz. The broken line in the second panel is the derivative of the GOES 1.5-10 keV profile, a proxy for the hard X-rays during the first flare when no hard X–ray data were available.

Figure 2 shows the 17 GHz and SXT images of the pre–flare and flare emission and their coalignment. The first and second row of panels are preflare times, except that the second row has been preflare–subtracted to show the noise level associated with the subtraction process. The source structure of 17 GHz emission is simple and compact (i.e., unresolved at less than 15″). The SXT source (although at a later time) is aligned almost perfectly with the 17 GHz source. A hard X–ray image made with photons accumulated over the whole flare (not shown) is also perfectly coincident with the 17 GHz and soft X–ray sources. The soft x–ray source is remarkable for its very small size and the fact that it shows no expansion as it evolves.

4 Event of May 29

The May 29 burst at 17 GHz occurred at 06:00 UT when no HXT data are available (Fig. 3). The only data available are a single long-exposure full-disk SXT through the Dagwood filter at 06:00:50. All active regions are saturated in this and other similar images, but it shows one very narrow vertical saturation spike which could be coincidental with the Nobeyama radio source. The 3-25 keV GOES data show no rise in SXR emission at all at this time. There is a suggestion of a small rise in the 1.5 - 10 keV soft x-rays but it is superimposed on a gradual rise which started much earlier. Both SXT and HXT had impulsive emission at 06:09 UT when there was no obvious 17 GHz emission in the correlation plot (Fig. 3). However, the images do show that there is 17 GHz emission at 06:09 UT corresponding to the impulsive HXR and SXR emission: it appears clearly on 17 GHz contour plots (Fig. 4). Other sources present are steady, and are associated with the fact that subtraction was carried out with an image made almost 1 hr earlier, and there has been some evolution in the region. Like the 25 May event, this burst at 06:00 UT is also polarized (-20%) during the implusive phase. The burst source is simple and compact at its impulsive phase (Fig. 4). The SXR source at 06:09 UT is more complex, having several component sources, of which one component has an impulsive rise and decay in correspondence with the HXR time profile. The radio data for the second burst do not show an impulsive profile, and are only very weakly polarized (-2%).

5 Discussion

The 17 GHz event of 1993 May 29 (06:00 UT) is remarkable for the apparent lack of associated x-ray emission (Figure 3). In the absence of SXT or HXT data, we can attempt to estimate the hard x-ray profile associated with this event by differentiating the soft x- ray time profile and invoking the x-ray "Neupert effect" (e.g., Dennis & Zarro [1]). Clearly, differentiating the 3 - 25 keV soft x-ray light curve implies a very low level of hard x-ray emission. The small event at 06:09 UT, by contrast, shows a significant enhancement in soft x-rays but negligible radio emission.

Kosugi, Dennis & Kai ([2]) have shown that the ratio of the peak hard x-ray flux to the peak 17 GHz flux varies by only a factor of 3 over a very large sample of flares. The low microwave flux observed in the 06:09 UT event is consistent

1993 May 29

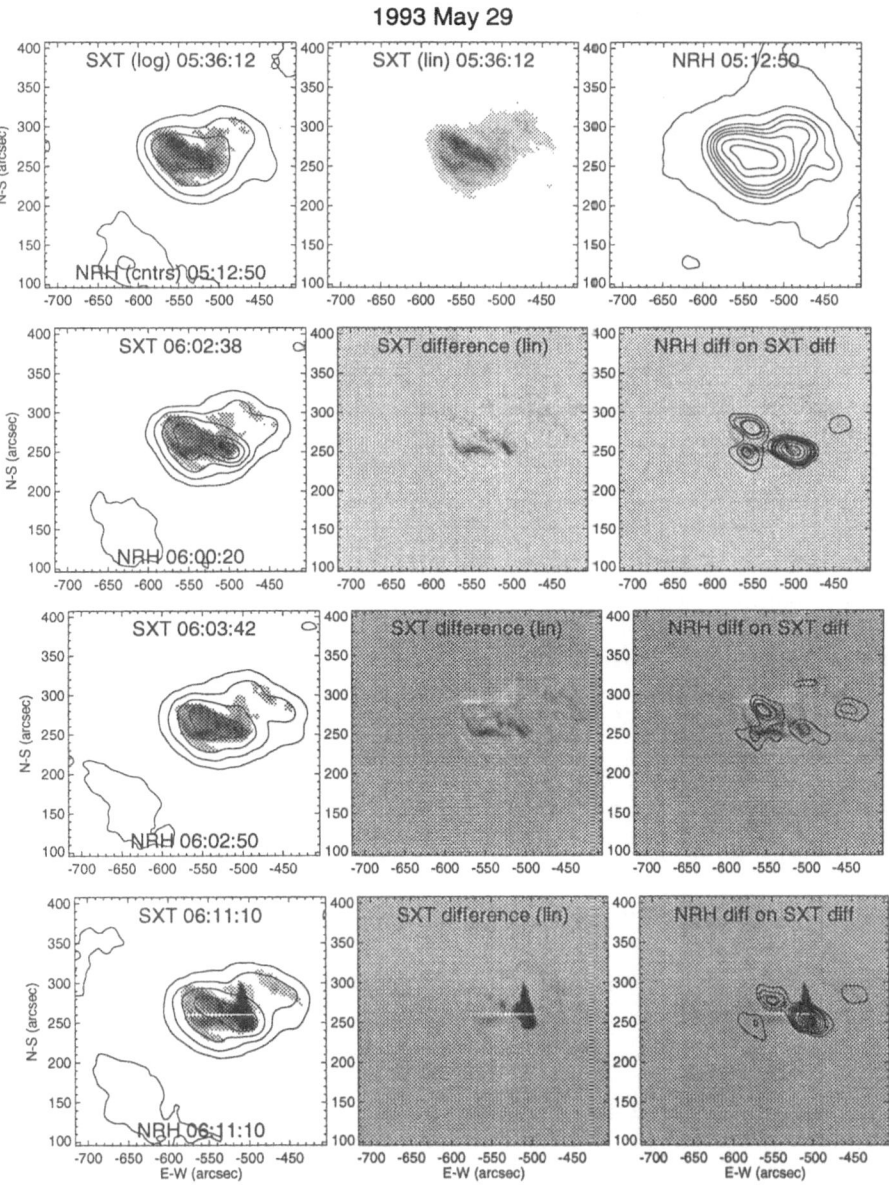

Fig. 4. Images of the 1993 May 29 06:00 UT flare in soft X–rays and at 17 GHz at selected times. In each row the left panel shows the 17 GHz contours overlaid on the SXT image at the corresponding time. The middle panels show the SXT images with a preflare image subtracted, while the right panels show the same preflare–subtracted SXT image with contours of a preflare–subtracted 17 GHz image overlaid.

with this relationship, and with the small size of the event in x-rays (GOES class B7). However, the 06:00 UT radio burst is clearly not consistent with this relationship.

Remarkably, we know of one other event which seems to be inconsistent with the Kosugi et al. relationship, and it shares several other properties consistent with the 06:00 UT event presented here. The previous event (from 1989 June 23) was analyzed by White et al. [4]: it showed a linear rise to a sharp peak in 6 seconds (for this event, 6-7 seconds), followed by an exponential decay with an e-folding decay time of 18 seconds ($\sim$ 14 seconds for this event). It reached a peak flux of 1.7 sfu at 8 GHz, 15 GHz and 86 GHz (coincidentally the May 29 burst has a similar flux at 17 GHz). The GOES data showed no enhancement in soft x-rays in the 1989 event. The earlier event showed some other peculiarities not necessarily shared by this event: in that case, the event was not detected at 5 GHz with an upper limit of order 0.1 sfu implying a very sharp low frequency cut-off as well as a flat spectrum above 8 GHz. On 1993 May 29 the 3.8 GHz microwave patrol data unfortunately have a data gap at the time of the event, so we do not have other information on its spectrum. The impulsive nature and weak or absent thermal signatures leave little doubt that it represents acceleration of energetic electrons without the usual conversion of a significant amount of energy to hot thermal plasma as is usual in a flare. Such radio–rich events are not yet understood: White et al. [4] discussed a number of models for the earlier events.

In the case of the 93 May 25 flare, this event was mostly remarkable for the short rise and fall of the 17 GHz emission: a rise time of 4 seconds and a decay time of about 6 seconds. It has a time profile similar to those observed by White [6] at 3mm. It has a very simple source structure: compact and coincident in position in SXR, HXR and 17 GHz radio. It is polarized around its peak. However, unlike the May 29 event, it is a simple but ordinary flare in the sense that it has SXR associated with it.

Acknowledgements. This work was partially supported by NSF grant ATM 93–16972, by NASA grant NAG W–1541 and NASA/CGRO grant NAG 5–1450. The use of BIMA for scientific research is supported by NSF grant AST 93–14847.

References

[1]   Dennis, B. R., & Zarro, D. M.: *Solar Phys.*, **146**, (1993) 177
[2]   Kosugi, T., Dennis, B. R., & Kai, K.: *Astrophys. J.*, **324**, (1988) 1118
[3]   Kundu, M. R., White, S. M., Gopalswamy, N., & Lim, J.: *Astrophys. J. Supp.*, **90**, (1994) 599
[4]   White, S. M., Kundu, M. R., Bastian, T. S., Gary, D. E., Hurford, G. J., Kucera, T., & Bieging, J. H.: *Astrophys. J.*, **384**, (1992) 656
[5]   White, S. M., & Kundu, M. R.: *Solar Phys.*, **141** (1992), 347
[6]   White, S. M.: in *High Energy Solar Phenomena: A New Era of Spacecraft Measurements*, (eds.) J. Ryan & W. T. Vestrand, AIP Conf. Proc. **294**, (1994) 199
[7]   White, S. M., & Kundu, M. R.: in preparation (1994).

Retrieving Information from Digital Solar Radio Spectrograms

André Csillaghy

Institute for Astronomy, ETH-Zentrum, CH-8092 Zurich, Switzerland

Abstract. The quantity of observational data recorded by high resolution spectrometers is in constant increase and must be managed efficiently, in order to be best exploited. More precisely, operations for searching, comparing and counting events in the spectrogram archive must be available. However, the large, variable size of spectrograms, as well as their low signal to noise ratio, prevent the use of conventional methods to implement these tasks. We propose an alternative method, more adapted to the kind of data we manage. Spectrograms are reduced to smaller representations, called image icons. They are built from data density considerations in a higher-dimensional space. An image size reduction allows browsing among icons rather than among original images, a much faster operation. Ordering the archive by its content, which facilitates searching and comparing tasks, is investigated using self-organizing maps. Such a map conserves the topological order of the input space. This property may be used to automatically distinguish dissimilar emission types.

1 Exploiting large datasets requires efficient management operations

Digital, high resolution spectrometers — as well as many other scientific measuring instruments — increase their capacity to record observational data faster than workstation decrease their secondary memory (disk) access time. For instance, the solar radio spectrometers Phoenix and Ikarus at ETH have recorded roughly "only" 1 GByte per year, but the datataking rate will eventually increase by a factor of 100. This growth is required for investigations of relatively low-flux, narrowband structures like microbursts (Benz, 1994). Moreover, observation programs running on more than one instrument necessitate continuous observation, a memory expensive process. The exploitation of the data is therefore slowed down by relatively inefficient software possibilities to browse or to access it.

The data recorded in the past and future clearly need an improvement of retrieval techniques. More precisely, the following, typical operations must be available:

- find specific events (corresponding to a known date, time or type);
- search for similar structures;
- browse interactively among images in the archive;
- count events belonging to a given class.

We propose to (1) transform images into *icons*, which outlines the information content of the spectrogram in a compressed form — allowing fast browsing and (2) investigate a method to sort the archive by its content (*content-based indexing*) — admitting searches of similar structure, as well as counting operations.

2 Spectrograms: an unconventional kind of images

Images are commonly used to depict a scene containing geometrical, spatially distributed objects. In this usual case, the axes x and y of the image represent spatial coordinates of pixels (*picture elements*), which are equally distant in both x- and y-dimension.

However, images can be used to represent other kind of information. Spectrograms are examples of images which do not contain spatial information. The axes, representing time and frequency, have mostly an uneven frequency distribution. This representation is extremely vivid in solar radio astronomy, where the emission frequency of solar radio bursts is correlated with altitude in the solar corona.

The beginning of a type IV event and a diffuse continuum, extracted from the catalogue published by Isliker and Benz (1994), both recorded by the Phoenix spectrometer of ETH Zurich (Benz et al., 1991) are shown in Fig. 1.

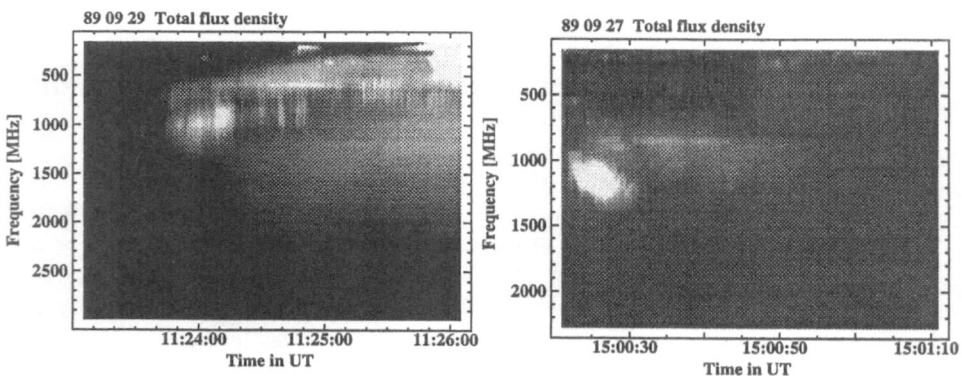

Fig. 1. Left: A very broadband type IV event, drifting to lower frequencies, recorded on September 29, 1989. Right: two long-duration diffuse continua (patches), slowly drifting, recorded on September 27, 1989.

In this article, we do not consider physical and semantic aspects of the spectral information. Following this rather technical point of view, we list common properties of spectrograms:

- Multidimensionality; each data item, (alternatively called *record*), is described by a three-dimensional tuple $< t, \nu, \phi >$ where t is the time, ν the frequency and ϕ the flux density. This record will eventually be extended to include polarization.
- Varying number of sweeps; in opposite to most spatial images, spectrometers usually record a variable amount of sweeps, ranging from a few to thousands. Therefore, some images are relatively small (short duration events), while other can even not be displayed in full resolution on the screen (type IV events).
- Widely varying information content; in spikes or microbursts, the enhanced flux — carrying the information — consists of relatively few pixels. In other structures, like type IVs, the majority of pixels in an image carry information on the structure (since such events are broadband and long duration).
- Large number of spectrograms; the total number of events recorded in Zurich since 1978 is estimated at 300'000, although only about 1 % can be accessed directly.

3 Building image icons

Most existing methods to compress images (Storer and Cohn, 1992) suffer from the fact that the compressed image cannot be used for display purposes before being decompressed first. Other ways of transformation which could produce small size, symbolic representations, like Voronoi Skeletons (Ogniewicz, 1993), assume an optimal signal to noise ratio. The (lossy) compression algorithm presented in section 3.1 avoids both shortcomings.

The properties an image icon must possess are summarized below:

- Compression: the icon must be much smaller in size than the original image, in order to be displayed fast and use economically disk space.
- Direct visualization: the icon is displayed without a decompression stage.
- Structure preservation: the structures formed in the original image by enhanced emission should also be visible in the icon.

3.1 Image representation in its attribute space

A spectrogram (and generally an image) need not to be stored in a bitmap. An alternative is a set of three-dimensional records, each consisting of three values <time, frequency, flux>. These values are called the record's *attributes*. The set spawn a three-dimensional Cartesian product space, the *attribute space*. In the latter, similar data points may form clusters, as sketched in Figure 2(a) for a simpler, two-dimensional example. In a spectrogram, a cluster is formed by a plateau, i.e. a region of similar flux in time and frequency.

The attribute space is partitioned into regions containing no more than a fixed number of records. (In other words, an adaptive grid is fitted to the

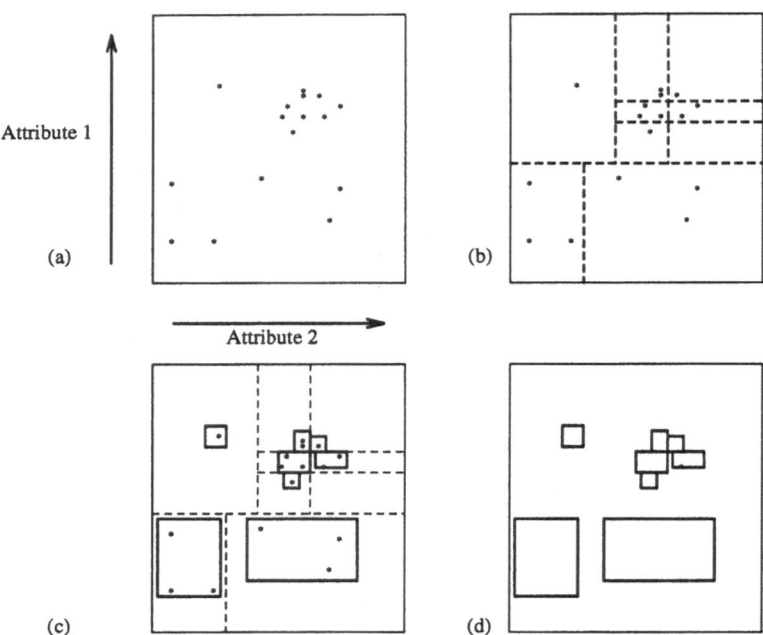

Fig. 2. A two-dimensional data set is represented in attribute space. (a) Similar data points, or records, form clusters. (b) The space is partitioned into regions containing no more than a given number of points, here 3. (c) For each region, a "box" is adjusted to the data depending on the records distribution in the region. (d) The real data is "forgotten" and only the regions are considered.

three-dimensional data space.) This representation is interesting because similar records, nearby in the attribute space, will generate regions of small volumes, while less dense data will generate larger volumes. Therefore, the shape of such a geometric region correspond to the data distribution in attribute space. This situation is shown graphically in Figure 2 (b).

By considering the region's geometry (if the partitioning has been done suitably), a lossy, compressed representation of the original image is produced (see Figure 2d).

3.2 Visualizing the attribute space

The attribute space is visualized by computing the data's mean and standard deviation for each attribute i and each region

$$\bar{x}^i = \frac{1}{N} \sum_{k=1}^{N} x_k^i \qquad i = 1, 2, 3 \tag{1}$$

$$\sigma^i = \frac{1}{N}\sqrt{\sum_{k=1}^{N}(x_k^i - \bar{x}^i)^2} \qquad i = 1,2,3 \qquad (2)$$

where n is the number of records in a region. The tuple $< \bar{x}^1, \bar{x}^2, \bar{x}^3, \sigma^1, \sigma^2, \sigma^3 >$ is named a *3-box*. It is displayed on the screen with a rectangle centered in $< \bar{x}^1, \bar{x}^2 >$, with an extent given by σ^1 and σ^2. The colour of the box is inferred from the value given by $\bar{x}^3$.

Therefore, the attribute space is visualized by displaying a list of 3-boxes. Two examples are shown in Figure 3, and may be compared with the original image shown in Figure 1. As one can notice, not all regions are displayed. This topic will be discussed in the next subsection.

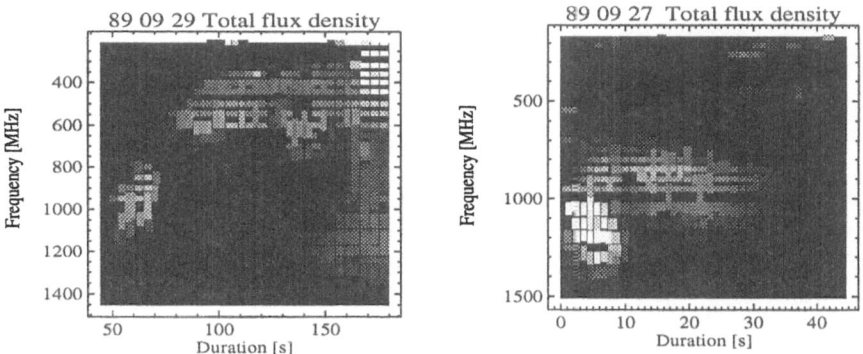

Fig. 3. The icons corresponding to the type IV and patch events shown in Figure 1. The whole attribute space has been divided into 2426 and 2371 regions, respectively. From them, only 384, respectively 363 are displayed. Since the icons are packed in 8640 bytes independently of the original size, the compression factor, defined by the ratio of the original image size to the icon size is 57.33 and 13, respectively.

3.3 Selecting boxes

The image compression is not only achieved by considering regions instead of data points, but also by selecting a subset of boxes. In fact, structural information is available already in a (small) number of boxes. Most regions can therefore be eliminated. The selection process is application dependent. Hereafter we present selection criteria used for visualizing and compressing solar radio spectrograms.

Because of their simple geometric shape, the selection process is done quite easily. For the case of spectrograms, the following criteria may be used:

- Effective number of records in a region. Only regions with more than 40 % of the maximal record number in a region are selected;
- Region's extension. The variance in time and frequency of a box, σ^1, σ^2, must be less than 2.5 % of the total range (the data in large regions is too much spread to bring information about structures).
- High flux. The 400 regions with highest mean flux, μ^3, which passed the former selection, are displayed.

More criteria may be applied, until a satisfactory selection is done. Moreover, threshold values given here are not sensitive and can vary roughly. They have lead to acceptable icons. However, new selection criteria, as well as threshold values will be tested more systematically in future work.

3.4 Compression factor and display time

As have been stated before, the number of regions is fixed (≈ 400), independently of the original size of the spectrogram. The compression factor is therefore

$$c = \frac{\text{original size}}{8640} \quad . \tag{3}$$

It increases linearly with the image size. In practice, the image icon is written in three FITS records (Wells et al., 1981), totalizing 8640 bytes. The compression is best for large images. For 1 MB images (which are common in radio astronomy), we get a factor of 115.7.

As the icon have always the same size, the unique display time, (reading from the disk included) is 1.47 second for a DEC Station 5000 running with a single user.

4 Using boxes to classify structures

Searching information in an image archive is clearly facilitated if the latter can be accessed (primarily) by its content, and not by an arbitrary search key, as the observation date and time (used today in many archives). A content-indexed archive gives the researcher the possibility to carry out comparisons between many similar images. Furthermore, counting images of a given class may lead to statistical information, as e.g. the fraction of type III bursts with spikes over a long period (more than a solar cycle).

Conventional cluster analysis such as the K-Means Algorithm (MacQueen, 1967), are not appropriate as method to build a content-based order in the spectrograms archive, for the following reasons. First, although belonging to a same class of emission, spectrograms can look quite different. Second, more than one emission process can appear at the same time as, for example, spikes and type III bursts. Finally, it is not always clear what emission processes underlie an emission type, hence the concept of "class" must be somewhat flexible.

4.1 Self-organization and adaptive processes

We are currently investigating an adaptive system to handle classification of spectrograms. Kohonen (1990) developed such a system, called *self- organizing map* (SOM) and relying on concepts from the artificial neural network technology. A detailed discussion of the self-organizing map and its underlying theory is found in the book of Kohonen (1989). A comprehensive introduction to the theory of neural computation is given by Herz et al (1991).

A simple SOM is shown in Figure 4. It is basically a two-dimensional artificial

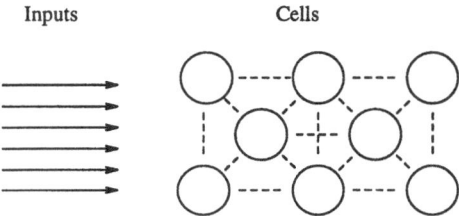

Fig. 4. A self-organizing map consists of an array of interconnected cells. A class of input records is associated to each cell of the network. The number of input is arbitrary.

neural network, where each *cell* is tuned to react to an input pattern class. Therefore, the spatial location of a cell reaction in the map corresponds to a specific region in the input space. The tuning of the map is done by learning from a set of records, called *learning set*. During the learning phase, the map has cells i, characterized by weight vectors m_i, updated after each input, following the formula

$$m_i(t+1) = \begin{cases} m_i(t) + \alpha(t)[x(t) - m_i(t)] & \text{if } i \in N(t) \\ m_i(t) & \text{otherwise} \end{cases} \quad (4)$$

where t is the (discrete) updating time and $x(t)$ is the current input vector. $N(t)$ describes a *neighbourhood* function determined by the application of the SOM. At last, $\alpha(t)$ is called the *adaptation gain*, $0 < \alpha(t) < 1$. After the learning phase, the weight vectors are left unchanged and determine which node will react to an (arbitrary) input.

Since the choice of vectors to be updated depends on a neighbourhood function, the topological structure of the input space is conserved in the map. Therefore, nearby cells react also to nearby input classes.

A simple example is shown in Figure 5. After the cells of a map have learned to react to the geometry of a set of cubes, two samples are "presented" to the network. The one with a small volume make a node react, say, at the bottom left of the map while the box with a large volume has a reaction placed at the top right. Moreover, the reaction of neighbouring cells corresponds to only slightly different boxes.

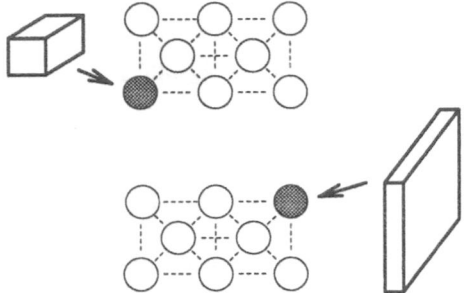

Fig. 5. A simple functional example of the self-organizing map is shown. Assume that the map is trained to react to boxes (characterized by 6 values). It will organize itself so that different shapes of boxes will make react different regions of the map.

4.2 Self-organizing maps, spectrograms and boxes

The properties of spectrograms, listed in section 2, make them inappropriate to serve directly as input to a SOM. Nevertheless, the icons are more adapted to this process, because using many small 6-dimensional boxes as input instead of a single, large bitmap is more flexible. The ability of the map to "recognize" different structures is hence larger. In addition, we teach the map to react only to selected boxes, and "forget" the boxes consisting of noise, unrealistic geometric shape, etc.

For a given icon, the cell reactions are recorded for every box of the icon. The number of times a cell has reacted is then computed, and a 20 × 10 map is produced. Results are shown in Figure 6. It is clear that different sets of boxes show a different pattern of cells reactions. Type IV emission made (for this specific learning set) cells react along a line starting at location < 11, 3 > and extending to the upper right corner. The reactions generated by the patch icon are mainly localized in the lines Nr. 0 and 9, and also the last column.

However, these are preliminary results, since it is currently not possible to link classes of emission to a specific part of the map. For instance, spikes have shown reactions located in many different parts of the map. We expect to solve this problem by calibrating the map more adequately.

5 The framework: an interactive system

Icons are useful only in an interactive environment. They are the foundations, with the content-indexed archive, of a system allowing their use in a user-friendly frame. Such a system includes a selection of spectrograms through a graphical icons menu interface. Instead of choosing a name from a list, the user searches, by browsing among events, for the icon roughly corresponding to his wishes. He then eventually selects it (e.g. by "pointing and clicking"). Only at this time, the real image is loaded for analysis.

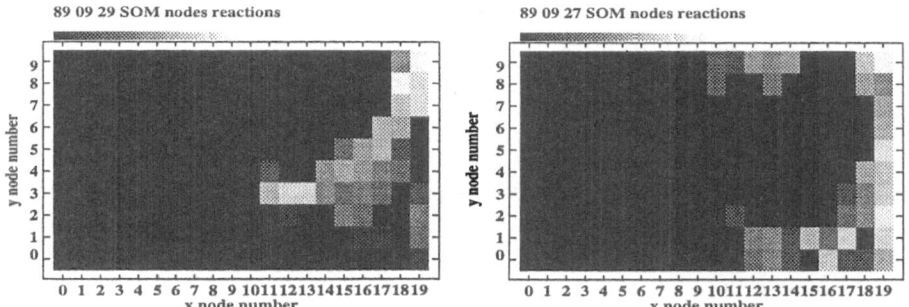

Fig. 6. Self-organized maps produced for the two events of Figure 1. The difference in the location of the cells reaction is clearly visible. Type IV emission have generated a pattern following a nearly diagonal line, whereas the diffuse continua correspond to reactions in the two lowest and the highest lines, and the 9th column of the map.

A specified domain of the archive can be selected. As the latter is content-indexed, the events of the considered domain are similar. The domain is thus the most likely to search for the specified kind of data.

6 Conclusions, future work

Undoubtedly, large amounts of data requires efficient, computerized data management techniques. For many recording instruments, and particularly in radioastronomy, image reduction (for browsing) and content-indexing (for searching) are two keywords for their implementation.

Considering data distribution in the attribute space make possible a transformation of the image into a structure consisting of a set of boxes. The geometry of such boxes reflects the original data density. Applying selection conditions on boxes to build an icon allows to compress the image by a factor determined by the application. For radio spectrograms, it varies linearly with the size of the image. Therefore, the larger the image, the better the compression factor. As icons have a fixed size, they are displayed in a given time (1.47 s for a DEC Station 5000), independent of the original size.

The simplicity of boxes makes them easy to use as input for a self-organizing map. The latter help to build a content-indexed archive, where images are sorted by their content and not by an arbitrary search key, as e.g. their date/time of observation.

The content-indexed icons are in the center of an interactive system currently developed for managing solar radio observations done by the Phoenix and Ikarus spectrometers. It will allow fast information retrieval, and, for example the search for narrowband, low flux structures in large data amounts.

However, many problems are still to be resolved. First, the selection of boxes is still empirical. We need to analyze more carefully the best values to find the selection thresholds. In the work of classification, the self-organizing map must be better calibrated, and the most effective number of cells to use has to be found. This is an experimental work which needs many trials.

Acknowledgements

The author acknowledges helpful discussions with A.O. Benz and H. Hinterberger. The spectral radio observations are carried out by all members of the Radio Astronomy Group. This work is partly supported by the Swiss Science National Foundation, Grants Nr. 20-40336.94 and Nr. 20-36395.92.

References

Benz, A. O., 1994, *this volume*
Benz, A. O., Güdel, M., Isliker, H., Miszkowicz, S., and Stehling, W., 1991, *Sol. Phys.* **133**, 385
Herz, J., Krogh, A., and Palmer, R. G., 1991, *Introduction to the theory of neural computation*, Addison Wesley
Isliker, H. and Benz, A., 1994, *A&AS* **104**, 145
Kohonen, T., 1989, *Self-Organization and associative memory*, Springer, Berlin (etc.), third edition, Springer Series in Information Sciences
Kohonen, T., 1990, *Proceedings of the IEEE* **78(9)**, 1464
MacQueen, J., 1967, in *Proceedings of the 5th Berkeley Symp. on Probability and Statistics*, University of California Press, Berkeley
Ogniewicz, R. L., 1993, *Ph.D. thesis*, Swiss Federal Institute of Technology, Zurich
Storer, J. A. and Cohn, M. (eds.), 1992, *Proceedings of the Data Compression Conference*, Snowbird, Utah, IEEE Computer Society, IEEE Computer Society Press
Wells, D. C., Greisen, E. W., and Harten, R. H., 1981, *A&AS* **44**, 363

Coupled Magnetohydrodynamic and Kinetic Development of Current Sheets in the Solar Corona

Bernhard Kliem

Astrophysical Institute Potsdam, An der Sternwarte 16, D – 14482 Potsdam, Germany

Abstract. This review considers (1) the stabilization of current sheets by shear flows, (2) the dynamical behaviour of coronal current sheets due to the interaction of magnetohydrodynamic (i.e., tearing and coalescence) and kinetic current-driven instabilities, (3) the resulting spatial and temporal scales of energy release, (4) the acceleration of particles in such an environment. The dynamical current sheet model naturally explains fragmentary energy release, and it may lead to a better understanding of particle acceleration and of the cross-field propagation of the flare disturbance.

1 Introduction

Current sheets are expected to form in magnetized plasmas with high electrical conductivity where magnetic diffusion is nearly negligible in comparison to the convective evolution of the system. Since diffusion is weak, the sheets possess small transverse scales in general and are directly observable only in the laboratory, planetary magnetospheres, and the solar wind. One exception is the Helmet streamer structure in the solar corona in which a large-scale metastable current sheet is embedded, but the scales of dynamic processes within this sheet are apparently less than the resolution limit. One can also trace the foot of the heliospheric current sheet in the lower solar atmosphere using magnetograms or soft X-ray images, the sheet itself cannot be observed on the Sun. Nevertheless, current sheets are supposed to occur in many places in the Sun as well as in the majority of magnetized cosmical plasmas. This assumption is based on two observational facts and it has also theoretical foundations. First, highly conducting magnetized plasmas show a tendency to develop a cellular structure with sharp boundaries instead of diffuse transitions. This can be seen most clearly in near-earth space. Second, such plasmas frequently show dynamical energy release events. These require dynamical dissipation of currents which, in turn, requires sufficient current densities and – for given magnetic field strength – correspondingly small spatial scales. As an alternative to sheets, the current in energy release regions might be concentrated into filaments, but, as will be discussed below, current filaments and current sheets are in fact closely related, filaments (or flux bundles) leading to sheets and vice versa. Third, ideal plasmas slowly driven at the boundaries pass through a sequence of equilibria but may reach a point at which no neighbouring equilibrium exist. Further perturbations lead to the formation of singularities in ideal MHD, i.e., current sheets for non vanishing resistivity. Such behaviour is in fact observed in numerical experiments.

The energy stored in current sheets may be released by a variety of processes. There are four classes of instabilities:

- ideal MHD instabilities,
- resistive MHD instabilities,
- kinetic current-driven instabilities,
- radiative/thermal instabilities.

Ideal instabilities release part of the free energy by rearranging the current in space, kinetic instabilities release energy by dissipating the current, resistive instabilities are a hybrid of these, and radiative/thermal instabilities can support the MHD instabilities by changing the resistivity. Steady-state energy release in a Petschek-type reconnection geometry appears to be irrelevant for the high Lundqvist numbers and long current sheets in the solar corona, due to the occurrence of instabilities. This was shown by Leboeuf et al. (1982), Lee & Fu (1986), and Yin & Ip (1991) who obtained formation of multiple X-lines as soon as the ratio of sheet length to width exceeded ~ 10 and the Lundqvist number exceeded $\sim 10^3$. There is no coherent picture of the energy release in current sheets. This review is concerned with the coupling of kinetic and magnetohydrodynamic instabilities which lead to rapid and intermittent energy release in current sheets. Characteristic scales of these processes are derived, supporting the hypothesis that current sheet dynamics is the underlying process for the production of energetic particles, which lead to radio, hard X-ray, and gamma-ray bursts.

Let us quantify the second argument above for the case of solar flares. Images at UV to X-ray wavelengths show that flares occur in magnetic loops or arcades of loops at spatial scales $\gtrsim 10^9$ cm (e.g., Shimizu et al. 1994). On the other hand, the hard X-ray (HXR) emission of flares occasionally shows temporal fine structure with rise times of single spikes within complex events down to 20 ms (Kiplinger et al. 1983, 1984). The mildly relativistic electrons which produce the HXR emission are known to carry a substantial fraction ($\sim$ 20 %) of the total energy released in a flare (Duijveman et al. 1982). The HXR fine structures must therefore reveal the temporal structure of the energy release itself. The observed timescale $\tau_{HXR} \sim 20$ ms, combined with the signal propagation velocity in MHD, the Alfvén velocity V_A, leads to spatial scales of the energy release in the lower corona $< 10^7$ cm in these extreme cases. If we require anomalous effects to come into play during the dynamic energy release, then the currents must possess spatial scales of order $10^4 - 10^5$ cm (see below, Eq. [9]).

The spiky appearance of the HXR and radio emissions supports the hypothesis that the flare energy release proceeds in a fragmentary manner. Since the characteristic electron energies of $10^4 - 10^5$ eV are reached in single HXR spikes, the flare could be composed of a big number of *separate* and possible largely independent energy release events. The flare as a whole shows the signatures of an instability: sudden onset and exponential rise of the emissions in the impulsive flare phase after a quiet period of energy storage. This suggests that an instability at a large scale ($10^9 - 10^{10}$ cm) drives instabilities at small scales ($< 10^7$ cm) in a big number of different places within the flare volume, and that the *small scale* instabilities produce the nonthermal particles. In this context the occurrence of short, electron-dominated events at gamma-rays is significant. In one event the electron bremsstrahlung spectrum was observed to extend up to the 40-65 MeV range, with a duration of the spike less than the instrumental intergation time of 2 s (Rieger & Marschhäuser 1990). This shows that the small scale energy release must be able to accelerate electrons to $\sim 5 \times 10^7$ eV.

Since only about 10 per cent of all HXR emitting flares show time structure below 1 second (Kiplinger et al. 1983), one might doubt the concept of fragmentary energy release. However, the radio emission of the impulsive flare phase *generally* carries signatures of fragmentation. The majority of flares emits type III radio bursts during the impulsive phase, big flares generally emitting *groups* of type III bursts. Each type III burst is the signature of a *separate* electron beam, and a few times 10^2 such bursts can be counted in some flares (Aschwanden et al. 1990). With increasing sensitivity and resolution of HXR instrumentation in recent years, the correlation between type III radio bursts and HXR spikes is seen to increase (with clear one to one correspondence in some cases; Aschwanden et al. 1993), suggesting that also the type III radio bursts reveal the history of the energy release process itself. The decimetric spike bursts, also frequently seen during the impulsive phase of compact flares, suggest an even stronger fragmentation (Benz 1985). Up to 10^4 short ($20 - 100$ ms), narrow band ($\Delta f/f < 0.02$) spikes occur in groups at various locations within a certain frequency interval, which suggests various source heights in the inhomogeneous corona. The bandwidth indicates source dimensions $< 2 \times 10^7$ cm. It is, however, possible that the decimetric spikes do not directly reveal the fragmentation of the flare energy release, since they occur with a high correlation but a time delay of several seconds to the HXR emission (Aschwanden & Güdel 1992).

One of the driving factors for the interest in current sheet dynamics is their property to release energy in a discontinuous manner for parameters typical of the solar corona, i.e., long sheet extensions and high Lundqvist (magnetic Reynolds) numbers, which resembles the fragmentary energy release in flares. (Compressibility, requiring low plasma-β, which is also typical of the low corona, further enhances the impulsiveness of the discrete energy release events in current sheets.) This behaviour is brought about in two different manners. The first is based on the occurrence of the coalescence instability during the nonlinear stage of the tearing instability (Leboeuf et al. 1982, Tajima et al. 1987, Priest 1985, Kliem 1988) and will be the main topic of this review. The second route to the occurrence of repeated discontinuous energy release events was explored in a series of papers by Lee and coworkers (Lee & Fu 1986, Shi et al. 1991 and references therein) and by Yin & Ip (1991). Their simulations considered a current sheet driven by a broad profile of plasma inflow towards the sheet in the

incompressible and compressible cases. The plasma is then squeezed out along
the sheet with a velocity which increases with distance and tears blobs of plasma
off the ends of the sheet. For magnetic Reynolds numbers exceeding a few 10^2
in the incompressible calculations of Shi et al. and ~ 2000 in the compressible
calculations of Yin & Ip, discrete blobs with closed magnetic field lines are ejected
from the ends of the sheet, and the reconnected flux and the induced electric field
show discrete maxima at the times of magnetic island formation. Although this
mechanism of discrete energy release is interesting and deserves further study
for solar coronal parameters, it will not be considered here any further since it
appears to induce less strong electric fields than the other mechanism and it does
not require the coupling of kinetic and magnetohydrodynamic instabilities. In
the following, we will concentrate on the interaction of the tearing, coalescence,
and kinetic current-driven instabilities, respectively, in coronal current sheets.
It will be demonstrated that such interactions can be expected for conditions
of the solar corona and that they naturally lead to fragmentary energy release
which peaks during repeated "coalescence events". Characteristic scales of this
so-called "dynamic" or "impulsive-bursty" regime of current sheet behaviour
are shown to correspond to scales of millisecond spikes in the radio and HXR
emission of flares.

The interaction of the instabilities will be considered for a single, isolated
current sheet, which alone is able to lead to repeated energy release events. This
does not contradict the notion of a 'statistical' flare (Lu & Hamilton 1991, Lu
et al. 1993, Vlahos 1995). These authors argue that the occurrence of flares,
in particular their occurrence distribution as a function of HXR flux, may be
described within a formalism of complex system dynamics, based on a model of
solar active regions which contain a multitude of potential energy release sites
distributed at random within the volume of energy storage. This approach is
promising but needs to be supplemented by an information about the concrete
plasmaphysical mechanisms of the energy release at the elementary scale, which
might be gained from a consideration of current sheets.

In the following sections, I will briefly discuss ways of formation of current
sheets in the solar corona, characterize the presumably most important instabilities in current sheets, present a model for "dynamic current sheet development"
based on the interaction of these instabilities, which reproduces the scales of
HXR and radio spike emissions, and demonstrate that such dynamic current
sheets are efficient sites of particle acceleration.

2 Formation of current sheets in the corona

The ideas on the formation of current sheets are based on the existence of *independent magnetic flux systems* in the low-β corona, which are rooted in the
high-β convective zone where turbulent as well as organized motions are imposed
at the footpoints of the coronal loops. Compared with coronal scales, these motions are sub-Alfvénic. The partly turbulent nature of the photospheric velocity
field suggests that individual flux systems are driven in an independent manner

by these motions, but this could not be established definitely since the photospheric velocity field cannot be measured completely. For example, if a flux system is seen to move in the corona, it is not known to what extend the moving flux system pushes the preexisting flux system(s) and to what extend the preexisting flux system is driven in a similar manner as the new flux system by organized motions in the convection zone.

Shearing motions at the scales of the sunspots exist without doubt and their effect on the coronal magnetic configuration has been studied in a number of numerical simulations. Since the driving velocities at the (photospheric) boundary are sub-Alfvénic, the system goes through a sequence of equilibrium configurations, with generally increasing but diffuse currents. For sufficient shearing the system may reach a state at which no neighbouring equilibrium exists and further shearing at the boundary leads to the occurrence of singularities (in ideal MHD). For finite resistivity, these singular surfaces are current sheets. The numerical simulations show a topology change at this point (Mikic et al. 1988, Inhester et al. 1992). Although there is a debate about the validity of the assumption of strict photospheric line tying and the requirement of field nulls in the initial configuration (Karpen et al. 1990) and whether a *singular* state is accessible in *finite* time due to *smooth* photospheric motions (Wang & Bhattacharjee 1994), it is clear that current sheets of small, but possibly nonvanishing width will form at least if field nulls are initially present. Possibly this occurs also if there are no field nulls initially. The existence of nulls in the coronal magnetic field is an open question, being probable, however, at least for the proposed complex magnetic configurations, as for example in Vlahos (1995). Also the small scale turbulent motions in the photosphere introduce a shear component of the coronal magnetic field, which presumably leads to a big number of small scale current sheets throughout the corona and might be responsible for the smallest flares and the heating of the corona (Parker 1988).

Also the relative motion of separate flux systems toward each other is supposed to lead to the formation of current sheets. This case is realized during the emergence of new flux into the atmosphere (Matsumoto et al. 1993) and during coronal mass ejections. The creation of *thin* current sheets in this situation is an assumtion, which may be justified *a posteriori* through the ability of the supposed current sheet to explain the energy release which is frequently observed in connection with newly emerging flux. For example, the flare described by Kiplinger et al. (1984), which showed a large number of subsecond spikes, occurred in connection with newly emerging flux of reversed polarity, and the flare which was particularly rich in decimetric radio spikes (Benz 1985) was triggered by magnetic flux intrusion of reversed polarity between two preexisting spots. The current sheets in this case can be considered as being continuously driven by inflowing plasma. The following section will present theoretical arguments that such sheets in fact shrink down to very small widths. Typical inflow velocities, as derived from the observations of arch filament systems, are of order $\lesssim 10^6$ $\mathrm{cm\,s^{-1}}$.

Relative motion of separate flux systems toward each other can also result from a magnetohydrodynamic instability of individual flux systems. A typical case is the kink instability of a flux system, induced by supercritical shear. Flux tubes become kink unstable if the field line twist exceeds $\sim 4.8\pi$ (Mikic et al. 1990). Since the motion is caused by an ideal instability, the flux system pushes against the surrounding flux with a velocity $\lesssim V_A$. At the boundary a current sheet with a strong inflow will be induced.

3 Instabilities in current sheets

A uniform current sheet is stable in ideal MHD, hence dynamic evolution may be initiated by a resisitive magnetohydrodynamic instability involving changes of the topology or by a kinetic current-driven instability.

3.1 Tearing mode instability

The basic resistive instability of a current sheet is the tearing mode, see the comprehensive review by Priest (1985). If the plasma is at rest, the tearing mode has a threshold current density orders of magnitude below the threshold of kinetic current-driven instabilities. Growth of tearing perturbations occurs at wavenumbers $k < 0.64\, l_{CS}^{-1}$, where l_{CS} is the current sheet half width. This requires that the ratio of sheet width to length satisfies $l_{CS}/L_{CS} < 0.1$. The length of the current sheet, $2L_{CS}$, is set by the external conditions which led to the formation of the sheet, e.g., the scale of newly emerging flux ($< 5 \times 10^{10}$ cm). For given magnetic field strength, Ampere's law then relates the maximum sheet width for tearing instability to a minimum current density:

$$j = \frac{c}{4\pi}\nabla \times \mathbf{B} \sim \frac{c}{4\pi}\frac{B}{l_{CS}} \tag{1}$$

which turns out to be four orders of magnitude below the kinetic instability threshold for typical coronal parameters. The growth rate of tearing modes at a given wavenumber is $\gamma_{TI} = \tau_{TI}^{-1} \approx (kl_{CS})^{-2/5}S^{-3/5}\tau_A^{-1}$, where $S = \tau_r/\tau_A = 4\pi\sigma l_{CS}V_A c^{-2}$ is the Lundqvist number, τ_r and $\tau_A = l_{CS}/V_A$ are the resistive and Alfvén time scales, respectively, $V_A = B(4\pi N m_i)^{-1/2}$ the Alfvén velocity, N the particle density and σ the electrical conductivity. Maximum growth occurs at a wavelength $\lambda_{max} \approx 4.5S^{1/4}l_{CS}$ and the minimum growth time has the well known scaling

$$\tau_{TI} \sim S^{1/2}\tau_A \tag{2}$$

We remark that numerical simulations of a *turbulent* plasma have shown growth of the tearing mode at wavelengths close to the minimum of $\sim 10l_{CS}$, i.e., much smaller than $S^{1/4}l_{CS}$ (Biskamp & Welter 1989). Let us evaluate τ_{TI} for typical parameters of the lower solar corona. Throughout this paper, $N = 10^{10}$ cm^{-3}, $T = 2.5 \times 10^6$ K $= 215$ eV, $B = 200$ G will be used for the purposes of numerical estimates. These values imply $\beta = 16\pi NTB^{-2} = 4.3 \times 10^{-3}$, $V_A = 4.4 \times 10^8$ cm s^{-1}, a plasma frequency $f_{pe} = 900$ MHz, and a cyclotron frequency $f_{ce} = 560$ MHz. The smallest current sheet half width derived in Section 3.2 for these parameters, $l_{CS} \approx 7 \times 10^3$ cm, gives $\tau_{TI} = 0.4$ s, which exceeds the rise time of the millisecond spikes. Consequently, the tearing instability based on the classical coronal resistivity cannot explain the shortest timescales of coronal energy release.

A shear flow along the current sheet, inevitable in current sheets which are driven by inflow of plasma, has a strong stabilizing effect on tearing modes. A simple estimate is obtained by requiring that the growth time of the magnetic islands, τ_{TI}, be smaller than the time the shear flow takes to distort the islands substantially. The latter time will be of order λ_{max}/u_{out} where u_{out} is the outflow velocity along the sheet. Mass conservation gives a relation between the inflow and outflow velocities:

$$u_i L_{CS} \sim \xi u_{out} l_{CS} \qquad (3)$$

where ξ is the compression factor. In the compressible case, i.e., low plasma-β and weak guide field component in the sheet, we have $\xi \sim \beta^{-1}$. We find that inflow velocities $u_i/V_A < 4.5(l_{CS}/L_{CS})\beta^{-1}S^{-1/4} \sim 6.6 l_{CS}/L_{CS}$ stabilize tearing modes. Bulanov et al. (1978) considered the problem quantitatively for the Harris equilibrium

$$\mathbf{B} = B_0 \tanh(y/l_{CS})\,\mathbf{e}_x \qquad (4)$$

and the following shear flow profile

$$\mathbf{u} = h\,x\,\mathbf{e}_x \qquad (5)$$

which was found to stabilize tearing modes at wavenumbers

$$k_x \geq S\,l_{CS}^{-1}\,(h\tau_r)^{-5/2} \qquad (6)$$

Using $k \geq 2\pi/L_{CS}$ we obtain the condition for the existence of a nonvanishing k-interval

$$M_A = \frac{u_i}{V_A} < 0.2\,\xi^{6/7}\,S^{-3/7} < 0.2 \begin{cases} \beta^{-6/7}S^{-3/7} & \text{for compressible plasma} \\ \\ S^{-3/7} & \text{for incompressible plasma} \end{cases} \qquad (7)$$

where Eq. (3) and the expression $h = 2M_A/(\xi\tau_A)$, which follows from $u(x = L_{CS}/2) = u_{out}$ and Eq. (3), have been used. Inequality (7) is equivalent to

$$l_{CS} < 0.2\,\frac{c^2}{4\pi\sigma V_A}\,M_A^{-7/3}\beta^{-2} \quad \text{(compressible case)} \qquad (8)$$

Using the coronal parameter values given above, we find for the compressible case

$$l_{CS} < \begin{cases} 4 \times 10^7 \text{cm} & \text{for } u_i = 1 \text{ km s}^{-1} \\ 2 \times 10^5 \text{cm} & \text{for } u_i = 10 \text{ km s}^{-1} \\ 8 \times 10^2 \text{cm} & \text{for } u_i = 100 \text{ km s}^{-1} \end{cases}$$

It should be noticed that Inequality (8) depends very sensitively upon the field strength (the r.h.s. is $\propto B^{16/3}$), slightly higher values of B relaxing the condition on l_{CS} significantly, slightly lower values strengthening this accordingly. For incompressible plasma, the requirement on the current sheet half width increases by a factor $\beta^{-2} \sim 5 \times 10^4$. Incompressibility is given if $\beta \gtrsim 1$ (which is not true in the lower corona) or if a nonnegligible guide field component, $B_z \not\ll B_{0x}$, is present in the current sheet. The latter condition is frequently satisfied in the corona, particularly for those current sheets which form due to increasing shear or due to instability of preexisting embedded flux tubes, but certainly not always: new flux may emerge in the photosphere with polarity opposite to the surrounding preexisting flux. Just those cases of new flux emergence appear to involve a higher flare probability than average, but also emerging flux systems with non-reversed polarity produce flares. One should also notice that Eq. (3), based on a one-dimensional simplification of the geometry, is only an order of magnitude estimate. Although a definite conclusion cannot yet be drawn from this consideration of tearing mode stabilization for the case $B_z = 0$, it appears probable that stabilization occurs for the majority, if not all, of the current sheets driven by inflow. This strong effect of the inflow is due to the high conductivity of the coronal plasma, which determines tearing mode growth (Eq. [2]). The inflow or continued shearing motions may thus be able to reduce the width of the current sheet until the threshold of a kinetic instability is reached.

3.2 Kinetic current-driven instability

There is a multitude of kinetic current-driven instabilities. These have been reviewed, among others, by Papadopoulos (1979), Spicer & Brown (1981), Norman & Smith (1978). Depending on the ratio of the guide field component and antiparallel field component, B_{0z}/B_{0x}, parallel or cross-field instabilities are relevant. We are interested here in order of magnitude estimates of the threshold current density for these two cases. This threshold or critical current density is conveniently expressed via a critical electron-ion drift velocity, $j_{cr} = Nev_{cr}$.

The lower hybrid drift (LHD) instability, which belongs to the group of cross-field instabilities, appears to have the lowest threshold: Huba et al. (1978) find $v_{cr\,LHD} \approx 0.2[1.0]v_{ti}$ for $\beta \leq 1.0[2.5]$ and $f_{pe}^2/f_{ce}^2 \gg 1$ (cf. their Figure 3). Here $v_{ti} = (T_i/m_i)^{1/2}$ is the ion thermal velocity. This instability is quenched by high values of β which occurs in the center of the sheet for $B_{0z} = 0$ (at the current density maximum), but, since it has the lowest threshold even for $\beta \sim 1$, it occurs quite close to the current sheet center. The anomalous diffusion of the magnetic field caused by the instability then creates steepening magnetic field gradients which propagate jointly with the instability region right into the center of the sheet (Drake et al. 1981) and lead to anomalous dissipation in a manner similar to the other kinetic instabilities. Although a nonvanishing guide field component has a stabilizing influence, the instability continues to exist with relatively low threshold even for substantial B_{0z}, Huba et al. (1982) find $v_{cr\,LHD} \sim v_{ti}$ for $B_{0z}/B_{0x} \sim 1$ (the maximum value they considered).

For currents flowing parallel to the magnetic field, i.e., highly sheared current sheets, the ion acoustic instability and the electrostatic ion cyclotron instabilitiy may have the lowest threshold. The critical drift velocity of the ion acoustic instability is (Papadopoulos 1977)

$$v_{cr\,IA} = \left(\frac{T_e}{m_i}\right)^{1/2} \left[1 + \left(\frac{m_i T_e}{m_e T_i}\right)^{1/2} \exp\left(-\frac{3}{2} - \frac{T_e}{2T_i}\right)\right]$$

This instability requires $T_e > T_i$ so that $v_{cr\,IA}$ always exceeds the ion thermal velocity. The ion cyclotron instability has a critical drift (Benford 1983)

$$v_{cr\,IC} = 0.3\,\frac{T_i}{T_e}\,v_{ti}$$

which is generally below the ion thermal velocity.

Depending on the values of B_{0z}/B_{0x} and T_e/T_i, each of these instabilities may be relevant, but generally the lower hybrid drift instability or the ion cyclotron instability will have the lowest threshold. In both cases, $v_{cr} \lesssim v_{ti}$. In the following, v_{ti} will be used as a conservative estimate of the critical drift. From Eq. (1), the corresponding critical current sheet half width then is

$$l_{cr} \sim 4\beta^{-1} r_{ci} \qquad (9)$$

where $r_{ci} = v_{ti}/\omega_{ci}$ is the ion cyclotron radius. For our coronal parameter values, $l_{cr} = 35 \times B[\text{Gauss}]$ cm. The corresponding Lundqvist number is $S = 6.5 \times 10^8$. The growth time of the kinetic instabilities is of the order of several ω_{LH}^{-1} or ω_{ci}^{-1}, respectively, i.e., instantaneous on magnetohydrodynamic time scales. Here $\omega_{LH} = (\omega_{ce}\omega_{ci})^{1/2}$ is the lower hybrid frequency. The instabilities lead to plasma heating, they may cause radio emission (Benz & Wentzel 1981) and particle acceleration (Wu et al. 1981), and there is a resonant coupling to modified (oscillating) tearing modes (Sen & Sundaram 1981), but their main effect on the macroscopic development of the current sheet is the anomalous resistivity resulting from the wave turbulence. The anomalous resistivity exceeds the classical value by several orders of magnitude for all current-driven instabilities. For the LHD instability the anomalous collision frequency is $\nu_{an} \sim \omega_{LH}/2$ (Papadopoulos 1979) which yields the anomalous resistivity

$$\sigma_{an\,LHD}^{-1} = \frac{4\pi\nu_{an}}{\omega_{pe}^2} = \frac{v_{ti}}{c}\beta^{-1/2}\frac{\Lambda}{\ln\Lambda}\sigma^{-1} \qquad (10)$$

where Λ is the number of particles per Debye sphere and σ^{-1} is the classical resistivity. The inverse relation holds between the anomalous and the classical Lundqvist number, and our coronal parameters yield $S_{an}/S = 4.5 \times 10^{-5}$ for the case of LHD instability. Based on S_{an}, tearing perturbations grow much faster and condition (7) on the inflow velocity is relaxed by nearly two orders of magnitude. (In tokamak plasmas the opposite can be true: suppression of the tearing mode by turbulent fluctuations [Esarey et al. 1986], but this effect is based on stochastic electron orbits in overlapping magnetic islands and is not relevant to extended cosmic current sheets.)

Since the kinetic current-driven instabilities grow practically instantaneously on MHD time scales, one cannot expect that they occur all along an extended sheet (as has been supposed, e.g., in Heyvaerts et al. [1977]; we have $L_{CS} < 10^{10}$ cm and $l_{CS} \sim l_{cr} \sim 7 \times 10^3$ cm) and over the full width of the sheet. Instead, due to fluctuations in the plasma parameters and the inflow velocity profile, onset of kinetic instabilities will occur at a few points distributed along the sheet and be restricted to the current density maximum, i.e., the sheet center. These points will be separated by at least several mean free path lengths ($\lambda_{mfp} \sim 5 \times 10^6$ cm). Once the threshold is reached at a point in the current sheet, the ensuing rapid diffusion and fieldline reconnection leads to local inequilibrium and sets up a flow pattern consistent with the tearing mode, i.e., in x direction away from the newly created X-line, which sucks in plasma and field lines towards the sheet in y direction (Ugai 1984, Scholer & Roth 1987). This leads to rapid field line reconnection over a y range broader than the range where anomalous resistivity is excited, and it sustains the supercritical current density at the X-line for several Alfvén times (Kliem & Schumacher 1995). A rapid filamentation of the current sheet similar to the usual tearing instability results, leading to irregularly sized current filaments (magnetic islands), which we refer to as *induced tearing*. Since the resistive part of this evolution (at the X-lines) is governed by σ_{an}, we expect that induced tearing proceeds at the time scale

$$\tau_{TI\,an} \sim S_{an}^{1/2} \tau_A$$

which is supported by simulations of Hoshino (1991). Due to the large separation of the X-lines that are created by spontaneous kinetic instability, the induced tearing creates long thin sections of high current density in place of the X-lines. These sections between the newly created magnetic islands may become unstable with respect to so-called secondary tearing (Scholer & Roth 1987), which forms current filaments of smaller size in x direction.

The coupling of the tearing instability with kinetic instabilities leads to much more rapid growth than the coupling with the radiative instability, which can increase the tearing mode growth rate over the classical value, Eq. (2), by reducing the temperature in the vicinity of the X-lines (Van Hoven et al. 1984) and would be relevant in the absence of kinetic instabilities.

3.3 Coalescence instability

Having found that tearing or induced tearing leads to filamentation of the current sheet, we are now interested in the further development of the current filament chain. In the absence of further perturbations, the tearing mode enters a phase of algebraic (reduced) growth and saturates when the width of the magnetic islands approaches l_{CS} (e.g., White 1986). However, in a disturbed and dynamic environment the configuration is immediately unstable with respect to the approach of neighbouring current filaments (Pritchett & Wu 1979). If nonvanishing resistivity is taken into account, this instability leads to complete coalescence of neighbouring current filaments. The tearing and the coalescence instabilities are driven by the same force, the attraction between currents flowing into the same direction. The coalescence instability may thus be viewed as the completion of the collapse of the current sheet (or of sections within the sheet), which is initiated and enabled by the tearing instability. It is therefore not surprising that it is the coalescence instability which releases the main part of the free energy in the current sheet (Leboeuf et al. 1982). This energy release occurs in a discrete manner by successive coalescence of neighbouring islands. Such *coalescence events* may correspond to individual fragments of the flare energy release.

There is no complete analytical description of the coalescence instability whose properties have to be inferred mainly from numerical experiments (Pritchett & Wu 1979, Biskamp & Welter 1980, 1989, Leboeuf et al. 1982, Tajima et al. 1982, 1987, Bhattacharjee et al. 1983, Richard et al. 1989). The coalescence instability has two phases. Initially it is an ideal magnetohydrodynamic instability that leads to a chain of pairs of current filaments lying side by side. This is followed by a resistive phase of reconnection between the approaching filaments and merging into single filaments. These two phases of the coalescence instability, which are more clearly distinguishable for high Lundqvist numbers, will be denoted by CI_1 and CI_2, respectively. The characteristic time scale of the ideal phase is essentially the Alfvénic time scale based on the distance λ_{CI} between approaching current filaments

$$\tau_{CI_1} = \zeta^{-1} \frac{\lambda_{CI}}{V_A}, \qquad \zeta = \frac{u}{V_A} \sim 0.1 - 1 \tag{11}$$

Here u is the velocity of filament approach. The value of ζ is strongly dependent upon the degree of current concentration at the magnetic O-lines, it approaches zero if the current is uniformly distributed in the sheet (no filamentation). The range of ζ quoted here is representative of moderately to strongly peaked current density profiles within the filaments. The scale λ_{CI} is set by the current filament formation process, it ranges from several l_{CS} to $\lambda_{max\,TI}$ or several λ_{mfp}, respectively, for classical or induced tearing (Sec. 3.1, 3.2). For this whole range we have $\tau_{CI_1} \ll \tau_{TI}$ for typical values of S ($> 10^8$). The typical time scale of the resistive coalescence phase, τ_{CI_2}, depends strongly on the plasma parameters and is not well known for high S; since the rapidly approaching current filaments enforce reconnection during the CI in general more strongly than external flows do so during the TI, τ_{CI_2} lies between τ_{TI} and τ_{CI_1}. In fact, in the favourable case of strongly peaked current density and compressible plasma, $\tau_{CI_2} \approx \tau_{CI_1} \approx \lambda_{CI}/V_A$ has been observed in simulations (Tajima et al. 1982, Bhattacharjee et al. 1983).

In a three-dimensional consideration, the coalescence of a current filament pair is supposed to start at a certain location and then to propagate along the filament pair with approximately Alfvén velocity. At any given point the process has the duration $\tau_{CI_1} + \tau_{CI_2}$, but the total duration of the coalescence as seen by a remote observer is of order $\tau_{\parallel} = L_{\parallel}/V_A$. Here $L_{\parallel}$, the extent of the coalescence process along the filaments, is the quantity which should correspond to observational source sizes, see Fig. 1. For $L_{\parallel} = 2 \times 10^7$ cm (the estimated size

of the shortest HXR and decimetric spike bursts sources) our parameter values
yield $\eta_\| = 25$ ms.

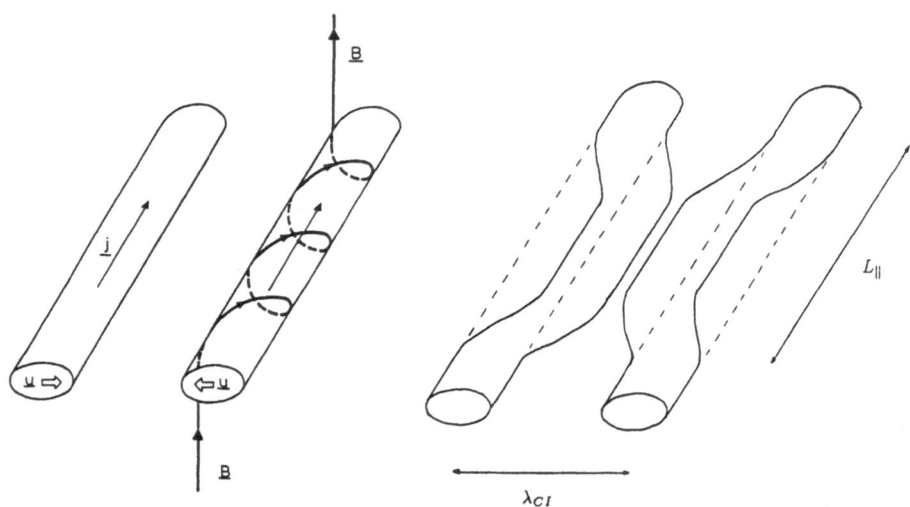

Fig. 1. Sketch of current filament coalescence.

Due to its rapid development, the coalescence instability is likely to lead locally and temporarily to supercritical current densities and the corresponding
kinetic instabilities. First, for the high Lundqvist numbers typical of the solar corona, reconnection of the field lines does not proceed rapidly enough and
magnetic flux will be piled up between the approaching filaments, especially in
the case of fast coalescence ($\zeta \to 1$). A study of this process has not yet been
performed but it was indicated by the simulations of Pritchett & Wu (1979).
Second, rapidly coalescing filaments create a strong flow $u_x \to V_A$ at their trail-
ing edge. This results in an elongated section of the sheet which is narrowed by
the ensuing inflow of plasma towards the sheet, such that $j > j_{cr}$ may occur.
This behaviour is well known in a single X-line configuration (Ugai 1984, Scholer
& Roth 1987), and it occurs in a multiple X-line configuration, too.

It has not yet been possible to follow the sketched sequence of instabilities in
a single numerical experiment, but each step is supported by certain simulations.
Figure 2 shows an example of induced tearing and subsequent coalescence.

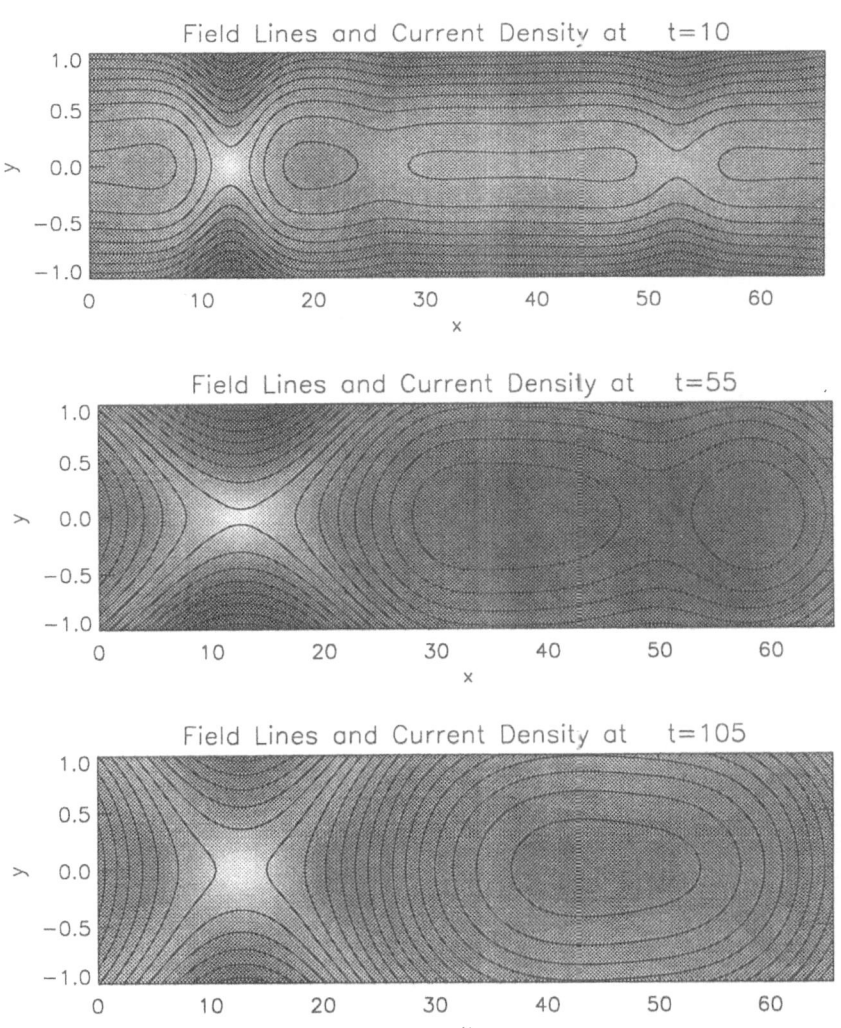

Fig. 2. Harris equilibrium (4) initially disturbed by setting anomalous resistivity (of different amplitude) at $x = 13$, 27, 53, $y = 0$, and $0 < t < 2$. Quantities are normalized to l_{CS} and τ_A, respectively. The box is periodic in x, symmetric in y and open at $y = 5$ (250×80 grid, compressible case, $S = 2000$, $\beta = 0.1$). Magnetic field lines and current density are overlayed. Induced tearing has produced three current filaments by $t = 10$, consistent with $\tau \sim S_{an}^{1/2} \tau_A$. Subsequently complete coalescence occurs with $\zeta \sim 0.3$. Grayscales are normalized to the current density peak (which is located at an X-line and decays during the run) in each panel separately (Kliem & Schumacher 1995).

4 Dynamical current sheet evolution and scales of fragmentary energy release

Our discussion of instabilities in current sheets leads to the scenario of a *dynamic current sheet*, also known as *impulsive bursty reconnection* (Leboeuf et al. 1982, Priest 1985, Tajima et al. 1987, Kliem 1988). First, tearing leads to filamentation, then rapid coalescence of the current filaments occurs. This creates elongated, nearly uniform sections in the current sheet between groups of coalescing filaments, these are again unstable with respect to (secondary) tearing. Consequently, the process may repeat in a cyclic manner in sections of long current sheets. This allows for a big number of separate energy release events which occur primarily as successive *coalescence events* of individual current filament pairs (Kliem 1988). This is sketched in Fig. 3. Recent simulations by Malara et al. (1992) support this heuristic scenario (Fig. 4).

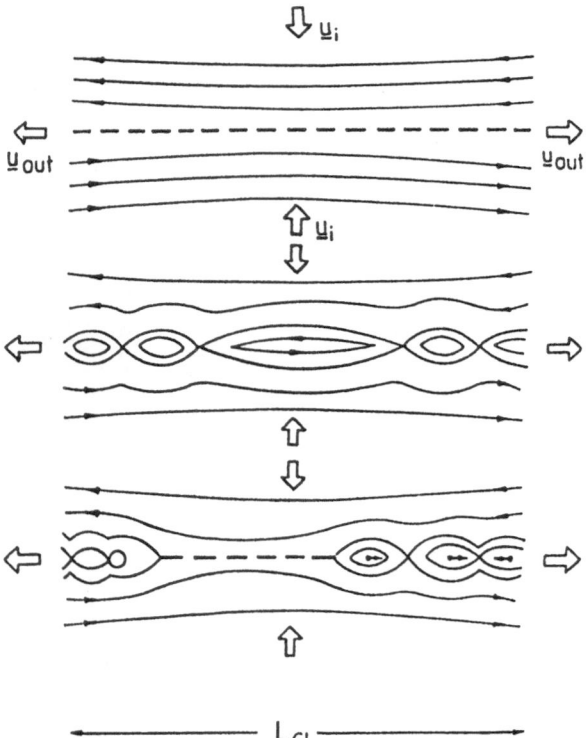

Fig. 3. Basic cycle of the dynamic current sheet model: current filaments are produced by tearing instability, the coalescence of filaments then re-creates sections of characteristic length L_{CI} free of filaments in which the process may repeat.

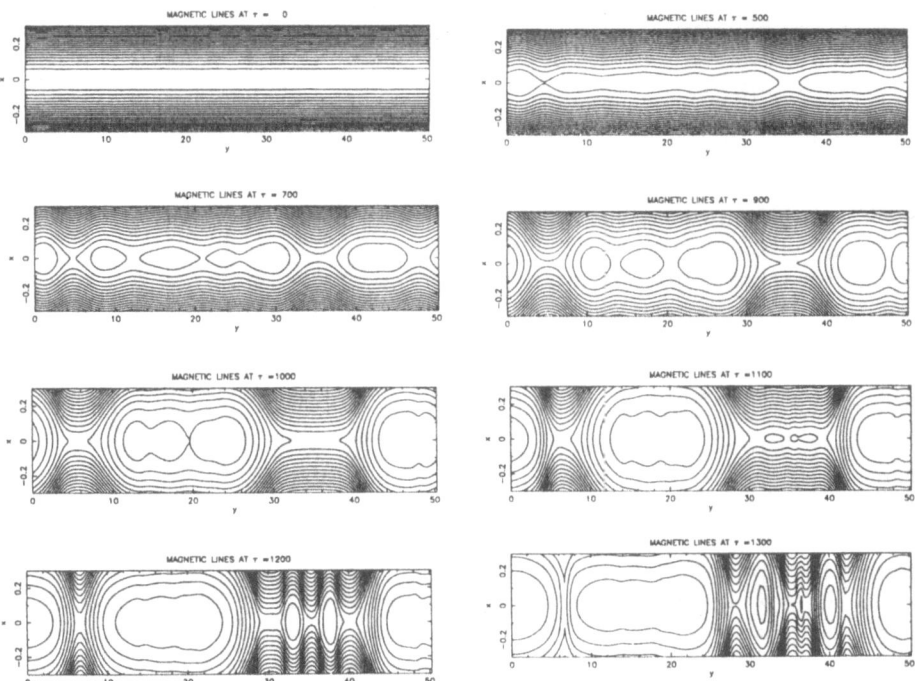

Fig. 4. Two-dimensional, incompressible, resistive MHD simulation of the evolution of a long current sheet ($L_{CS}/l_{CS} = 40$; only central part shown) at high Lundqvist number ($S = 2000$), with free boundary conditions in x, periodic boundary conditions in y, and uniform resistivity (from Malara et al. 1991). Time is in units of $\tau_a/2$ in this figure. The growth of tearing modes ($t = 0 - 700$) and a sequence of coalescence events on a shorter time scale ($t = 900 - 1100$ and $t > 1200$), leading to partial re-creation of the original current sheet configuration, can be clearly seen. The formation of structures in these sections at $t > 1100$ may be caused by secondary tearing.

A continuing inflow towards coronal current sheets occurs under various conditions, in particular in sheets which are driven by newly emerging flux or by an unstable coronal flux system (Sec. 2). Except for very small inflow velocities ($u_i < \sim 10 \text{ km s}^{-1}$), the flow will stabilize tearing modes that are based on the low classical resistivity of the corona, which leads to shrinking of the sheet width to the critical value for kinetic instability onset and consequently to induced tearing. Since the coalescence instability then rapidly sets in, the internal dynamics, being of order $\lesssim V_A$ (i.e., $\gg u_i$ in general), will dominate the flow in the neighbourhood of the current sheet, which effectively prevents further shrinking of the sheet width. Also under the action of the repeated cycles of tearing and coalescence, the current sheet half width is expected to stay close to the critical value, $l_{CS} \sim l_{cr} \sim 4\beta^{-1} r_{ci}$, where the narrowing of the sheet by the inflow and the broadening of the sheet by the flow connected with the instabilities and by the partial dissipation of the current layer balance on average. The tearing-coalescence cycles probably lead to intermittent occurrence of anomalous resistivity.

A characteristic scale, L_{CI}, of the tearing-coalescence cycles can be estimated from the condition that the total coalescence time of the current sheet section which shows cyclic behaviour, $L_{CI}/(\zeta V_A)$, is comparable with the time scale of new filament generation (secondary tearing) after stage 3 of Fig. 3, which is $\gtrsim \tau_{TI\,an} \sim S_{an}^{1/2}\tau_A$. (Scholer & Roth [1987] have numerically found about four times this value as the time scale of secondary tearing.) Roughly after this time has elapsed, the current sheet section emptied by filament coalescence will have formed new filaments, and the process may repeat. Then we have

$$L_{CI} \gtrsim \zeta S_{an}^{1/2} l_{cr} \tag{12}$$

which, with $\zeta \sim 1$, is of order $\gtrsim 10^6$ cm for our parameter values. It is even conceivable that several 'active' sections, which support independent tearing-coalescence cycles, exist in very long sheets where $L_{CS} \gg L_{CI}$.

We have suggested that the energy release occurs basically at the scales of the coalescence instability in a filamentary current sheet close to the threshold width for onset of kinetic current-driven instabilities. In the plane perpendicular to the current flow these scales are of order $l_{cr} \times \lambda_{CI}$ where λ_{CI} lies in the range $\sim (10 - 4.5S^{1/4})l_{cr}$ for the case of classical tearing and $\sim (10l_{cr} -$ several $\lambda_{mfp})$ for induced tearing, respectively (Sec. 3.1, 3.2). Numerically, $\lambda_{CI} \sim (7 \times 10^4$ cm$-$ several 5×10^6 cm). It is a requirement of the coalescence instability that the typical scale in the direction of current flow, $L_\parallel$, is much larger than the distance between filaments, λ_{CI}. In comparison, the sizes of the acceleration regions underlying the shortest HXR spikes, are of order 10^7 cm (Sec. 1). Typical time scales of coalescence events, $\tau_\parallel = L_\parallel/V_A$, are bounded by $L_\parallel/V_A \gg \lambda_{CI}/V_A > 7 \times 10^4$ cm$/V_A = 0.2$ ms. With a value of $L_\parallel$ somewhat smaller than typical loop lengths in the corona, $L_\parallel = 3 \times 10^8 - 10^9$ cm, we find an upper limit $\tau_\parallel < 0.7 - 2$ s.

Let us finally estimate the energy released by coalescence events (Kliem 1991). For coalescence of two current filaments with initial distance nearly twice their diameter, Tajima et al. (1982) have found that the released energy is about one sixth of the energy that would be released by complete approach of two current carrying wires of similar geometry. Incorporating this factor into the expression for the released energy during approach of n current carrying wires of diameter a $(\sim 2l_{CS})$, mutual distance R, and length $L_\parallel$, we obtain

$$W_{CI} = \frac{(n-1)^3 R^2 L_\parallel B^2}{24\pi^2 n} \ln\left(\frac{R}{a}\right) \tag{13}$$

Since $(n-1)R = L_{CI}$, this is a substantial fraction of the magnetic energy contained in the volume $L_{CI}^2 L_\parallel$ for moderate values of R/a (< 10). It is clear that a factor smaller than $\frac{1}{6}$, which has yet to be determined, must be applied for large values of R/a. Equation (13) is therefore only a very rough estimate, but it shows that W_{CI} can be substantial. With our parameter values and $a = 2l_{cr}$,

$R/a = 2$, $L_\parallel = 10^8$ cm, we find for coalescence of two current filaments $W_{CI}(n = 2) = 5 \times 10^{18}$ erg. The energy for coalescence of ~ 40 filaments, filling a range $\Delta x = L_{CI} = 10^6$ cm, is $W_{CI}(n = 40) = 1.3 \times 10^{22}$ erg — in the range of nanoflares. Coalescence over an observable lateral extent ($\sim 10^8$ cm) and a length $L_\parallel \sim 10^9$ cm would release a substantial fraction of 1.3×10^{29} erg.

5 Particle acceleration in dynamic current sheets

Besides the necessity of reproducing the observed spatial, temporal, and energetic scales, a crucial test of any energy release model is its ability to accelerate particles. The requirements for impulsive flare phase acceleration are substantial: ~ 20 % of the released energy goes to energetic particles; the particles must be accelerated to a large fraction out of the *thermal* distribution; up to $\sim 10^5$ eV; within $\lesssim 10^{-2}$ s; occasionally up to $\lesssim 10^8$ eV. Mainly electrons are accelerated. To investigate acceleration in a dynamic current sheet, test particle calculations have been performed in a field model representative of coalescence in a multiple X-line current sheet geometry (Kliem 1994). The results show that the strong electric field induced by the coalescence instability is indeed able to energize electrons and protons out of their thermal distributions up to $\sim 10^6 - 10^7$ eV and $\sim 10^7 - 10^8$ eV, respectively. This occurs at the coalescence time scale (Eq. [11]). The multiple X-line configuration is essential since it leads to trapping of a large fraction of the particles in the configuration, which enables sufficient acceleration — in contrast to a single X-line geometry where the particles generally quickly escape from the vicinity of the X-line and do not experience sufficiently strong acceleration in spite of the existence of substantial electric fields near the X-line. The following analytical model of the electromagnetic field was considered (Fig. 5):

$$B_x = -B_0 \frac{\sinh(y/l)}{\cosh(y/l) + \epsilon \, \cos(x/l)} \tag{14}$$

$$B_y = -B_0 \frac{\epsilon \, \sin(x/l)}{\cosh(y/l) + \epsilon \, \cos(x/l)} \tag{15}$$

This magnetic field, known as the Fadeev equilibrium, was used as initial condition in many simulations of the coalescence instability. The parameter ϵ ($0 \leq \epsilon < 1$) controls the current localization within the islands, its choice was $\epsilon = 0.3$, similar to, e.g., Pritchett & Wu (1979), which makes the island half width $z_h \approx l$. An electric field into the invariant direction was formally added to simulate the occurrence of convective electric fields due to filament motion during the coalescence instability. The asymmetry of magnetic islands about the O-lines during the coalescence process, particularly pronounced for compressible plasma, is taken into account by the electric field

$$E_z = E_{z0} \, \cos(\frac{x}{2l}) \, \cos^2(\frac{x}{4l}) \tag{16}$$

The configuration (14-16) is not self consistent (since $\nabla \times \mathbf{E} \neq 0$ implies $\partial \mathbf{B}/\partial t \neq 0$), but it is expected to reveal the particle orbit characteristics (meander motion and drift motion), on which the acceleration is based, qualitatively correctly.

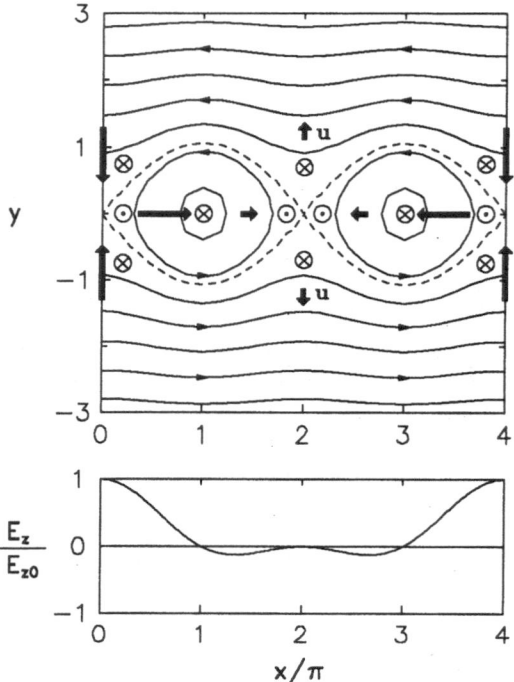

Fig. 5. Model configuration (periodic in x) for calculation of test particle trajectories. The separatrix is shown dashed. The flow u implied by $\mathbf{B}(x,y)$, $E_z(x)$ is indicated by heavy arrows. Arrows out of the plane give the direction of the ∇B drift for positive charges. Lengths are normalized to l $(= l_{CS})$ in this and subsequent figures (from Kliem 1994).

Furthermore, the orbit calculations show that the particles acquire a large portion of their final energy in a time interval $\ll \tau_{CI_1}$, much shorter than the time scale of magnetic field evolution.

This investigation was focused at the acceleration by a perpendicular electric field in order to show that the coalescence instability always involves strong particle acceleration (even without the occurrence of anomalous resistivity) due to the strong convective electric field connected with the plasma motions. The convective electric field is generally considered as being unable to contribute to particle energization, but there are two important exceptions to the rule, relevant to the case under study:

(1) at magnetic X- and O-lines (similar to the situation in homogeneous current sheets) particles perform a meander-like motion and decouple from the fluid,
(2) particles drift across the magnetic field due to inhomogeneities in $\mathbf{B}$; those particles that enter the region where an $\mathbf{E}_{conv}$ is induced in an inhomoge-

neous $\mathbf{B}$ from a region with a different local flow velocity $\mathbf{u}$ and are collisionless at the scales of the $\mathbf{u}$ and $\mathbf{B}$ variations, experience energy gain or
loss, respectively, depending on the sign of $\mathbf{E}_{conv}$.

The calculations show that the majority of particles is energized during (repeated) phases of ∇B drift motion along the strong electric field in the vicinity
of X-lines. Those particles which are already sufficiently superthermal at $t = 0$
can enter a meander orbit at one of the O-lines and reach the highest final
energies there.

In a typical run the relativistic equations of motion were integrated for 10^3
test particles and a time interval $\Delta t = \tau_{CI_1}$ ($= 2\pi l/\zeta V_A$ for the Fadeev equilibrium). The test particle were initially randomly distributed over the area of
one magnetic island, $0 \leq x_0/l \leq 2\pi$, $-1 \leq y_0/l \leq 1$, $z_0 = 0$, and the surface of
a sphere in velocity space (all particles had the same initial energy). Figure 6
shows the final positions and energy distribution of 10^3 protons at $t = \tau_{CI_1}$. It is
seen that the majority of particles has gained energy and is trapped in the configuration. Particles with orbits close to the separatrix surfaces travel distances
of many island lengths in the sheet, especially the more energetic particles. This
lateral motion (essentially cross-field if a strong guide field component exists)
can reach velocities $v_\perp \sim 3V_A$ and trigger activity in other places of the sheet.

Characteristic maximum particle energies are shown in Fig. 7a as a function
of the initial particle energy and the electric field strength. Since the actual
maximum energy in a test particle distribution sensitively depends upon the
details of the closest approach of one of the particles to an X-line, "characteristic"
maximum energies (obtained by disregarding the uppermost 0.5, respectively 1.0,
per cent of the final test particles) are given. The electric field is normalized to the
maximum convective field $E_0 = V_A B/c$ such that it is equal to ζ (cf. Eq. [11]).

For $r_{ci}(0)/l \lesssim 10^{-2}$ and $E_{z0}/E_0 = \zeta \lesssim 10^{-1/2}$ the acceleration due to the ∇B
drift in the vicinity of X-lines dominates. In this range there is only a weak
dependence upon the initial particle energy. This shows that the acceleration
is effective for the bulk of the thermal particles (the thermal energy of coronal
particles lies between the first and second column of data points in Fig. 7a),
which is a great advantage of the multiple X-line current sheet acceleration
over acceleration at shock waves or in turbulent wave fields. The characteristic
maximum particle energy scales roughly as

$$W_{max} \sim e \frac{L_{CI}}{2} E_{z0} = \frac{u}{V_A} e \frac{L_{CI}}{2} \frac{V_A B_0}{c} \tag{17}$$

which is indicated by arrows in Fig. 7a.

For reasons of computational economy, proton orbits were calculated in the
majority of cases. The dependence of the maximum particle energy upon mass
is shown in Fig. 7b for two characteristic cases. It scales approximately proportional to $(m/m_i)^{1/4}$ (the acceleration is less efficient for particles with smaller
cyclotron radius). Hence, at this level of description more energy goes to protons, opposite to the observations. Inclusion of further effects such as electron
preheating may change this result by increasing the mean free path of electrons

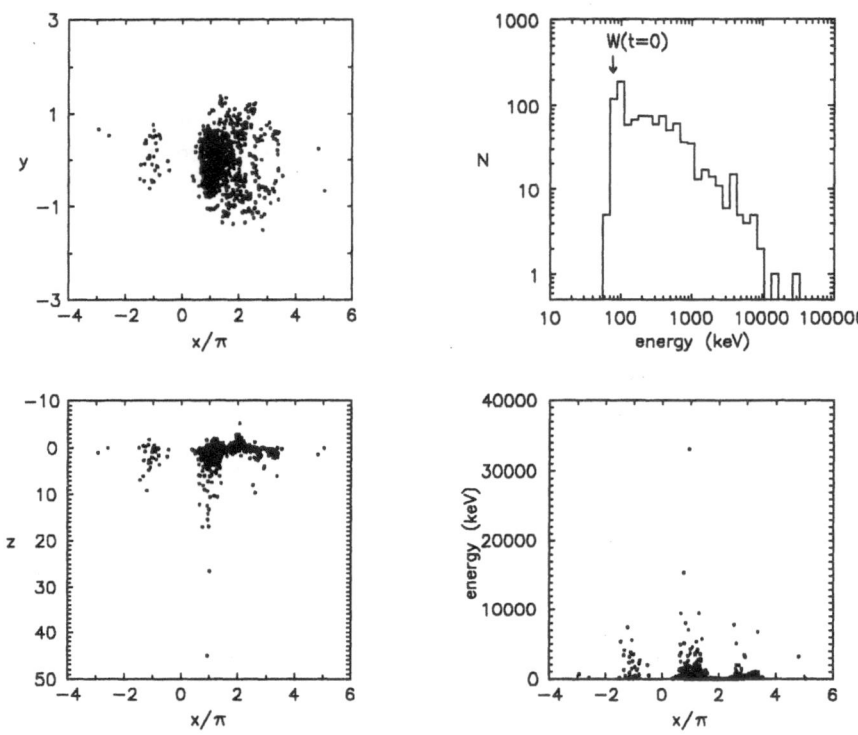

Fig. 6. Final positions, energies, and the corresponding energy distribution of 10^3 test particles (protons), which were initially monoenergetic and randomly distributed over the area of one island, after one coalescence time τ_{CI_1} (Kliem 1994). Parameters are $E_{z0}/E_0 = 0.1$ and $r_c(0)/l = 0.01$ (i.e., initial energy $W(0) \approx 77$ keV for the chosen value of $l = 3l_{cr} = 2 \times 10^4$ cm).

over that of the ions, rendering electron acceleration more easily possible. The accelerated particles are nearly isotropically distributed over pitch angle.

The presented model can account for several important properties of particle acceleration during the impulsive phase of solar flares, viz. bulk acceleration, fragmentation of the acceleration, time scales in the millisecond range, maximum energies in the MeV range. In comparison with a single X-line configuration, a filamentary current sheet is far superior with respect to particle acceleration because near Alfvénic velocities and correspondingly strong electric fields occur near the X-lines, many particles have rapid access to an X-line, there are repeated drift acceleration phases at the X-lines due to trapping of the particles in the configuration, and a small population of highly energetic particles can be produced by O-line meander acceleration.

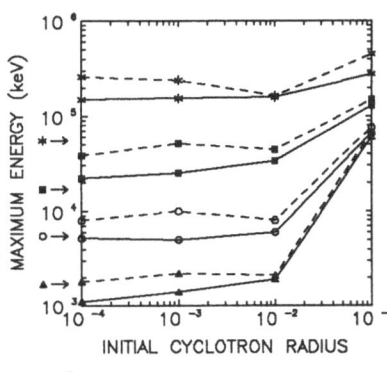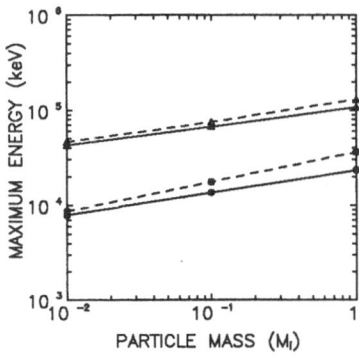

Fig. 7. (a) Characteristic maximum particle energies over initial cyclotron radius, normalized to l (Kliem 1994). The corresponding initial energies range from 7.7 eV to 7.7 MeV. Solid lines: energy reached by 1 per cent of the test particles; dashed lines: energy reached by 0.5 per cent of the test particles. Values of the applied electric field strength are $E_{z0}/E_0 = 10^0$ (*asterisks*), $10^{-0.5}$ (*squares*), 10^{-1} (*circles*), and $10^{-1.5}$ (*triangles*). Arrows indicate the approximate values from Eq. (17). (b) Dependence of the characteristic maximum particle energy on mass. Parameters are $E_{z0}/E_0 = 10^{-0.5}$ and $r_c(0)/l = 10^{-4}$ (*dots*), 10^{-1} (*triangles*). A good fit to the numerical values is given by $W_{max} \sim (m/m_i)^\delta$ with $\delta \approx 1/4$.

References

Aschwanden M.J., Güdel M., 1992, ApJ 401, 736

Aschwanden M.J., Benz A.O., Schwartz R.A., Lin R.P., Pelling R.M., Stehling W., 1990, Solar Phys. 130, 39

Aschwanden M.J., Benz A.O., Schwartz R.A., 1993, ApJ 417, 790

Benz A.O., 1985, Solar Phys. 96, 357

Benz A.O., Wentzel D.G., 1981, A&A 94, 100

Benford G., 1983, ApJ 269, 690

Bhattacharjee A., Brunel F., Tajima T., 1983, Phys. Fluids 26, 3332

Biskamp D., Welter H., 1979, Phys. Rev. Lett. 44, 1069

Biskamp D., Welter H., 1989, Phys. Fluids B1, 1964

Drake J.F., Gladd N.T., Huba J.D., 1981, Phys. Fluids 24, 78

Duijveman A., Hoyng P., Machado M., 1982, Solar Phys. 81, 137

Esarey E., Freidberg J.P., Molvig K., Beasley, Jr. C.O., VanRij W.I., 1986, Phys. Fluids 29, 200

Heyvaerts J., Priest E.R., Rust D.M., 1977, ApJ 216, 123

Hoshino M., 1991, J. Geophys. Res. 96, 11 555

Huba J.D., Gladd N.T., Papadopoulos K., 1978, J. Geophys. Res. 83, 5217

Huba J.D., Gladd N.T., Drake J.F., 1982, J. Geophys. Res. 87, 1697

Inhester B., Birn J., Hesse M., 1992, Solar Phys. 138, 257

Karpen J.T., Antiochos S.K., DeVore C.R., 1990, ApJ 356, L67

Kiplinger A.L., Dennis B.R., Emslie A.G., Frost K.J., Orwig L.E., 1983, ApJ 265, L99

Kiplinger A.L., Dennis B.R., Frost K.J., Orwig L.E., 1984, ApJ 287, L105

Kliem B., 1988, ESA SP-285, 117

Kliem B., 1991, in B. Schmieder and E.R. Priest (eds.), Flares 22 Workshop, Dynamics of Solar Flares, Obs. de Paris, p. 119

Kliem B., 1994, ApJ Suppl. Ser. 90, 719

Kliem B., Schumacher J., 1995, in preparation

Leboeuf J.N., Tajima T., Dawson J.M., 1982, Phys. Fluids 25, 784

Lee L.C., Fu Z.F., 1986, J. Geophys. Res. 91, 6807

Lu E.T., Hamilton R.J., 1991, ApJ 380, L89

Lu E.T., Hamilton R.J., McTiernan J.M., Bromund K.R., 1993, ApJ 412, 841

Malara F., Veltri P., Carbone V., 1992, Phys. Fluids B4, 3070

Matsumoto R., Tajima T., Shibata K., Kaisig M., 1993, ApJ 414, 357

Mikic Z., Barnes D.C., Schnack D.D., 1988, ApJ 328, 830

Mikic Z., Schnack D.D., Van Hoven G., 1990, ApJ 361, 690

Norman C.A., Smith R.A., 1978, A&A 68, 145

Papadopoulos K., 1977, Rev. Geophys. Space Phys. 15, 113

Papadopoulos K., 1979, in S.-I. Akasofu (ed.), Dynamics of the Magnetosphere, Reidel, Dordrecht, p. 289

Parker E.N., 1988, ApJ 330, 474

Priest E.R., 1985, Rep. Prog. Phys. 48, 955

Pritchett P.L., Wu C.C., 1979, Phys. Fluids 22, 2140

Richard R.L., Walker R.J., Sydora R.D., Ashour-Abdalla M., 1989, J. Geophys. Res. 94, 2471

Rieger E., Marschhäuser H., 1990, in R.M. Winglee and A.L. Kiplinger (eds.), Max '91 Workshop # 3, Max '91/SMM Solar Flares: Observations and Theory, Univ. Colorado, Boulder, p. 68

Scholer M., Roth D., 1987, J. Geophys. Res. 92, 3223

Sen A., Sundaram A.K., 1981, Phys. Fluids 24, 1303

Shi Y., Wu C.C., Lee L.C., 1991, J. Geophys. Res. 96, 17 627

Shimizu T., Tsuneta S., Acton L.W., Lemen J.R., Ogawara Y., Uchida Y., 1994, ApJ 422, 906

Spicer D.E., Brown J.C., 1981, in S. Jordan (ed.), The Sun as a Star, NASA SP-450, Washington, D.C., p. 413

Van Hoven G., Tachi T., Steinolfson R.S., 1984, ApJ 280, 391

Tajima T., Brunel F., Sakai J., 1982, ApJ 258, L45

Tajima T., Sakai J., Nakajima H., Kosugi T., Brunel F., Kundu M.R., 1987, ApJ 321, 1031

Ugai M., 1984, Plasma Phys. Controlled Fusion 26, 1549

Vlahos L., 1995, these proceedings

Wang X., Bhattacharjee A., 1994, ApJ 420, 415

White R.B., 1986, Rev. Mod. Phys. 58, 183

Wu C.S., Gaffey, Jr. J.D., Liberman B., 1981, J. Plasma Phys. 25, 391

Yin S.-P., Ip W.-H., 1991, Phys. Fluids B3, 1927

Acceleration and Radiation from a Complex Active Region

Loukas Vlahos

Department of Physics, University of Thessaloniki,
54006 Thessaloniki, GREECE

Abstract. Active regions are treated in this review as a "paradigm" of a **complex dynamical system**. Active regions are formed by magnetic fibers escaping from the turbulent convection zone. Random movements of the feet of the fibers in the photosphere and emergence of new magnetic flux from the convection zone are responsible for the formation of neutral sheets and magnetic discontinuities inside the active region, which are the sites for magnetic dissipation. A simple model, based on the scenario of self organised criticality, reproduces many of the known flare characteristics. The same model is used to provide a large number of nanoflares, which can heat the corona. We also show that this complex, inhomogeneous active region, is an efficient accelerator and reproduce the observed dm spikes and type III bursts.

1 Introduction

In this review an attempt is made to analyze the evolution of active regions using concepts from the theory of complex systems. It is well known that magnetic fields, formed in the base of the convection zone, have a filamentary structure (Leighton 1963; Sheeley 1967; Spruit and Zwaan 1981). Their evolution and propagation inside the turbulent convection zone is a topic of intense discussion (Bogdan 1992). Magnetic fields emerged in the solar surface are continuously re-arranged from the random motion of the footpoints at the photosphere and the emergence or submergence of fibers. It is then obvious that the observed activity in the photosphere and corona is the outcome of the evolution not only of isolated magnetic configuration inside the active region but the dynamic readjustment of a system with at least three main components (a) dynamo dynamics (b) convection zone dynamics and (c) active region evolution. The active region follows the evolution of the drivers below the photosphere and develops spontaneously current sheets and waves to dissipate the stresses developed. We proceed by giving simple definitions of the "complex system" and suggest that active regions should be treated as complex systems.

1.1. What is a complex system?

We define as "complex system" a dynamical system composed of a large number of different interacting elements. Traditionally, we use statistical mechanics to study the evolution of systems with large number of components. Ideal gases and solids are systems with very large number of components but in both cases two fundamental assumptions are made

(1) The components are identical

(2) The interactions are either very weak and can be ignored (like the ideal gas) or we use linearization methods to simplify the problem (like in the case of solids).

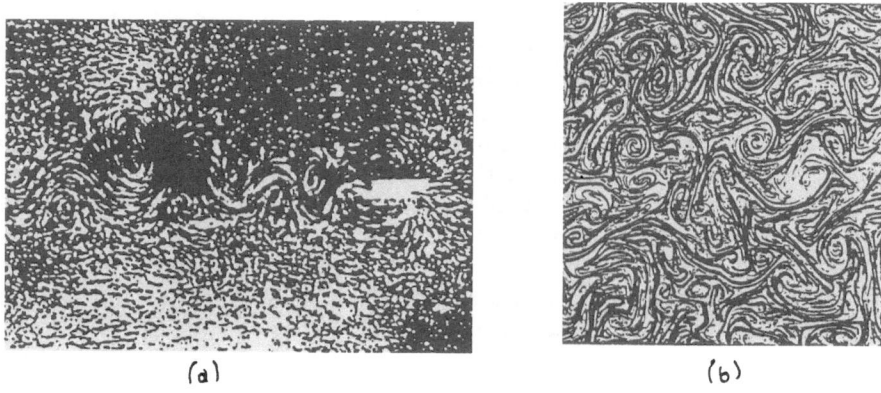

(a)   (b)

Fig. 1. (a) The "vortex street" in the wake of a circular cylinder moving steadily (from Clutter, Smith and Brazier 1959) (b) Numerical simulations from 2-D MHD turbulence (from Biskamp 1993). The formation of structures (vortices in fluids and irregular current sheets) is obvious.

These systems are not considered as complex. Systems with many different interacting components (with strong or weak interaction) will be called complex systems (the human brain, the turbulent flow, computer system with different electronic components etc.). We choose, in this approach, to oversimplify the components of the system whose global behaviour we would like to model. These simplifications enable us to apply rigorous methods and obtain rigorous results. Our approach is dynamical: we start from the local description of the system, in terms of short-term state changes of the components as a result of their interactions. We expect the global description of the system from the method, that is to say the long term behaviour of the system as a whole. The global behaviour of the system is not an addition of the behaviour of the components of the system but new properties appear, which are not predicted from the properties of the components (Weisbuch 1991).

Systems away from equilibrium are full of complex structures and new forms. In Fig. 1a we present a nearly turbulent flow and in Fig. 1b simulations from a turbulent MHD system. The striking characteristic in both systems is the appearance of structures at many different scales. Naturally, we arrive at two fundamental questions: (1) Is the active region a complex system? (2) Can we use methods from complex system theory to study the evolution of active regions?

1.2. Are active regions complex systems?

In this review we consider active regions as a collection of fibers of all sizes and magnetic strengths. This highly inhomogeneous topology appears to have a large scale "organization" due to the overall velocity fields in the convection zone and photosphere. The "components" of our system are the fibers. These fibers, studied as isolated systems (as it has been done so far), have a limited contribution on the evolution and dynamics of the whole active region. The interaction of hundreds or millions of fibers produces many new characteristics. Fibers are partially emerged, a large portion is still submerged in the turbulent convection zone. It is then obvious that viewing the active region as a collection of interacting fibers with turbulent boundary conditions is the best way to follow its dynamic reconstruction. This enormous task can be done either by $3D$ numerical simulation (Galsgaard and Nordlund 1994) or using methods developed in the theory of complex systems dynamics (Lu and Hamilton 1991). We will assume here that active regions are complex systems so in section 2 we outline a method, borrowed from the theory of complex dynamical systems, for treating the energy dissipation of a complex active region, and in sections 3 and 4 we propose a new model for particle acceleration, dm spikes and type III bursts, based on the complexity of active regions.

2 The statistical flare

Research on magnetic energy release remains one of the unsolved problems in astrophysics. In simple topologies (single loop, single arcade, etc.) two forms of energy dissipation have been studied extensively (1) transport and dissipation of waves in the stellar atmosphere (Zirker 1993) and (2) magnetic storage and sudden release through the formation and dynamics of current sheet(s) (Priest 1992). The first type of dissipation is considered as the best candidate for coronal heating and the second as the most probable mechanism for flares. This analysis is correct for simple topologies while in complex magnetic field topologies (collection of fibers) new opportunities for dissipation emerge, e.g. interaction of fibers. In complex active regions wave propagation and dissipation and current sheet formation and evolution are non-linearly coupled (Sudan and Longcope 1992). Lin et al. (1984) reported, using very sensitive instruments, that many small scale releases were detected and the peak-luminosity distribution of flares ($N(L)$) followed a power-law with exponent $\simeq -1.8$ (Fig. 2). This important discovery was confirmed by several authors later (Dennis 1985; Crosby et al. 1992;

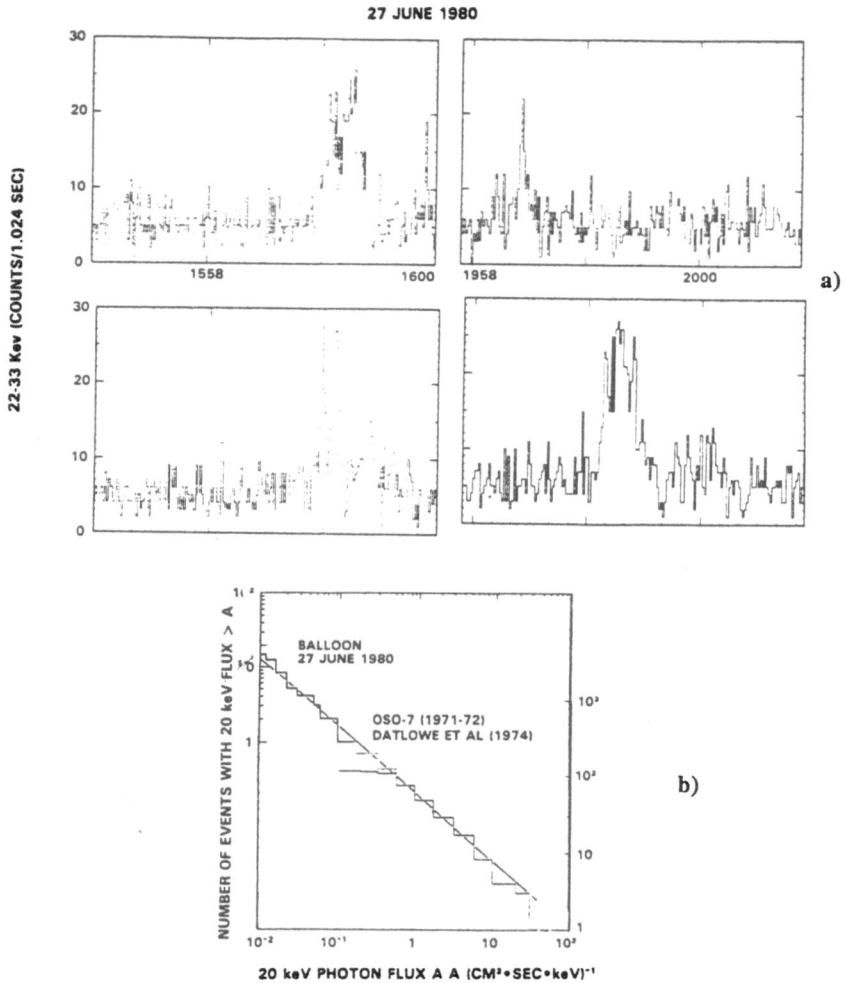

Fig. 2. (a) The four largest hard X-ray microflares are shown at 1.024 sec resolution (b) The distribution of the integral number of events versus peak 20 keV photon flux for the solar hard X-ray microflares observed (from Lin et al. 1984).

Pearce et al. 1993). Approaching the solar flare not as a unique phenomenon but rather studying the statistical properties of the behaviour of active region emission naturally changes the scope of the research from the local properties (dynamics of a current sheet, etc.) to the global characteristics. The analysis of flares as a time series demands new types of models for the global evolution of the active region.

Studying the flare distribution we naturally arrive at the conclusion that many still unobserved small flares (named nanoflares by Parker (1988)) are present and if their contribution is significant, they may be able to heat the corona. Nanoflares, according to Parker, are the outcome of the evolution and reconstruction of complex active regions. Hudson (1991) studied this problem in detail and proved that unless nanoflares have much steeper power law index for the peak luminosity distribution $\sim$ (-4) from the one observed for flares, the energy content on these mini-flares is not enough to heat the corona.

2.1. Models for the evolution of a complex active region

There are currently two ways the global modelling of the active region can be done: (1) $3D$ MHD simulation and/or (2) using methods from the complex system theory. We are going to use the second method here and follow the work first proposed by Lu and Hamilton (1991) and Lu et al. (1993). They assumed that the emergence as well as changes of magnetic flux in the photosphere place new flux of random magnitude and direction at **randomly** located points inside the active region. Simple rules for the redistribution of magnetic field and the release of magnetic energy in the vinicity of a discontinuity are then applied. This model is able to predict a power-law behaviour for the number of release events formed versus the peak luminosity which has an index similar with the one observed. Specifically, they constructed a $3D$ cubical lattice with vector field $\mathbf{F}_i$ associated to every grid point i. They defined as "slope" the quantity

$$d\mathbf{F}_i = \mathbf{F}_i - \sum \omega_j \mathbf{F}_{i+j} \tag{1}$$

where ω_j is an arbitrary weighting function and $\mathbf{F}_{i+j}$ is the field of the six first-order neighbours of the grid point i. A critical threshold F_{cr} is introduced and defines an *instability criterion*, $|d\mathbf{F}_i| > F_{\mathrm{cr}}$. An algorithm is constructed, which consists of selecting randomly a grid point, adding a small increment $\delta\mathbf{F}$ with $|\delta\mathbf{F}| << F_{\mathrm{cr}}$, and searching the grid for possible instabilities. If a grid point is found to be unstable, then itself and its neighbours are automatically readjusted in such a way that the field is conserved and redistributed **isotropically**. At the same time an amount of energy, depending on $d\mathbf{F}_i$ is released. The grid is then again scanned, and if new instabilities are found, the same procedure of readjusting the field is applied. One keeps scanning the grid and readjusting the field until the system has relaxed into a state in which no grid point is unstable. Then again a small increment $\delta\mathbf{F}$ is added to a randomly selected grid point, and the same procedure of scanning and readjusting is applied. This model predicts the prevalence of a power-law behaviour in the occurrence distributions of relaxation events vs. total energy, peak luminosity and the duration of events. The consistency with observations leads to the conclusion that one might gain in this way an insight into the physics underlying the flaring mechanism.

Vlahos et al. (1994) extended the work of Lu and Hamilton (1991) in two important points: (1) they have extended the instability criterion to the *second order neighbours* assuming that neighbours of an unstable point have relaxed criteria which depend on the strength of the instability of the initial point. The steeper the gradient the easier the triggering of the neighbour is. (2) They

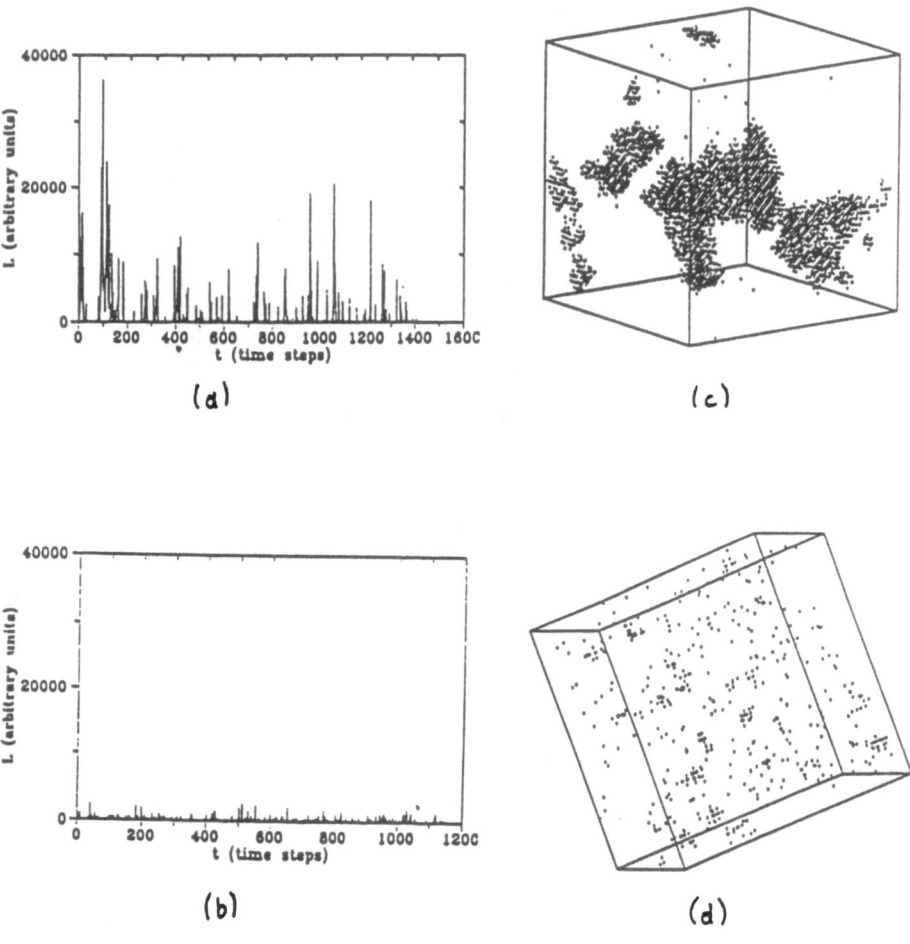

Fig. 3. (a) Typical time series from an isotropic model, (b) for an anisotropic model, where the nanoflares dominate, (c) $50 \times 50 \times 50$ simulation box from an isotropic model, (d) the same simulation from an anisotropic model.

construct an anisotropic model. The relaxation procedure provides directionality towards the neighbour or the neighbours that need to be readjusted according to the instability criterion. In the anisotropic model, each grid point is characterized by six slopes instead of one (as it is the case with eq. (1) used in the isotropic model). These slopes at the grid point i are defined by

$$dB_{ij} = B_i - B_j \qquad (2)$$

where B_{ij} is the field of the first order neighbours and the new stability criterion is $dB_{ij} > B_{cr}$. The results are remarkably different between the two models (see Fig. 3). The anisotropic model is dominated by small isolated events, distributed everywhere inside the simulation box, while the isotropic model shows more clustering of discontinuities and much larger outbursts (Fig. 3).

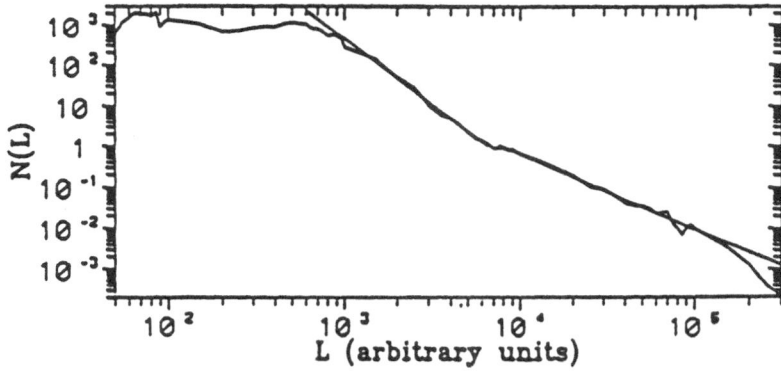

Fig. 4. The combination of isotropic and anisotropic models provides two regions of power law behaviour for the frequency distribution vs the peak luminosity.

The anisotropic model provides also a steeper peak-luminosity distribution, with a slope around -3.5. The events are relatively small and reminescent of nanoflares. The extension of the instabilities to the second order neighbours on the other hand, produce large and small clusters which are responsible for flares of all sizes (this is why we call this type of flaring, the statistical flare). Vlahos et al. (1994) showed that the combination of an isotropic and an anisotropic model provides two regions of power-law behaviour: a low energy region (nanoflares) with index ~ -3.5 and a high energy region with index ~ -1.8 (see Fig. 4). This result is in accordance with the assertion of Hudson (1991), who stated that a peak luminosity power-law index ~ -4 is needed to account for coronal heating and agrees with the observations reported above.

3 Particle acceleration from a complex active region

A complex active region, as described above, will have an enormous impact on the methods used for particle acceleration and transport of energetic particles. It is well known that a large fraction of the energy released during solar flares is deposited to the energetic particles. The energy spectrum of the accelerated particles during solar flares can be found either by in situ measurements at 1 AU (e.g. Lin, Mewaldt and van Hollebeke 1982) or by the spectral index of the emitted radiation at several wavelengths (X-rays, γ-rays and radio emission).

There is a long list of acceleration mechanisms known (e.g. MHD waves, shock waves, DC electric fields and coherent radiation) but seldomly these processes are connected to the dynamics of the energy release process. The two subjects, energy release and particle acceleration, are discussed and reviewed separately. We have already presented above a model for energy release in complex active regions, in this section we develop a model for particle acceleration and transport in complex magnetic topologies.

Anastasiadis and Vlahos (1991, 1994) assumed that particle acceleration is not confined inside a single non-linear structure (current sheet, shock wave or Double layer) but receives several "kicks" from many structures before escaping in the upper atmosphere or precipitate in the low corona. An avalanche of current sheets will confine the particle in space so its final energy, when it escapes from a small or big cluster, will be the outcome of N-interactions. They assumed that slow shocks escapping from current sheets will propagate inside a highly inhomogeneous plasma and develop (temporarly) fast shocks (with Mach number M_A between $1-3$). The interaction of an ensemble of shock waves with electrons and ions from the tail of the ambient maxwellian distribution was studied in detail. They simulated an active region with a box having characteristic length $L = 10^9$ cm and mean ambient magnetic field $B_0 = 10^2$ G. The ambient density used was $n_0 = 10^{10}$ cm^{-3}, the plasma parameter beta is $\beta = 5$ and the Alfvén velocity is $V_A = 2.18 \times 10^8$ cm s^{-1}. The number of shock waves (N_S) formed inside a complex active region is randomly placed in a restricted part of the simulation box (for convenience we place the shocks at the middle of the box. Particles (electrons and ions) are accelerated when there is a particle shock interaction by the shock drift mechanism. The shock drift mechanism was chosen for its simplicity. In addition to the acceleration process, particles will be subjected to loss due to Coulomb collisions. The kinetic energy loss rate in eV sec^{-1}, for the case of electrons is (Bai and Ramaty 1979)

$$\frac{dE_e}{dt} = \begin{cases} -1.55 \times 10^{-7} n_0 E_e^{-0.5} & if \quad E_e \leq 160 keV, \\ -3.8 \times 10^{-7} n_0 & if \quad E_e > 160 keV \end{cases} \quad (3)$$

and for the case of ions is (Ryan 1986)

$$\frac{dE_i}{dt} = -7.7\ 10^{-9}\ n_o\ E_i^{-0.5} \quad for\ E_i \geq 10\ \ \text{MeV} \tag{4}$$

where $E_{i,e}$ is the kinetic energy of the particle in eV and n_0 is the density in cm^{-3}. Particles can also lose energy during their interactions with shock waves (Webb et al. 1983). The Coulomb losses are active continuously on the trajectory of the particles. On the other hand, the acceleration (or deceleration) can occur only at the position of the shock waves (localized process). Particles are trapped at the boundaries of the simulation box, and they introduced magnetic mirroring with ratio $R_m = 0.1$. The trapping affects the trajectories of the particles, since a few will return in the acceleration site by reflection at the boundaries and thus undergo more encounters with the shock waves. The rest will escape from the simulation box. The role of random position mirroring on the evolution of the energy distribution was also investigated. This small-scale mirroring is active downstream and upstream of the shocks. Such a process, has a similar effect on the trapping of the particles in the neighbourhood of shock fronts, which was studied recently by Decker (1993).

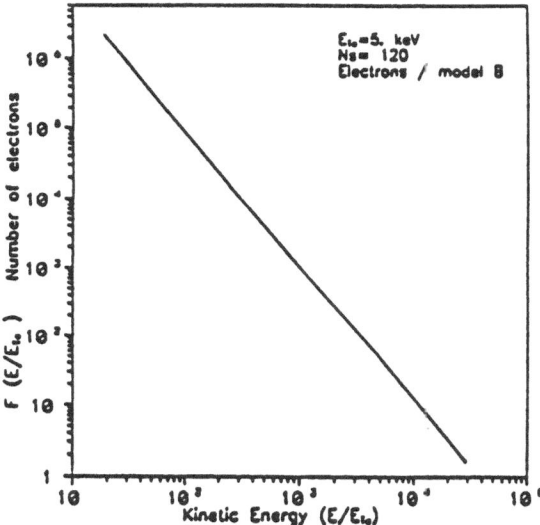

Fig. 5. The energy distribution of power law form, for the case of electrons for model B and $N_S = 120$ shock waves in active region.

The particle population which interacts with the N_S shock waves has initially a Maxwellian distribution in the velocity space, e.g.,

$$f_j(v_j) = \frac{n_0}{(2\pi)^{3/2} V_{Tj}^3} \exp(-\frac{v_j^2}{2V_{Tj}^2}) \qquad j = i, e \tag{5}$$

where v_j and V_{Tj} are the particle velocity, and the thermal velocity respectively. The kinetic energy range, we are interested in, is 2 MeV$\leq E_i \leq$ 13 MeV for the ions and 20 keV$\leq E_e \leq$ 200 keV for the electrons. The above energy ranges correspond to particles with $2 \leq (v_j/V_{Tj}) \leq 5$ for $V_{Te} = 4.19 \times 10^9$ cm sec^{-1} and $V_{Ti} = 9.79 \times 10^8$ cm sec^{-1}. The initial distribution was normalised in such a way that $f_j(5 \, V_{Tj}) = 1$.

Anastasiadis and Vlahos (1994) considered two different models, denoted A and B. In model A the particles are trapped only by magnetic mirroring at the boundaries of the simulation box. On the other hand, in model B, particles are subject to additional trapping by a random position mirroring inside the acceleration volume.

The resulting energy distribution of the particles (electrons and ions) can be fitted well, with correlation coefficient $R \geq 0.8$, by a power law, e.g.,

$$F_j(\frac{E_j}{E_{Tj}}) \approx K(\frac{E_j}{E_{Tj}})^{-s}, \qquad j = i, e \tag{6}$$

where E_j is the kinetic energy of the particles, E_{Tj} is the thermal kinetic energy ($E_{Te} = 5$ keV for the electrons, $E_{Ti} = 500$ keV for the ions), K is a constant, and s is the power-law index. In Fig. 5 we present the energy distribution for the case of electrons interacting with $N_S = 120$ shock waves for model B, having power law index $s = 1.92$.

For both models, the maximum energy E_m increases as the number of shock waves being present in the active region increases. In the case of model B the particles reach higher energies, compared to model A, since they interact several times with the active shock waves. For model A, the acceleration time t_a remains almost constant, for ions ($t_a \approx 0.5$ s) and for electrons ($t_a \approx 0.3$ s) for $N_S \geq 80$, but increases when $N_S < 80$, since the particles are not accelerated efficiently and spend more time inside the box. For model B the behavior of t_a is different. When N_S increases the acceleration time decreases for the ions. On the other hand, for electrons t_a is higher than that of model A, since they are trapped more efficiently.

The variation of the power law index s with the number of shock waves present, is presented in Fig. 6a for the case of ions and in Fig. 6b for the case of electrons. Notice that the variations of the power-law index for the case of ions are not significant. The mean value for model A is 3.5 and for model B is 4.5 with dispersion 0.4. On the other hand, the power-law index of the electrons for the model A is decreasing as the number of shocks is increasing up to $N_S = 80$. This is due to the fact that the environment is not an efficient accelerator and the energy distribution has softer power-law index. In the case of model B we have smaller variation of s around the mean value 1.95 with dispersion 0.37.

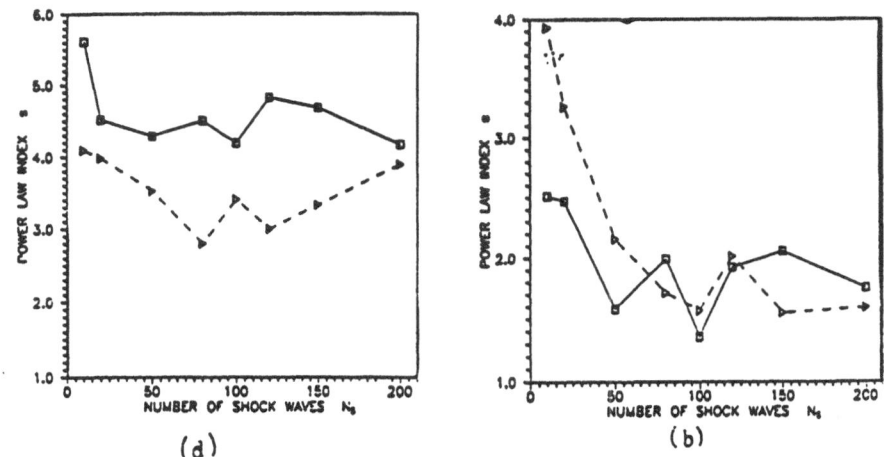

Fig. 6. (a) The variation of the power-law index s with the number of shock waves N_S for ions, for model A($\triangle$) and model B(o) (b) the same for electrons.

The energy density per shock is $W_S \simeq 4.12 \times 10^3$ ergs cm^{-3}. If we assume that $N_S = 200$ shock waves are present, then a large percentage of particles ($n_t \geq 10^{-2} n_0$), with velocities $\geq 2 \times V_T$, will be "energized". Under these assumptions the energy density of ions for model B is, $W_i \simeq n_t E_m \simeq 6.9 \times 10^4$ ergs cm^{-3}. It is clear that the fraction of the total energy carried by the shocks and transferred to the energetic ions is

$$P_i = \frac{W_i}{N_S \times W_S} \simeq 5.9 \times 10^{-2}. \tag{7}$$

Using the same parameters, the energy given to the acceleration of electrons can be estimated:

$$P_e = \frac{W_e}{N_S \times W_S} \simeq 0.5 \times 10^{-2}. \tag{8}$$

It is obvious that less than 10% of the total energy carried by the shock waves goes to the acceleration of the particles (ions and electrons). The rest of the energy released goes to heating.

Complex active regions with randomly placed acceleration sites (shocks) will accelerate electrons and ions with power law index $< s > \sim 4$ for ions and $< s > \sim 2$ for electrons. These numbers are not sensitive to the number of acceleration sites (shocks). The acceleration time can be very fast $t_\alpha < 1$ sec but trapping of a small population of electrons and ions for longer times is also possible. The maximum energy increases with the number of shocks present, so large clusters will not only produce large flares but will accelerate electrons and ions to high energies very fast. Only a small fraction of the energy carried by the shocks will be deposited to electrons and ions at the tail of the local Maxwellian ($E \geq 50$ keV).

Further research on more complex accelerators will be necessary to understand the characteristics of acceleration and transport from a collection of discontinuities, as those shown in Fig. 3c.

4 Type III bursts and dm spikes from a complex active region

The type III bursts are among the best studied solar radio bursts. Their interpretation is that the burst is caused by a fast stream of electrons with mean velocity one third of the speed of light. Modeling and interpretation of type III bursts was concentrated so far on the problem of stable propagation of a beam of electrons.

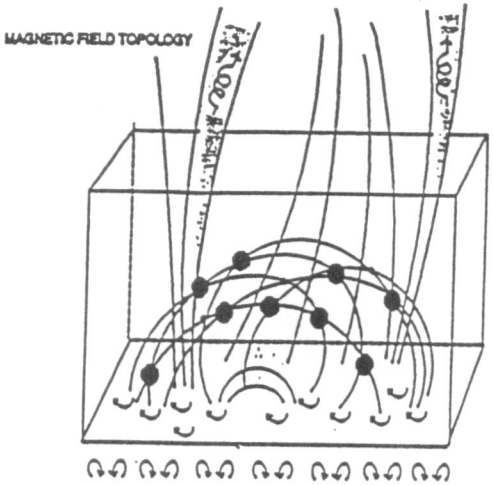

Fig. 7. A complex active region with open and closed fibers. A collection of N beams propagate in open fibers.

The model for type III bursts based on complex active regions differs substantially from the models used so far since the single flux tube should be replaced by many independent injections on different flux tubes. Vlahos and Raoult (1994) studied the formation of type III bursts from a complex active region (Fig. 7). They assumed (1) that the fast energy release around clusters of current sheets will develop a super-hot component in the low corona and the fast particles will form a positive slope at a distant z away from the energized volume, (2) the beam is stabilised non-linearly suffering only from Coulomb and weak wave losses, (3) the interaction of plasma waves with ion-sound waves produces the observed emission at ω_p, (4) the magnetic field is split in fibers with random density and in some fibers we inject a beam with random characteristics.

It is well documented now that electron beams with energies $\leq 50 - 100$ keV, injected from the solar corona, travel large distances almost freely. Recent observations, at a distance as far as four A.U. (Buttighoffer et al. 1994), show the presence of electron beams with energy ≤ 40 keV that have propagated freely from the acceleration site. Former in situ observations at 1 A.U. (Lin et al. 1981) have detected secondary wave levels that favour nonlinear stabilisation, but the current accuracy of plasma wave measurements did not allow to determine which non linear process was going on.

The theory of strong turbulence was used by Vlahos and Raoult (1994) to follow the evolution of the beam since it allows an easy representation for the beam and because available literature on the subject provides relations between primary, secondary and ion acoustic wave levels which are needed to estimate the expected radio emission. We recall briefly the main results of the strong turbulence theory below.

4.1. Beam stabilization

A semi-quantative system of equations was used to study the evolution of the beam-plasma system (Papadopoulos 1975; Vlahos and Rowland 1984; Hillaris et al. 1988 1990). The system captures the main characteristics of the nonlinear evolution, which are as follows:

1. The beam excites linearly plasma waves with a rate $\gamma_b \simeq (n_b/n_0)(v_b/\Delta v_b)^2$ ω_e and wave number $k_0 = \omega_e/v_b$, where ω_e is the plasma frequency, v_b is the beam velocity, Δv_b the velocity spread around v_b, n_0 the local ambient plasma density. The resonant waves excited by the beam have total energy W_1 (all quantities are normalised, $W \equiv W/n_0 T_e$ etc.).

2. When the W_1 reaches a critical W_c the resonant waves are transferred nonlinearly to the secondary plasma waves with energy W_2 with a rate

$$\gamma_{mod} \simeq \begin{cases} 0, & W_1 < W_c \simeq (v_e/v_b)^2 \\ \sqrt{\frac{W_1}{3} \cdot W_c} & W_1 > W_c \end{cases}$$

and the ion acoustic waves W_s with a rate $\alpha_{sc} \simeq \frac{\delta W_s}{q^2}$ where γ_{mod} and α_{sc} are normalised to ω_e, the wave number of the ion acoustic waves is $q \simeq \sqrt{W_{1max}^3} \simeq 0.2$ and δW_s describes the part of the ion acoustic oscillations that is not trapped inside the cavities (Vlahos and Rowland 1984).

3. The secondary plasma waves and the ion acoustic waves are damped with rate $\nu_L \simeq q^{-3} \exp(-1/2q^2)$, for the plasma waves and $\nu_1 \simeq 6 \times 10^{-4} q$, for the ion acoustic waves. The above processes are presented well by the system of equations

$$\frac{\partial W_1}{\partial t} = 2\gamma_{\mathrm{b}} W_1 - 2\gamma_{\mathrm{mod}} W_2 - \alpha_{\mathrm{sc}} W_1 \tag{9}$$

$$\frac{\partial W_2}{\partial t} = 2\gamma_{\mathrm{mod}} W_2 - \alpha_{\mathrm{sc}}(W_1 - W_2) - \nu_{\mathrm{L}} W_2 \tag{10}$$

$$\frac{\partial W_{\mathrm{s}}}{\partial t} = 2\gamma_{\mathrm{mod}} W_2 - \nu_1 \delta W_{\mathrm{s}} \tag{11}$$

Solving the system of equations we found that the primary waves grow first in each height and when $W_1 > W_c$ the daughter waves start to grow. The 3-wave system will reach a critical stable point with amplitudes $W_1^{\mathrm{s}}, W_2^{\mathrm{s}}$ and $W_{\mathrm{s}}^{\mathrm{s}}$. The non-linearly saturated values depend strongly on the characteristics of the beam (density, speed etc.) (Hillaris 1992).

The total energy absorbed by the waves is a negligible fraction of the free energy of the beam, and the beam evolution follows closely the free-streaming equation:

$$\frac{\partial f_{\mathrm{b}}}{\partial t} + v_{\mathrm{b}} \frac{\partial f_{\mathrm{b}}}{\partial z} = -\nu_{\mathrm{eff}} f_{\mathrm{b}} + f_0 \tag{12}$$

where f_0 is the injection distribution and in the term ν_{eff} we have kept only the influence of Coulomb collisions. Coulomb collisions will not be negligible since sometimes the beams are injected in overdense fibers, according to observational results deduced from radio and white light measurements. Coulomb collisions will slow down the lowest energy particles of the beam, which would suffer quasilinear relaxation quicker than strong turbulence stabilisation and would be poorly represented by a free-streaming equation.

The free streaming equation (Eq. (12)) will be appropriate to represent the beam escape out of the injection site as long as the cumulative energy losses remain small compared to the initial beam energy. If the cumulative wave losses happen to be comparable (say 1/3) to the initial beam energy, the beam has probably suffered distortions that cannot be approximated by a free streaming equation. Future studies should replace Eq. (12) with a diffusion equation for a better representation of the beam evolution in the outer corona and the interplanetary space.

4.2. A model for a random collection of beams

We have mentioned already that a complex active region is modeled here as a collection of fibers, where beams are randomly injected when a cluster happens to occur in the vicinity of an open fiber (Fig. 7). Each fiber, denoted with j, develops its own "atmosphere". The conditions in temperature and density are different but they follow the standard Saito model

$$n(z) = \alpha_j n_0 e^{-z/\mathrm{H}}$$

where α_j is a random number chosen between one and twenty, $n_0 = 4.7 \times 10^{23}$ Km^{-3}, $H = 96055$ Km, to approximate the standard Saito model up to $0.5R_\odot$.

The corona is assumed isothermal. At the base of j-th fiber, a proportion β_j of the plasma is heated impulsively to a temperature T_j (mean speed v_j, in a mean time τ_j, peaking at t_j, over a slice of thickness L_j). We choose β_j randomly between 10^{-7} and 10^{-5}, τ_j at random between 0.1 and 1 seconds, t_j at random in a given time interval, (between 0-15 sec in this study) L_j at random between 1000 and 10000 Km, and v_j is chosen at random between 10 and 30 times the ambient thermal speed.

We thus model a sporadic heating of a small portion in the base of each fiber by the following injection distribution:

$$f_{0j}(z,v,t) = \frac{N_{0j}}{\sqrt{2\pi}v_j} e^{-\frac{1}{2}(\frac{v}{v_j})^2} e^{-(z/L_j)^2} e^{-(t-t_j)^2/\tau_j^2}$$

where N_{0j} is proportional to $\beta_j \alpha_j n_0$. This injection of electrons expands in the fiber according to the equation (12), where $\nu_{\text{eff}} = k n_j(z)/\hat{v}^3$, where k is constant, giving the analytical solution

$$f_j(z,v,t) = g(v) \int_{-\infty}^{t} e^{-a_1} e^{-a_2} e^{-a_3} dt$$

with $g(v) = N_{0j} e^{-v^2/(2v_j^2)}/(\sqrt{2\pi}v_j)$ and

$$a_1 = (kHn_j(z)/v^4)(e^{v(t-t')/H} - 1),$$
$$a_2 = [(z - vt + vt')/L_j]^2,$$
$$a_3 = [(t' - t_j)/\tau_j]^2.$$

The time step for computation is fixed to 0.1 sec. In each fiber we pre-determine the altitude steps that correspond to the radio frequencies that we want to "observe" on the final estimated radio spectrum (every 25 MHz from 800 to 500 MHz, every 50 MHz from 500 to 150 MHz). In each fiber we check if the injected beam will stop within the computed range of frequencies by comparing the cumulative wave losses with the initial energy of the beam. We compute the distribution function in each fiber, at each altitude step and time step. From the shape of the computed distribution function we estimated the growth rate, then the primary and secondary wave levels and finally the emission at the fundamental

$$P(\omega = \omega_e) \approx 2\omega_e W_s W_1 (\frac{v_b}{c})^2.$$

At each step of computation we check that the secondary ion acoustic waves do not induce anomalous resistivity that would overtake Coulomb collisions, and that the growth rate does not exceed $10^{-4}\omega_e$ (see Vlahos and Raoult, 1994 for more details). The range of parameters used in this study is such that the checks mentioned above remain valid in the 800-150 MHz domain. We have

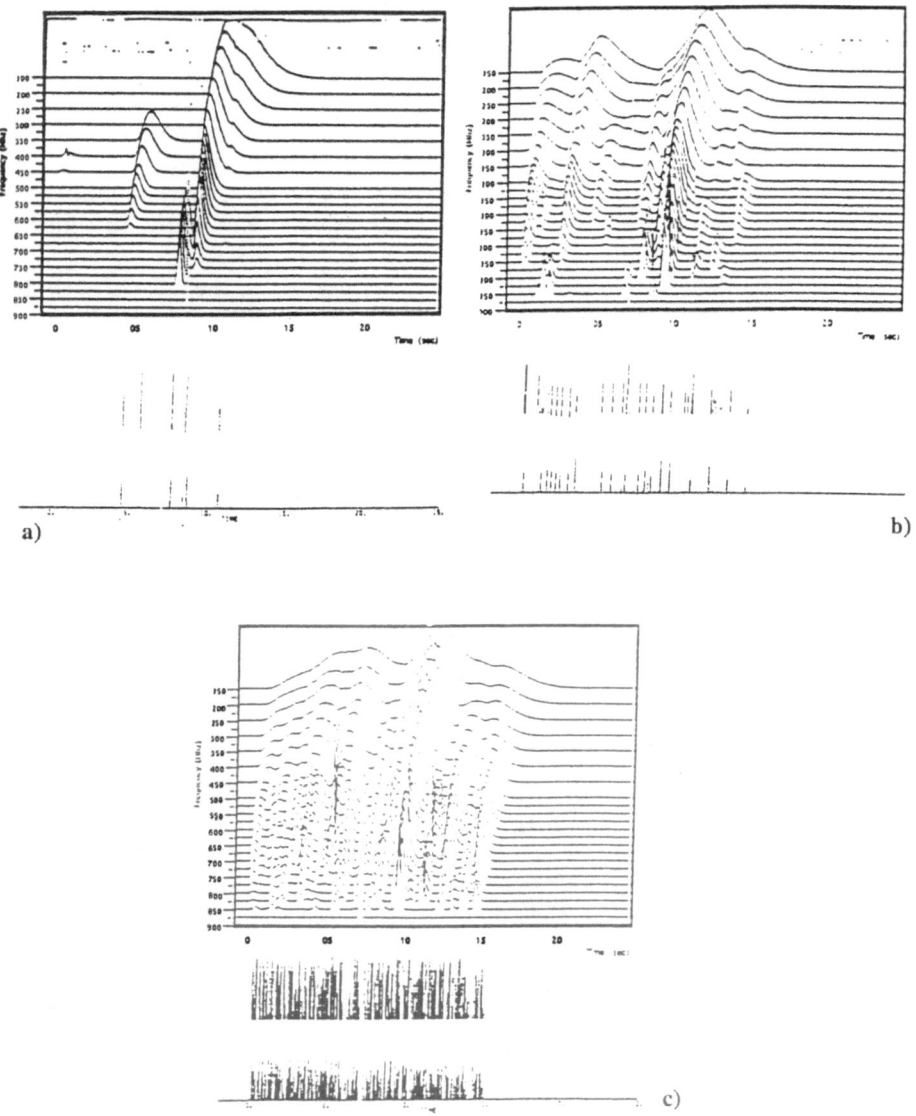

Fig. 8. Simulations from the injection of beams in different independent fibers (a) the injection of 10 beams, (b) the ejection of one hundred beams and (c) the injection of five hundred beams. All beams are injected between 1 to 15 sec. The marks in the bottom of each picture indicate the density variability of the fibers chosen to inject a beam and the second line the peak time of the injected beam. The length is proportional to the energetic of the beam.

already mentioned that the computation presented here cannot be extended to lower frequencies (< 150 MHz) since we need a more sophisticated beam propagation physics to follow the evolution at larger distances. Finally we add the contribution from the N fibers and form the simulated radio spectrum.

In Fig. 8 we plot the results obtained by using a flaring interval of 15 seconds and short injections with duration fixed at 0.1 seconds. The parameters are chosen randomly as was explained above. Only fundamental radiation is computed. The marks under the simulated spectrum indicate the relative density of the fiber used for the injected beam and the second row indicates the time and the relative strength of the injected beam.

Some beams stop abruptly close to the injection point due to fast energy depletion. These beams seem to simulate closely the decimetric spikes (Benz 1986). They obviously stop more abruptly in the computation than in reality, since we stop those beams when free-propagation cannot model the beam any further. Some beams propagate over long distances and merge, creating complex type III time-profiles, as on the observed spectra (see Güdel et al. 1991). Though every injection occurs around the same altitudes, the starting frequencies of bursts vary, since the fibers have different densities. Fig. 8a shows the simulated spectrum when we inject ten beams in different fibers. The result models very closely an "isolated" type III event. Fig. 8b shows the simulated spectrum for one hundred beams injected in different fibers. The result simulates a typical type III group and in Fig. 8c we inject 500 beams which simulates a type III event mixed with continuum emission. These computations suggest that beam propagation in a fibrous corona with fragmented energy release can account for various forms of radio emission in the decimetric-metric range.

5 Summary

Let me review the successes from modelling the active region as a complex system: (1) Using techniques developed in complex system theory, we showed that **random** loading of magnetic fields, inside the active region, depending on the instability criterion (isotropic or anisotropic), will release energy in the form of a time series which resembles closely the observed emission profile. This model predicts the observed peak-luminosity power-law and suggests that nanoflares (still unobservable) may be able to heat the corona. (2) Flares, microflares and nanoflares are clusters of current sheets with different size. (3) **Randomly** placed shock waves inside the active region will be able to accelerate the tail ($E \geq 50$ keV) of the local Maxwellian to high energies in very short time scales. (4) **Random** injection of beams in open fibers will reproduce the observed dm spikes and type III bursts.

We believe that a model for complex active regions is now ready to be applied in more realistic environments and help to understand several still unsolved problems like energy release, particle acceleration and electromagnetic emission from an evolving collection of fibers.

Acknowledgements: I would like to thank the organizers and the participants of this stimulating workshop. I thank Drs. A. Anastasiadis, R. Kluiving,
A. Raoult and Mr. M. Georgoulis for many stimulating discussions. Part of this
work was supported by the EEC program SCIENCE on "Coherent Radiation
and Particle Acceleration in Natural Plasmas".

References

1. Anastasiadis, A. & Vlahos, L. (1991): A&A **245**, 271
2. Anastasiadis, A. & Vlahos, L. (1994): ApJ **428**, 819
3. Bai, T. & Ramaty, R. (1979): ApJ **227**, 1072
4. Biskamp, D. (1993):" Nonlinear Magnetohydrodynamics", Cambridge University
   Press, p. 237
5. Benz, A.O. (1986): Solar Phys. **104**, 99
6. Bogdan, T.J. (1992):"Electromechanical coupling of the solar atmosphere", eds.
   D.S. Spicer & P. MacNeice, AIP, p. 79
7. Buttighoffer, A., Pick, M., Roelof, E.C., Hoang, S., Mangeney, A., Lanzerotti,
   L.S., Forsyth, R.S. and Phille, S.L. (1994): J. Geophys, Res., in press
8. Clutter, D.W., Smith, A.M.O. and Brazier, J.G. (1959): Douglas Aircraft Com-
   pany Report, no. ES29075
9. Crosby, N.B., Aschwanden, M.J. & Dennis, B.R. (1992): Solar Phys. **143**, 275
10. Decker, R.B. (1993): J. Geophys. Res. **98**, 33
11. Dennis, B.R. (1985): Solar Phys. **100**, 465
12. Galsgaard, K. and Nordlund, A. (1994): Space Sci. Rev. **68**, 75
13. Güdel, M., Aschwanden, M.J. & Benz, A.O. (1991): A&A **251**, 285
14. Hillaris, A., Alissandrakis, C.E. & Vlahos, L. (1988): A&A **195**, 301
15. Hillaris, A., Alissandrakis, C.E., Caroubalos, C. & Bougeret, J.-L. (1990): A&A
    229, 216
16. Hillaris, A. (1992): Ph.D. thesis, Univ. of Athens
17. Hudson, H.S. (1991): Solar Phys. **133**, 357
18. Leighton, R.B. (1963): Ann. Rev. Astr. Ap **1**, 19
19. Lin, R.P., Potter, D.W., Gurnet, D.A. & Scarf, F.L. (1981): ApJ **251**, 364
20. Lin, R.P., Meweldt, R.A. & van Hollebeke, M.A.I. (1982): ApJ **253**, 949
21. Lin, R.P., Schwartz, R.A., Kane, S.R., Pelling, R.M. & Hurley, C.C. (1984): ApJ
    285, 421
22. Lu, E.T., Hamilton, R.J. (1991): ApJ **380**, L89
23. Lu, E.T., Hamilton, R.J., McTierman, J.M. & Bromund, K.R. (1993): ApJ **417**,
    841
24. Papadopoulos, K. (1975): Phys. Fluids **18**, 1769
25. Parker, E.N. (1988): ApJ **330**, 474
26. Pearce, J., Rowe, A.K. & Yeung, J. (1993): Ap&SS **208**, 99
27. Priest, E. (1992): "Eruptive solar flares", Z. Švestka, B.V. Jackson and M.E.
    Machado (eds), (Berlin, Springer Verlag), p. 15
28. Ryan, J.M. (1986): Solar Phys. **105**, 365
29. Sheeley, N.R. (1967): Solar Phys. **1**, 171
30. Spruit, H.C. and Zwaan, C. (1981): Solar Phys. **70**, 207
31. Sudan, R.N. and Longcope, D.W. (1992): "Electromechanical coupling of the solar
    atmosphere", eds. D.S. Spicer and P. MacNeice (New York, AIP), p. 100

32. Vlahos, L. & Rowland, H. (1984): A&A **139**, 263
33. Vlahos, L. & Raoult, A. (1994): A&A, in press
34. Vlahos, L., Georgoulis, M., Kluiving, R. and Paschos, P. (1994): A&A, submitted
35. Webb, G.M., Axford, W.I. and Terasawa, T. (1983): "ApJ **270**, 537
36. Weisbuch, G. (1991): "Complex Systems Dynamics" (Redwood, Addison-Wesley)
37. Zirker, J.B. (1993): Solar Phys. **148**, 43

Flares in Accretion Disks

Jan Kuijpers[1,2,3]

[1] Astronomical Institute, Utrecht University, P.O. Box 80 000, 3508 TA Utrecht,
The Netherlands
[2] Faculty of Science, University of Nijmegen, Toernooiveld 1, 6525 ED Nijmegen
[3] CHEAF (Center for High Energy Astrophysics), P.O. Box 41882, 1009 DB
Amsterdam

Abstract. Present understanding of solar flares suggests that violent flare phenomena are common place in coronae of accretion disks. In particular near magnetized compact objects accretion can be modified by repeated magnetic energy storage and release. Further also in the absence of a central magnetosphere, such as around galactic and extragalactic black holes, magnetic flaring activity is possible in a corona on both sides of the disk.

Moreover strong observational evidence exists for the occurrence of magnetic flares in accretion disks in the form of X-ray and radio bursts. We point out similarities and differences of accretion disk flares with solar flares.

Finally coronal magnetic fields and magnetic flares can play an important role in providing a large effective "anomalous" viscosity. We estimate the equivalent alpha-parameter in case of a filling factor f of the disk surface with magnetic flux tubes to be $\alpha_B = (f^2 r)(6H)^{-1}$, where r is the central distance and H the disk half-thickness. It is suggested that observed bimodal behaviour in disk accretion is related to a threshold in magnetic filling factor.

1 Introduction

Violent releases of stored magnetic energy – magnetic flares – are predicted to be quite common in magnetized coronae, sandwiching accretion disks. If our present understanding of solar flares is correct flares result from kinematic distortion of coronal magnetic flux tubes by subphotospheric flows. Similar explosions are therefore expected in flux tubes in accretion disks when they become distorted by sufficiently powerful fluid motions. Indeed such magnetic effects may be responsible for the observed eruptions, the highly variable X-ray emission and radio synchrotron bursts from nuclei of active galaxies, X-ray binaries and protostellar disks. As conditions in accretion disks around compact objects differ greatly from the solar environment interesting new flare physics appear. Conversely the study of magnetic flares in accretion disks is useful for solar flare research by clarifying fundamental flare physics.

In Section 2 I will summarize the theoretical evidence for magnetic coronae enveloping accretion disks. Particular attention is paid to the problem of angular momentum transport in accretion disks in Section 3. Section 4 contains a brief

summary of the conditions for magnetic explosions to occur and outlines the
solar connection. In Section 5 I present a selection of disk observations which
can be understood in the framework of magnetic flares and examples are given
of new flare physics. Characteristic examples pertain to disks around magnetized neutron stars and white dwarfs, around stellar mass black holes, and in
Active Galactic Nuclei (AGN). Finally I show how magnetic flares may solve the
problem of anomalous viscosity in disks.

2 Thin Disks with Magnetic Coronae

I will confine myself to *thin* accretion disks (Fig. 1) around compact objects in
binaries and in AGN (for magnetic activity in thick disks see e.g. [CDS94]). If
the inward radial component of the accreting flow is small with respect to the

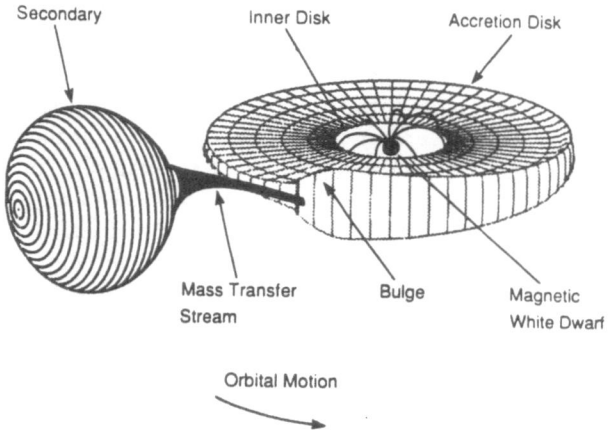

Fig. 1. Sketch of a thin accretion disk around a magnetic white dwarf fed by Roche
lobe overflow of a low-mass star [MRH88].

Keplerian speed the disk is centrifugally supported in the radial direction and
the condition for a disk to be thin follows from vertical hydrostatic equilibrium
[Pri81]:

$$H/r \approx c_s/v_K \ll 1, \tag{1}$$

where $H(r)$ is the disk half-thickness at radius r, $c_s(r)$ the sound speed and
$v_K(r) = \Omega_K(r)r = (GM/r)^{0.5}$ is the Keplerian speed. Typical numbers used for
thin disks are $H/r = 10^{-1}$ to 10^{-3} which are therefore highly supersonic with
Mach numbers $M = 10 - 10^3$.

Lynden-Bell [Lyn69] and Galeev et al. [GaR79] were among the first to suggest that magnetic flares operate in and near accretion disks (see also [LiP77,

KupI85, Kup90]). The observational evidence for the existence of magnetic activity in thin accretion disks is given in Section 5. Here I summarize our present knowledge on *dynamos* in disks as based on theoretical results (see also [Hey91, Bla92, Hey92, Pri92, Sci94, vaO94]). At first sight it is questionable if disks are magnetized since near- Keplerian shear flow is known to be linearly stable against the excitation of Kelvin-Helmholtz instabilities. Also there are no indications for convection in disks (except perhaps if an appreciable dust component is present or a hydrogen ionization zone [Roz94]). However there is an important magnetic instability – the *Balbus-Hawley instability* [BaH91, Haw92, BaG94, Cha61] –which operates under quite general conditions on vertical seed fields in Keplerian disks (Fig. 2). The instability can be simply understood by considering the

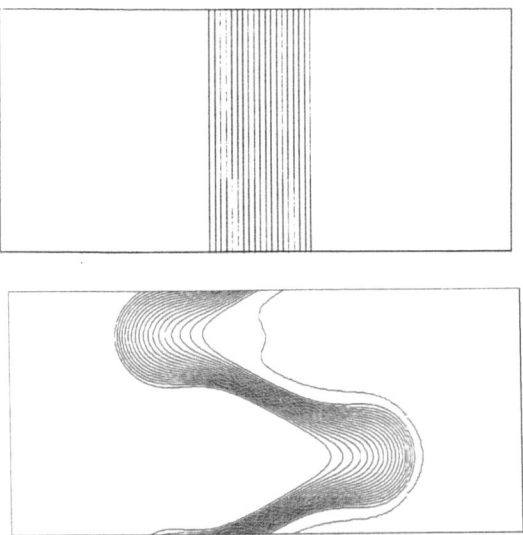

Fig. 2. Calculated distortion of an initially vertical, weak and thin magnetic field strand by [BaH91]. The vertical direction in the figure is the vertical (z-) direction in the disk, while the horizontal direction corresponds to the radial direction in the disk.

projection of a field strand onto the disk plane. As soon as the strand bulges out in the radial direction it transports angular momentum from the fast revolving inner gas to the slower outer gas. As a result the inner gas tends to fall inward while the outer gas which is sped up is swung outward and the strand is stretched even further. The instability operates when $k_z^2 v_{zA}^2 < -d\Omega_K^2/d\ell nr$, that is when the effect of rotation is sufficiently important (k_z is the wave vector component in the vertical direction and $v_{zA} = B_z(\mu_0\rho)^{-0.5}$ the Alfvén speed based on the vertical magnetic field component B_z. At what field strength the instability saturates in the nonlinear regime and whether the instability survives in a global treatment [DuK93, Kum94] remains to be studied.

Using the Balbus-Hawley instability Tout and Pringle [Tou92] have constructed a *disk dynamo:* out of a vertical component the instability creates a radial magnetic field component, from which the Keplerian shear flow generates an azimuthal field, which itself finally is acted on by buoyancy to recreate a vertical component, as long as the tube is embedded in the disk and has the same temperature as its environment. Magnetic flux tubes can then be thrown off by reconnection and the cycle repeats.

It should be mentioned that strong *deviations from a Keplerian flow* may exist in thin disks. It is true that the Keplerian flow profile is linearly stable to axisymmetric and to Kelvin-Helmholtz instabilities since the Rayleigh criteria for stability are met (monotonic outward increase of the specific angular momentum as a function of distance [Cha61], $j(r) \equiv rv_\phi(r) \propto r^{0.5}$, and the absence of a local extremum of vorticity). However the flow may be unstable to finite perturbations [Zah90] and moreover the criterion is a local one and derived for inviscid, axially symmetric flow. For instance, the presence of an inner disk edge matching the rotation speed of the central star, introduces an inflexion point as does a stiff structure carried around by the disk. Also axial symmetry may be absent because of the impact of gas on the outer disk. Clearly these effects need further investigation.

The effect of *rotation* on eddies of diameter D becomes appreciable as soon as the *Rossby number* defined as

$$Ro \equiv v/(2\Omega D) \qquad (2)$$

becomes less than unity, $Ro < 1$. In thin disks this can happen for flat two-dimensional eddies with horizontal sizes larger than the disk thickness, $D > H$. *Two-dimensional turbulence* is very different from three-dimensional turbulence. In particular 2-D turbulence has an *inverse energy cascade* and creates *long-living large-scale coherent structures* [Has85, BrR74, McW84, Hop93, Mar93]. The commonly used *isotropic* 3-D turbulence inside a disk is merely an assumption to supply the observationally required strong angular momentum transport. The same is true for the conventional upper cut-off in the turbulent scales at an eddy size comparable to the disk half-thickness (of course turbulence cannot be isotropic anymore at larger cell sizes). Further, although in a Keplerian disk the horizontal velocity difference over a radial distance D increases as $\Delta v \approx 0.5c_s D/H$ so that differential speeds across such 2-D eddies become supersonic, this does not automatically imply shock formation or the irrelevance of 2-D turbulence to disks. Until recently such 2-D disk turbulence has been neglected (an exception is [Abr92]) but perhaps it can solve the angular momentum problem and cause the anomalous viscosity in disks. Moreover such 2-D long-lived structures may provide for the very inhomogeneities whose frequency beat with the magnetosphere cause the so-called Quasi Periodic Oscillations [AlS85], strong fluctuations observed in low mass X-ray binaries. Also X-ray observations of Active Galactic Nuclei point to the existence of disk structures [ABL91]. Finally such a disk is reminiscent of the picture for the formation of the solar system originally proposed by Von Weizsäcker [VoW48].

In the case of the earth' atmosphere, of the ocean and of the major planets, cyclones and large fluid vortices are known to be extremely *stable*. A cross-fertilization may be expected between astrophysics and oceanography where these 2-D vortices are an important object of study. Preliminary numerical calculations [Yec92] (Fig. 3) already suggest that non-axisymmetric supersonic flows can be embedded inside a disk without the formation of shocks. If 2-D turbu-

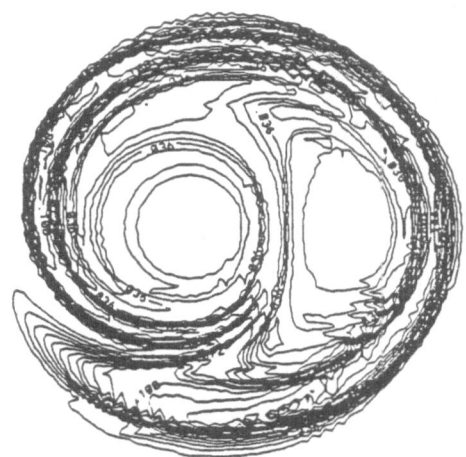

Fig. 3. Contours of constant surface density of a large-scale distortion of a shallow accretion flow [Yec92] suggesting the possible formation of 2-D structure without shocks.

lence does play a role in disks magnetic fields are still of importance. The reason is that the energy flow in the *inverse cascade* has to be dissipated. On earth this happens in thin boundary, so-called Eckman, layers at the bottom of the ocean. In accretion disks a solid boundary is absent and a different energy sink is needed. Large-scale magnetic fields sticking out into a corona may provide the answer and dissipate a large fraction of the turbulent kinetic energy by *coronal heating and flares*.

If magnetic fields are important in accretion disks the dynamical evolution of the disk depends critically on the spatial structure of the field inside the disk (the dynamo region). In the literature two classes of magnetized disks appear: those with spatially *smooth fields* [Bla76, BlP82, Dra83, Lo76, LoW87] and those with *intermittent flux tubes* [Cor85, Tor93, Maz93]. I take the point of view that disk fields have a *stochastic* instead of a regular appearance on the basis of observations of two well-studied magnetized objects: the solar magnetic field, which is anchored in discrete flux tubes [Zir88, Fou90] and the galactic magnetic field, which has a fluctuating component of a strength at least equal to the average component [Park79, Hei87, SFW86] and which has been observed to contain inhomogeneous threads or sheets [Wie93]. Note, however, that this pertains to the internal state of the disk. Outside the disk, in the corona, I shall assume that the magnetic field dominates the dynamics and is therefore *"space filling"*.

3 Angular Momentum Transport

In an accretion disk matter spirals inward faster as the outward rate of transport of angular momentum increases. For an effective scalar shear kinematic viscosity coefficient ν the inward radial speed in a nearly Keplerian disk is [Pri81]

$$v_r \approx -\nu/r. \tag{3}$$

The "molecular" viscosity, $\nu_{mol} \approx \ell v_{tr}/3$, falls short of the observed value of the effective "*Shakura-Sunyaev*" *viscosity* [SoF73], $\nu_{SS} = \alpha c_s H$, by a factor of $\tau_{ie}\Omega_K/(3\alpha) \approx 10^{-6}$. Here ℓ is the particle (e.g. proton) collisional mean free path, v_{tr} is the average particle momentum transport speed (e.g. the proton thermal speed), τ_{ie} is the proton-electron collision time, and α is a free model parameter in the Shakura-Sunyaev disk approximation and is observed from light variations in dwarf novae [FrK92] to be of order $0.1 - 1$. This deficiency constitutes the angular momentum problem in accretion disks. The nature of this *anomalous viscosity* ν_{SS} remains as yet unclear.

A number of different routes to anomalous viscosity have been suggested, each of which pertains to the dominance of a particular term in the expression for the force density in the azimuthal direction on a unit column of disk matter in a thin non-gravitating disk [AlyK90]

$$\frac{\partial}{\partial t}\Sigma v_\phi + \frac{\partial}{r^2\partial r}\left(r^2\Sigma v_\phi v_r - r^2 T_{r\phi} - r^2\int_{z_-}^{z_+}\frac{B_r B_\phi}{\mu_0}dz\right) + \frac{B_\phi B_z}{\mu_0}\Bigg|_{z_-}^{z_+}$$

$$= \frac{\partial}{r\partial\phi}\left(-\Sigma v_\phi^2 + T_{\phi\phi} + \int_{z_-}^{z_+}\frac{B_\phi^2 - B_z^2 - B_r^2}{2\mu_0}dz\right) - 2H\frac{\partial p}{\partial\phi}. \tag{4}$$

Here Σ is the surface density, $\mathbf{v}$ is the fluid velocity, $\mathbf{B}$ the magnetic field, p the gas pressure, $\mathbf{T}$ the viscous stress tensor, the suffices z, r and ϕ denote components in, respectively, the vertical, radial and azimuthal direction. The vertical bar denotes the difference of the function at z_+ and z_-, the upper and lower disk boundaries at position (r, ϕ). The various processes that can cause a torque are sketched in Fig. 4:

1. *Shocks and waves.* The pressure jump in an *oblique* shock or wave causes transport of angular momentum – the last term on the RHS – as has been known for a long time in galactic dynamics (e.g. [LyP74]). In accretion disks the case of shocks with narrow winding (Mach numbers not too different from unity) has been treated in [Mic84, LiS91, RSp93, Ch93]. The propagation of angular momentum by wave action in disks is discussed in [LuP93, DuV92] and in [Tag90] for the case of magnetosonic waves.

2. *Magnetized winds.* As is the case for stars magnetized winds can remove angular momentum efficiently by effectively placing the lever arm outside the stellar surface [Mes61, Sch62, WeD67, Mes68]. The winds are supported by centrifugal forces when magnetic fields open outward from the disk under a sufficiently large angle with respect to the rotation axis [BlP82, Dra83]. The corresponding torque

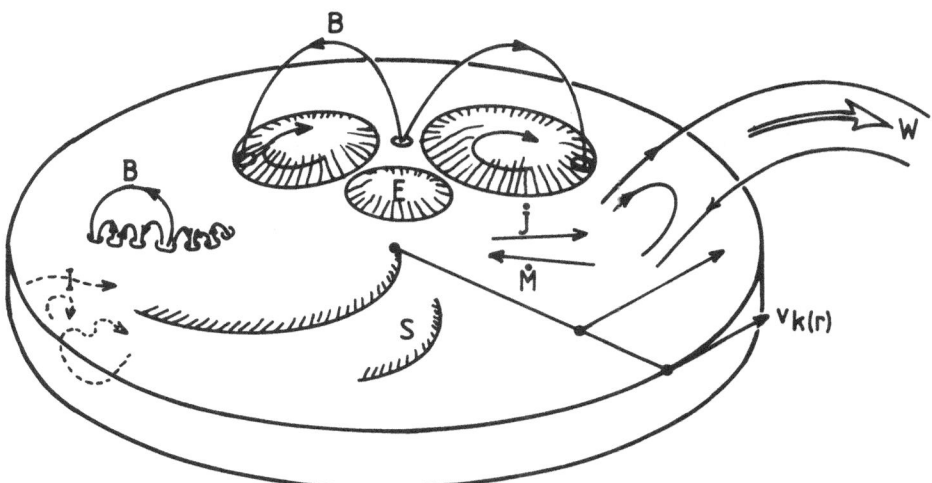

Fig. 4. Sketch of a number of processes causing anomalous viscosity in an accretion disk; B stands for magnetic field, S for spiral waves and shocks, E for 2-D eddies, W for winds, I for internal magnetic and velocity fields, $\dot{M}$ for the mass flow and $\dot{J}$ for the flow of angular momentum.

obtains from the last magnetic term on the LHS in eq.(4). Note that this term survives in case of axial symmetry whereas the previous effect does not.

3. *Isotropic fluid turbulence with and without magnetic turbulence inside the disk.* Isotropic fluid turbulence leads to a Reynolds stress via the second term on the LHS and is formally similar to the next term, the viscous stress tensor. Isotropic turbulence is assumed in the Shakura-Sunyaev alpha disk model [SoF73] and its presence is supported (but not proven) by the large value of the Reynolds number in an unmagnetized disk

$$ Re = \frac{v_\phi^2/r}{\mid \nu \boldsymbol{\nabla} \cdot \boldsymbol{\nabla} \mathbf{v} \mid_\phi} > 10^{14}. \tag{5} $$

If seed magnetic fields are added the turbulent velocity field leads to a strong enhancement of the angular momentum transport via the action of internal magnetic stresses (the first magnetic term on the LHS). The effect of space filling magnetic turbulence has been treated by [GeA92], whereas the effect of intermittent flux tubes inside the disk has been studied by Eardley & Lightman [EaL75] and by [Sck94, Scj94] for slender flux tubes. For internal circular flux tubes embedded inside the disk the most efficient transport of angular momentum appears to occur through aerodynamic friction by loops of intermediate field strength $1 \leq \beta_0 \leq 10$, and with a major radius comparable to the disk half-thickness, $0.5H \leq R_0 \leq 2H$, and a minor radius $a_0 = 0.1H$ ($\beta_0 = 2c_s^2/v_A^2$ is the ratio between the internal gas and magnetic field pressure). In this case the effective viscosity parameter alpha can reach a value $\alpha \approx 0.03 - 0.1$.

4. *Coronal loops.* Magnetic loops connecting different parts of the disk via a dilute corona are particularly effective in transporting angular momentum [Bur88, ?] and their effect appears in the second magnetic term on the LHS. For instance, comparing the coronal magnetic torque to the phenomenological stress in an alpha disk one finds for the local(!) ratio of torques

$$\frac{2B_z B_\phi r / \mu_0}{2\alpha p H} \approx \frac{r}{\alpha H(r)} \gg 1, \tag{6}$$

where we have assumed that $2B_\phi \approx B_z$ and that equipartition is reached in the footpoint regions $B^2(2\mu_0)^{-1} \approx p$. Also when compared to internal disk fields coronal loops are effective transmitters of angular momentum: Comparing the second magnetic term in eq. (4) on the left with the first term one finds a ratio

$$\frac{2B_\phi B_z / \mu_0}{B_\phi B_r 2H(r)/(\mu_0 r)} \approx \frac{r}{H(r)} \gg 1, \tag{7}$$

provided the various magnetic field components inside and outside the disk are all comparable. For loops of an extent comparable to the disk thickness an energy argument leads to the estimate $\alpha \approx 0.01$ assuming a filling factor for the disk area with magnetic fields of order unity [Bur88].

These coronal loops have to find their origin in flux tubes amplified by 3-D or by 2-D disk turbulence. Note that, contrary to conventional wisdom, even in the case of 3-D disk turbulence with a lower wave number determined by the disk thickness, the lengths of coronal loops need not by bounded by the disk thickness because reconnection in the corona tends to create larger loops. In Section 5.4 we will take into account the effect of magnetic flaring and demonstrate that tubes with a footpoint distance at least comparable to the disk half-thickness $L \geq H$ are the most effective coronal loops in transporting angular momentum, a result which in view of the low-beta assumption is surprisingly similar to the above result of [Sck94] for high-beta disk embedded tubes.

Apart from flux tubes anchored in the disk (Fig. 5a) magnetic links can exist between a magnetized central object and the disk if reconnection can take place (Fig. 5b). Because of the, in general, strong differences in rotation speeds between the footpoints such links will be shown to play a dramatic role in flaring activity and in the angular momentum budget.

5. *2-D eddies.* The study of shallow 2-D vortices in the highly supersonic disk regime has only just started. If such structures are important large-scale coronal magnetic fields may be required as the sites of energy dissipation at the end of the inverse cascade for 2-D turbulence. Note that their effect vanishes in case of axially symmetric fluid motion, just as the effect of axially symmetric shocks, while the other processes do not vanish for axial symmetry.

4 Magnetic Flares – The Solar Connection

A brief summary of the solar flare process is as follows:

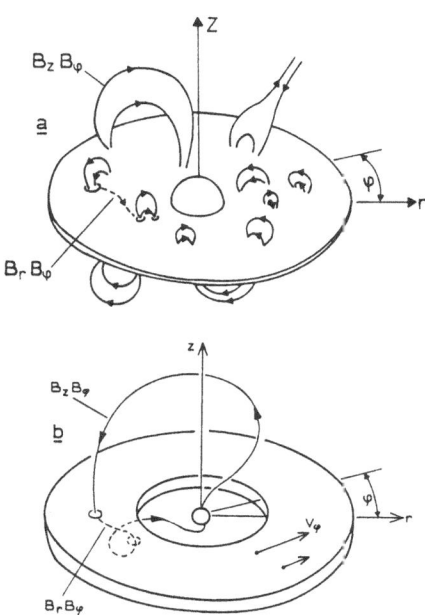

Fig. 5. Angular momentum is transported by magnetic fields embedded in the disk (stress $B_r B_\phi$) and by loops sticking out of the disk (stress $B_z B_\phi$). Relative footpoint motion causes disk flares in Fig. a, and a much larger magnetospheric flare in Fig. b.

1. A force-free magnetoplasma – the corona – is linked by magnetic fields to a dense dynamo – the subphotospheric convection zone. The corona is largely force-free in the sense that $2\mu_0 p B^{-2} \ll 1$ while the dense "driver" is characterized by $\max(2\mu_0 p B^{-2}, 2\mu_0 \rho v^2 B^{-2}) \gg 1$.

2. Kinetic energy in the driver is transformed into magnetic energy which is transferred to the corona by magnetic buoyancy and by electrodynamic coupling. This magnetic energy is not dissipated quickly but can be stored since the time scale for structural changes at the boundary between driver and corona, the *flow crossing time* $t_v \equiv \ell/\Delta v$, is less than the *Ohmic decay time* of the system $t_\eta \equiv \mu_0 \ell_B^2/\eta$, $t_v \ll t_\eta$ (the resistivity is $\eta = \sigma^{-1}$, σ is the conductivity).

3. When the travel time of magnetic signals across a particular coronal structure, the *Alfvén time* $t_A \equiv \ell/v_A$, is larger than the flow crossing time $t_A > t_v$, this structure evolves through a series of nearly force-free equilibria with, in general, increasing magnetic energy. It can be shown from the scalar magnetic virial theorem that the free magnetic energy of the distorted structure is at most comparable to the magnetic energy of the original potential (in the absence of electric currents) magnetic structure.

4. At some stage the slow evolution develops into a transient phase, as the result either of an mhd instability or of a catastrophic loss of equilibrium of the changing structure embedded inside the rest of the corona. Then the free mag-

netic energy is released during a few Alfvén crossing times [BiW89, Sch90, Rou94] (Figure 6) by *reconnection* [Bis93, Ber94]. At this stage resistivity in small vol-

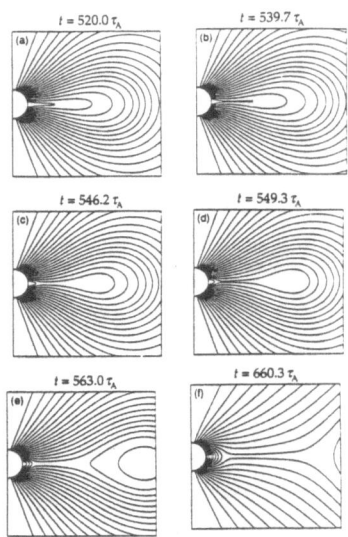

Fig. 6. Evolution of an axially symmetric magnetic coronal arcade as the initial dipolar configuration is sheared at the solar surface. The contours are field line projections onto the vertical plane at constant poloidal flux intervals [MiL94]. Times are expressed in the Alfvén time, defined as the ratio of the solar radius divided by a typical coronal Alfvén speed. Also see the figures in [BW89].

umes is essential and allows the fast transformation of part of the energy residing in coronal electric currents (the free magnetic energy) into particle acceleration, heating, plasma waves and mhd waves, and bulk motion. What determines the *partitioning* of the energy over these end products still forms an unsolved problem. The observations show that the energy release is *fragmented* and occurs in a multitude of thin sheets.

It would appear that similar magnetic flares occur on accretion disks when the above conditions are satisfied: We will therefore assume that the disk has a force-free magnetic corona which is anchored in the disk (and possibly in the central compact object) and that the Alfvén speed in the corona is suffiently large with respect to the Keplerian speed to allow a quasi-static evolution of a coronal structure in response to distortions at the boundary, $c_A \gg v_K$. The condition that the Ohmic decay time t_η is large in comparison to the flow crossing time (here a fraction of the Keplerian period) can be easily satisfied because of the large extent of astrophysical objects in general.

Finally the condition that coronal structures are distorted kinematically by motions of the footpoints implies that coronal magnetic stresses at the disk surface cannot bring the plasma lumps at both footpoints into corotation before a flare disrupts the connection. Since the average magnetic stress per unit surface area at the boundary of a force-free corona cannot exceed a value $B_z B_\phi (2\mu_0)^{-1} \leq B_z^2 (2\mu_0)^{-1}$ [AlyK90], flaring is restricted to loops with sufficiently large footpoint separations. Below we shall derive these limits separately for loops linking a disk to a central star (Section 5.1) and for loops anchored with both feet in a disk (Section 5.4). It will turn out that loops linking up a disk with a star cause powerful flaring activity, while an assembly of disk loops can cooperate to form a strong anomalous viscosity inside the disk.

5 Observations of Accretion Disk Flares – New Physics

Observations of a variety of accreting objects point to the importance of magnetic fields and magnetic flares in accretion disks:

1. Radio outbursts have been observed in a number of X-ray binaries and it is estimated that at least 10 % of them show strong and variable radio emission [Nel88]. The radio emission can usually be ascribed to the (gyro)synchrotron process and requires the production of energetic electrons and the existence of magnetic fields close to the accreting objects ($10^9 - 10^{12}$ m). Note that observations of small central radio sources in X-ray binaries are severely hampered by the accreting and expelled gas itself and may dependent crucially on the orientation of the observer with respect to the source (as is now known to be true for AGN). In particular for a fiducial spherical in- or outflow $\dot{M}_{-9}$ (in units 10^{-9} $M_\odot$ yr^{-1}) and a flow speed v_3 (in units of 10^3 km/s) radiation from a central distance r_{10} (in units 10^{10} m) has a low-frequency cut-off at the electron plasma frequency

$$f > 450(\dot{M}_{-9})^{0.5} v_3^{-0.5} r_{10} \text{MHz}. \tag{8}$$

Further the free-free absorption optical thickness at a frequency f from a central distance r_{10} outward and for a gas temperature T_4 (in units 10^4 K) is

$$\tau_{ff} = 35 \, \dot{M}_{-9}^2 \, v_3^{-2} f_9^{-2.1} T_4^{-1.35} r_{10}^{-3}, \tag{9}$$

and finally the image of a radio source in the galactic plane at 23 GHz is smeared out by scattering on electron density fluctuations, e.g. over a linear scale of 10^{12} m along a path of 10 kpc.

2. Observations of the emission lines from the chromospheres of quiescent accretion disks suggest a direct analogy with dynamo-generated chromospheres of rotating stars [Hor94].

3. The binary AE Aqr contains a white dwarf and shows optical "flares" [WHO93] and radio outbursts [BDC88] up to 1.3 mm wavelength [Aba93]. A gyro-synchrotron interpretation for the radio emission requires field strengths in expanding clouds of $2.5 \cdot 10^{-3}$ to 0.025 T. The white dwarf may have a surface magnetic field ($10 - 100$ T) and is the most rapidly spinning white dwarf known. Although an accretion disk may be absent [Horn94] and it has been proposed that the object is powered by a pulsar-like mechanism [DeJ94] some kind of magnetic relaxation is required to explain the eruptions.

4. The X-ray variability observed in AGN is often ascribed to magnetic flaring activity [FiR93]. Moreover a large part of the hard X-ray emission is inverse Compton radiation [Poz83] and therefore requires acceleration of electrons. Whereas in an X-ray binary magnetic flaring may be connected to the presence of a magnetosphere anchored in a star, for accreting black holes as in AGN any magnetic field must be anchored in an accretion disk.

5. The galactic center (Sag A) is a site of violent and episodic energy release possibly connected to the presence of an accreting black hole of mass 10^5 $M_\odot$ [Oor77]. The spectacular long (100 pc) and thin (< 0.1 pc) strands of magnetic fields discovered at radio wavelengths [Yus88] and extending away from the galactic center strongly suggest the eruption of a magnetic loop system from an accretion disk [HN89] or at least magnetic reconnection caused by fast moving clumps of matter [SeM94].

6. Pre-main sequence stars with protostellar disks [Bec94] or disk remnants are known to have large-scale magnetic fields [BaV92, HaN87] and show irregular light variations [Gah88, Gul94, CC93] some of which may be caused by magnetic interactions between disk and star. The relatively slow rotation of T Tauri stars ($\Omega_\star \approx 0.1 \Omega_K(R_\star)$) could be the result of magnetic wind breaking or magnetic interactions with an accretion disk [Har94]. Also large fluxes of chromospheric and transition region emission lines have been observed which require a magnetic emitting area larger than the stellar surface and are best placed in the disks [LeR92].

7. The highly collimated jets in Young Stellar Objects and T Tauri stars [Mun88, Mor94] and the presence of aligned ambient magnetic fields [StS86], the narrow jets in X-ray binaries [AcB83] and in AGN point to hydromagnetically driven outflows [Cam90, PuO94], magnetic confinement and the action of hoop stresses [LoB91, NeN92, Wa93].

5.1 Disks around Magnetized Neutron Stars and White Dwarfs

As a corollary I will use that a coronal magnetic structure with initial potential field energy $W_0 = B_0^2 V (2\mu_0)^{-1}$ (B_0 is the average potential field value, V is the volume) can store a maximum additional free energy of the same magnitude $W_f \approx W_0$ which is released in a flare during a few Alfvén crossing times t_A.

Magnetic accretion onto the stellar equator. Let us consider a magnetic flux tube connecting the magnetized star to the disk at radial position r and initially lying in a meridional plane. Let $\Omega_B(r) \equiv |\Omega_K(r) - \Omega_\star|$ be the beat frequency between the rotation frequencies of both footpoints. If the excess (deficit) angular momentum of the gas lump in the disk with respect to the stellar rotation rate is communicated to (from) the star before a flare occurs the magnetic link will reach corotation with the star. We will approximate the increase of the azimuthal magnetic field component by $B_\phi(r) = \Omega_B(r)t B_z(r)$. This relation only holds till a maximum time $t_f \approx \Omega_B^{-1}(r)$ when the distorted link is supposed to flare and relax to a poloidal configuration. However no flare occurs when the gas lump at the disk footpoint can be brought into corotation before that time. This will happen if the following condition is satisfied

$$B_z^2(2\mu_0)^{-1} < \rho c_s v_K (\Omega_B/\Omega_K)^2, \qquad (10)$$

while a flare will occur when the opposite is true. Assuming an equipartition field strength $B^2(r)(2\mu_0)^{-1} = p(r) = \rho(r)c_s^2(r)$ it follows that a gas lump embedded in the disk is brought into corotation only when it is located inside a narrow ring around the radius of corotation, $r_{co} \equiv (GM/\Omega_*^2)^{1/3}$, of width $2\Delta r \approx 1.3(H(r_{co})/r_{co})^{0.5}$. For other locations the lump gives rise to flares and it remains in the equatorial plane as long as a disk exists. In the inner region the steady disk is disrupted by the stellar field and eq.(10) can be used in combination with the given stellar field and conservation of mass and magnetic flux through the gas lump to determine what stellar field strength is needed to enforce corotation. For a dipolar field dependence near the neutron star and a field strength at the stellar surface substantially below 10^8 T it turns out that magnetically linked gas lumps stay in the equatorial plane and hit the star in a ring at the equator [AlyK90]. This has important observational consequences since one can now understand why weak field neutron stars such as occur in low mass X-ray binaries exhibit no X-ray pulses (which would arise if gas is forced to follow the stellar field onto the, in general, oblique magnetic poles) but instead show so-called Quasi-Periodic-Oscillations [vaK89].

Maximum flare energy from white dwarf and neutron star. The flare energy derives from relaxation of the magnetosphere of the compact object upon distortion by the accreting matter. Gravitational energy is converted and temporally stored as magnetic energy. It is therefore not surprising that the maximum flare power averaged over a long period is comparable to the accretion power, e.g. [AlyK90] find an average flare luminosity from one side of the disk

$$L_M \approx \pi \ell R_*^2 \left(\frac{R_*}{R}\right)^4 \Omega_B(R) \frac{B_*^2}{4\mu_0}$$
$$= 1.3 \cdot 10^{28} \frac{\ell}{10^4 \text{m}} \left(\frac{R_*}{10^4 \text{m}}\right)^6 \left(\frac{R}{10^5 \text{m}}\right)^{-4} \frac{\Omega_B}{200 \text{Hz}} \left(\frac{B_*}{10^6 \text{T}}\right)^2 \text{Watt}, \quad (11)$$

for a spherically symmetric magnetic umbrella of equatorial radius R and radial extent ℓ at the equator, a stellar dipole field of strength B_* at the stellar pole, a flaring rate $\Omega_B^{-1}(R)$ and a stellar radius R_*. This luminosity is in principle available for particle acceleration required to explain the observed radio synchrotron flares.

The maximum energy in one flare event is determined by the maximum free energy which can be stored in the stellar magnetosphere. Based on the virial theorem this energy is $W_f \approx r^3 B(r)^2 (2\mu_0)^{-1} \propto r^{-3}$, for a dipolar dependence of the stellar field with distance. We therefore expect the largest flares to occur close to compact, strongly magnetised objects and for high accretion rates so that the field structure near the star can be "loaded": max $W_f \approx R_*^3 B_*^2 (2\mu_0)^{-1}$ or $4 \cdot 10^{26}$ J for a white dwarf with a surface field of 1 T and $4 \cdot 10^{33}$ J for a neutron star with a surface field of 10^8 T. Such big concerted magnetic relaxations have been proposed to cause the bursts of accretion observed in X-rays from the rapid burster [KuK94].

Steady versus flaring accretion. There are two opposing views about the interaction between disk and magnetosphere: [GL79] assume that the magnetic field penetrates the disk over a large radial extent and remains stationary which implies that the field lines slip across the disk at a rate dictated by the relative motion between star and disk. Because of the high conductivity [AlyK90] on the other hand argue that the field links up to the disk over relatively small areas located in the magnetically compressed inner part of the disk and that the reconnection is essentially non- steady and of the flaring type. Typically these flares occur inside the pressure balance radius, where the external stellar field pressure dominates the original uncompressed gas pressure of the disk. Recently [vaB94] has computed a series of force-free equilibria. He shows that

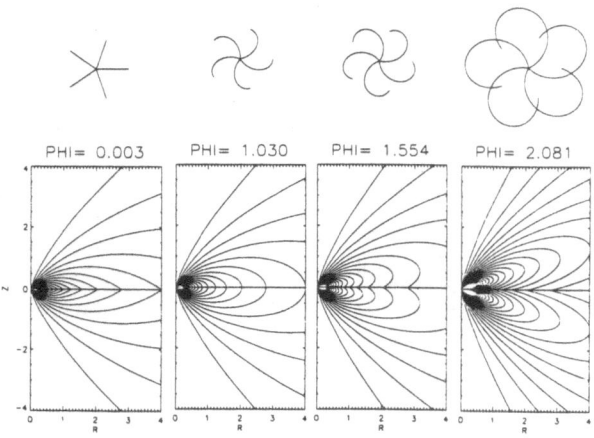

Fig. 7. The top row shows a sequence of projections of force-free field lines onto the plane of the disk and the bottom the projections in the meridional plane for different angular displacements PHI between the footpoints for a rigidly rotating disk and a vertical field component $B_{z,disk}(r) \propto r^{-2.5}$ [Bal94].

flare-like events must take place for differential rotation angles less than 3 radians assuming that stellar magnetic field can be pulled into the disk over a wide range of radii as a consequence of the non-potential nature of the field.

Compton resistivity. An important difference between the physical conditions in accretion disks and solar flares is the intensity of the radiation field. As a result in the inner disk flare accelerated electrons loose their energy by inverse Compton losses long before reaching the disk instead of causing "evaporation" of chromospheric material to coronal temperatures as in the sun [deVK92]. This Compton drag has been proposed by [vdO] to provide an anomalous resistivity for the dissipation of coronal currents in disk flares.

5.2 Disks around Stellar Mass Black Holes

Magnetic energy extraction from rotating black holes may play a role in the energy liberation in AGN [BlZ77, ThP88, KCa92, Par94]. Blandford and Znajek showed that a rotating hole generates a source voltage which drives an electric current if an axially symmetric force-free magnetic structure extends between the black hole and "infinity". Each magnetic surface rotates with a fixed speed and slips across the (faster rotating) black hole horizon on one end and across the (non-rotating) resistance at the other end. Maximum power can be extracted from the rotating hole if the load at infinity matches the impedance of the hole, in which case half of the "Ohmic" dissipation appears as useful energy at infinity. However [VOK93] have pointed out that the current may be too large for the force-free connection to permit a (quasi-)steady state. The reason is that in this case on one hand the current value is determined by $I = \oint (\mathbf{v} \wedge \mathbf{B}) \cdot d\mathbf{l}/R$ (R is the total resistance of the circuit), while on account of the magnetic virial theorem its energy, which is of order $W_M = \int B_\phi^2 (2\mu_0)^{-1} d^3\mathbf{r}$ where the toroidal magnetic field component B_ϕ is determined by the current through Ampére's law, should remain smaller than the original potential field energy $W_0 = \int B_0^2 (2\mu_0)^{-1} d^3\mathbf{r}$. Clearly for chosen values of the hole's mass, rotation rate and external magnetic field strength the value of the resistance at infinity determines if the current is suppressed to a sufficiently small value. If the load "at infinity" is formed by gas at the footpoint of a link in an accretion disk indeed a stationary current is not set up and the current system dissipates in the form of flares [VOK93]. Note that a closed current connecting to a black hole requires ingoing electrons on some field lines and ingoing positive charges on other field lines, and further continued infall of magnetized plasma is required as a (non-charged) black hole has no magnetic field of its own.

Electron-positron annihilation line. The black hole candidates 1E 1740.7-2942 and GRS 1124-68 (Nova Muscae) show a transient line-like feature in X-rays (gaussian hump at 480 keV) superimposed upon a persistent hard X-ray continuum (which can be modelled as a Comptonized disk component). The hump cannot simply be explained as a (gravitationally) redshifted electron-positron annihilation line since the strength of the line requires a rate of pair production far too high in relation to the observed weak high-energy end of the spectral continuum above 0.5 MeV. [vBO94] however show that the presence of a rotating stellar mass black hole makes an important difference and can lead to a sufficiently enhanced photon pair production rate as the infalling photons from the disk get blue-shifted and collide on a nearly circular orbit around the black hole so that a much larger amount of photons out of the hard X-ray continuum partake in pair creation.

Origin of the disk corona. [vaO94] points out that the density outside an accretion disk decreases very steeply with distance z from the midplane ($\propto exp(-z^2/L^2)$ for scale height L). A corona extending over a substantial altitude can therefore not be maintained by gas pressure. In fact similar to a pulsar magnetosphere [BGI93, Mic91] coronal gas can be produced and sustained by electric fields if magnetic fields are rooted in the accretion disk. The rotating magnetic fields produce electric fields with a component parallel to the magnetic field. For a magnetic loop with footpoint separation D the potential drop is of order $\Delta\Psi \approx \Omega_K(r)rDB(r)$. and it can be large enough to extract electrons from one footpoint and positive ions from the other footpoint. Depending on the potential drop and the geometry of the magnetic field even pair production is possible from curvature radiation.

5.3 Disks in Active Galactic Nuclei

Many of the peculiarities of various classes of AGN can be unified into the picture of a geometrically thin accretion disk orbiting a massive black hole ($M \approx 10^5 - 10^9 M_\odot$) and surrounded by a geometrically thick molecular dust torus which is looked upon from various angles [Bar89, Ant93]. In the absence of magnetic fields thermal accretion disks with the required accretion rate up to 1 $M_\odot$/yr would only emit blackbody radiation in the ultraviolet or at lower frequencies, in contrast with the observed X-ray, gamma-ray and non-thermal radio emission. Detailed modelling of the emission over the entire spectrum is now reaching a sophistication not too far from solar physics and rests on processes such as inverse Compton and synchrotron emission and irradiation of a disk from above, all of which require abundant particle acceleration, pair creation and magnetic fields in a corona [Bla91, RFa93, FiR93, Haa93, Sve94, SvZ94].

X-ray variability. The X-ray emission (2 - 10 keV) from some AGN shows variability which is well described by a power law, e.g. for NGC 5506 a power law f^{-1} describes the fluctuations between $3 \cdot 10^{-6} - 5 \cdot 10^{-3}$ Hz [PoM88, Mus93]. The variability can be explained by magnetic flares occurring outside 10 Schwarzschild radii if the flares dissipate by accelerating electrons in elementary bursts during an Alfvén time. The electrons transfer their energy to X-rays by inverse Compton losses in the disk radiation field and it is assumed that every flaring structure on the disk repeats the flaring cycle after ($(2\pi)^{-1}$ times) its Keplerian rotation period [deVK92].

Reconnection in jets. Astrophysical jets appear to be related to accretion disks [BuL93]. In the case of radio jets repeated injection of energetic particles or reacceleration is required to make up for the losses due to adiabatic expansion [Ver87, Ver93]. Such an energy injection may be caused by recurrent flaring of magnetized plasmoids as they travel outwards in the jet [USh85]. Possibly one starts with a linear force-free plasmoid which expands anisotropically so that the force-free field becomes non-linear and then relaxes according to Taylor's hypothesis [Tay74, Tay86, NoH83] to the linear force-free field with the same total helicity [Dix89, ChK86, deV93, VeP94]. Typically the available injection energy is a few per cent of the instantaneous total magnetic field energy [deV93].

5.4 Disk angular momentum transport revisited

If disks have magnetic coronae with flaring loops it is of interest to find out
how angular momentum transport depends on the radial footpoint separation of
the coronal loops [HnP89]. For loops with a *fixed* footpoint separation of $2H$ it
was found in [Bur88] that angular momentum is more effectively transported by
continuous current dissipation than by flaring. Here I will follow a different approach making direct estimates of the magnetic stresses and comparing loops of
different dimension. Consider an individual coronal flux tube as sketched in Fig.
8 extending over an interval (r_1, r_2). Depending on the separation two cases can
be distinguished: For very small footpoint separation the differential Keplerian
speed of the undisturbed flow is relatively small and one expects the magnetic
stress on the (non force-free) gas parcels at the footpoints to enforce solid body
rotation. However for sufficiently large separations the difference in angular momentum of the gas at the footpoints is too large to be transferred magnetically
before the flux tube is strongly distorted and disrupted by reconnection in a
flare.

I will assume that the coronal field can be considered force-free and that the
fluid shear builds up an azimuthal magnetic field component at each footpoint
out of an initially meridional field according to

$$B_\phi = B_z \Omega_B rt/\Delta r, \qquad (12)$$

until $B_\phi = B_z$ after a period $\tau = \Delta r/(\Omega_B r) = 2/(3\Omega_K)$ at which time the magnetic field is assumed to relax to a meridional configuration. Here $\Delta r \equiv r_2 - r_1 \ll$
r_2 is the radial footpoint separation of the coronal loop, $\Omega_B \equiv |\Omega_K(r_1) - \Omega_K(r_2)|$
is the *beat frequency* between the footpoints r_1, r_2 and eq.(12) approximates a
linear force-free arcade evolution within 30 % [Bur88]. For the average rate of
transport of angular momentum through a *flaring* loop into footpoint 2 at both
upper and lower sides of the disk we now find

$$\dot{J}_f \approx \frac{2A_2 B_{z2} r_2}{\mu_0 \tau} \int_0^\tau \frac{B_{z2} \Omega_B r_2 t}{r_2 - r_1} dt = \frac{A_2 B_{z2}^2 r_2}{\mu_0} = A_2 2\rho c_s^2 r_2, \qquad (13)$$

where A_j is the cross-section of the tube at the surface of the disk in footpoint
j. This result only applies to large and flaring loops.

For *small* loops eq. (12) still applies but corotation is reached at a time $\tau_s < \tau$.
A first condition for a small loop to reach rigid rotation is that the excess (deficit)
angular momentum of the gas at one footpoint be transported to the gas at the
other footpoint within a flaring time $\tau = 2(3\Omega_K)^{-1}$. Noting that rigid rotation
is obtained as soon as $\Omega_1 = \Omega_2 = \Omega_0$, where we approximate $r_0 = 0.5(r_1 + r_2)$,
and using eq. (12) we find the transfer time τ_s from equating half the difference
in angular momentum of the feet to the amount transported into the disk at
footpoint 2 during a time τ_s

$$A_2 2H_2 \rho_2 r_2 0.75(r_2 - r_1)\Omega_{K2} = A_2 r_2 B_{z2}^2 3\Omega_{K2}\tau_s^2(2\mu_0)^{-1}, \qquad (14)$$

resulting in $\tau_s^2 = (r_2 - r_1)/(2c_s \Omega_K)$. Secondly the condition that the transfer time of the excess momentum is less than the flaring time, $\tau_s < \tau$, leads to

$$r_2 - r_1 < H8/9. \tag{15}$$

One can verify that the magnetic tension in such small loops can oppose the excess (deficit) centrifugal acceleration of the outer (inner) foot. Whether the tension can also resist the azimuthal aerodynamic drag is more difficult to decide and depends on the details of the flow pattern and the magnetic filling factor of the annulus at radius r. We conclude that coronal loops with a footpoint separation larger than the disk half-thickness H undergo flaring while smaller loops could retain their identity. We now determine the average rate of transport of angular momentum into footpoint 2 for a small rigid loop

$$\dot{J}_s \approx A_2 r_2 B_{z2}^2 3(\Omega_{K2}(r_2 - r_1))^{0.5}(2c_s)^{-0.5}(2\mu_0)^{-1}, \tag{16}$$

and, comparing the result with (13), we find that the ratio of the average rate of transport of angular momentum by a flaring loop to that by a small rigid loop (keeping all other quantities the same) varies as

$$\frac{\dot{J}_f}{\dot{J}_s} = \frac{2^{1.5}H^{0.5}}{3(r_2 - r_1)_s^{0.5}} > 1. \tag{17}$$

This result implies that flaring loops with footpoint separations larger than the disk half-thickness, $r_2 - r_1 > H$, are more efficient in transporting angular momentum than small, $(r_2 - r_1)_s < H$, rigid loops and that the efficiency is independent of the footpoint separations of the flaring loops.

Finally we can make a formal comparison between the magnetic viscosity and the anomalous viscosity described by the alpha disk. Let us define a filling factor f for the coverage of the disk with closed magnetic fields, where $f = 1$ implies that the (upper and lower) surface is entirely covered with closed magnetic fields (half positive, half negative polarity on each side). We assume that the magnetic flux tubes overlap each other in radial extension as sketched in Fig. 8 so that the magnetic field leads to a "global" disk viscosity. The vertical coronal field component is related to the photospheric field component by flux conservation $B_{zc} = B_{zp}f$. The azimuthal coronal field component is limited on average to the vertical coronal field for a force-free corona on the basis of the virial theorem $B_{\phi c} \leq B_{zc}$. The magnetic stress per unit area is therefore proportional to $B_{zc}B_{\phi c} = B_{zp}^2 f^2$. If we now compare the average magnetic torque density per unit radial distance at radius r with that due to anomalous "alpha" viscosity we find a ratio

$$\frac{\dot{J}_B}{\dot{J}_\alpha} \approx \frac{f^2 r}{6\alpha H}, \tag{18}$$

from which it is found that magnetic fields with filling factors less than unity can form a physical realisation of the so-called alpha-parameter with an effective anomalous viscosity

$$\alpha_B = \frac{f^2 r}{6H}. \tag{19}$$

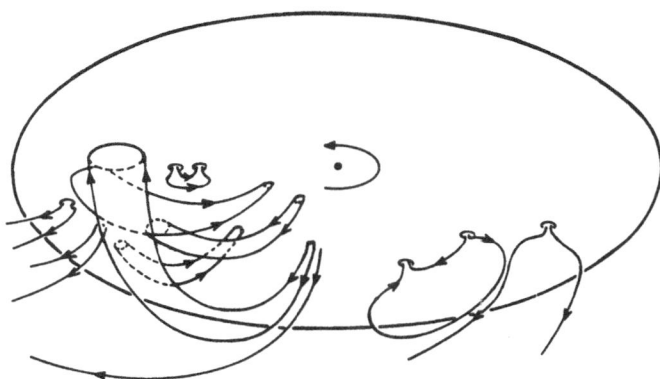

Fig. 8. Sketch of a radially overlapping system of coronal loops causing anomalous viscosity.

It should be noted however that we have assumed that the field strength at the disk surface corresponds to equipartition with the gas pressure inside the disk. Further we have assumed that the angular momentum is completely transported through closed flux tubes while in reality part of it can be lost through a wind in open tubes.

Magnetic loops in a disk only enhance the global transport of matter if they overlap each other in radial distance. For loops of a cross-section $\pi h^2/4$ at the photosphere this means that the magnetic surface filling factor should at least exceed a value

$$f > f_{min} = h(8r)^{-1} \approx H(8r)^{-1}, \tag{20}$$

where we have assumed that each ring with radial extent $2h$ is served by one (outgoing and one incoming) flux tube and that a flux tube at the photosphere has a typical diameter equal to the disk half-thickness H.

If indeed the effective viscosity in disks is formed by the concerted action of coronal magnetic loops we predict a bimodal [KZd94] behaviour in the accretion rate (and therefore in the X-ray luminosity): For sufficiently large filling factors (see eq.(20)) the radial inflow is fast while at times of smaller filling factor the inflow stagnates, the disk becomes strongly structured in the radial direction by the inhomogeneous distribution of flux tubes over the disk and accretion occurs in bursts. Perhaps the accretion shows a stochastic threshold as in tokamaks [Dub94] and the appearance of self-organized criticality [Vla94, Min94].

6 Conclusion

Observations of impulsive particle acceleration in accreting systems point to the existence of magnetic fields inside and above accretion disks. In this respect of particular interest are strongly magnetized white dwarfs and neutron stars in X-ray binaries, protostellar disks around bipolar sources in the infrared and T Tauri stars, and the magnetic formation of jets. This paper is a plea to the solar community to apply its knowledge to accretion disks and conversely to the X-ray community of compact objects to open itself to cross-fertilization from other disciplines. In particular the longstanding problem of energy partitioning –the relative importance of particle acceleration, heating, bulk motion and waves in magnetic explosions– will benefit from observations of flares in accretion disks.

Acknowledgement. It is a pleasure to thank the organizers and the participants for this stimulating workshop. I thank Roeland van Oss, George Schramkowski, Michael Raadu, Keith Horne, Michiel Nauta, Sjef Zimmerman, Loukas Vlahos and Marc Dubois for stimulating discussions. Part of this work was performed under EC-twinning Contract No. SCI*-CT91-0727 on Coherent Radiation and Particle Acceleration in Magnetized Plasmas.

References

[Aba93]   Abada-Simon, M., Lecacheux, A., Bastian, T.S., Bookbinder, J.A., Dulk, G.A.: 1993, ApJ 406, 692

[ABL91]   Abramowicz, M.A., Bao, G., Lanza, A., Zhang, X-H.: 1991, A&A 245, 454

[Abr92]   Abramowicz, M.A., Lanza, A., Spiegel, E.A., Szuszkiewicz, E.: 1992, Nature 356, 41

[AcB83]   Achterberg, A., Blandford, R.D., Goldreich, P.: 1983, Nature 304, 607

[AlS85]   Alpar, M.A., Shaham, J.: 1985, Nature 316, 239

[Aly80]   Aly, J.J.: 1980, A&A 86, 192

[AlyK90] Aly, J.J., Kuijpers, J.: 1990, A&A 227, 473

[Ant93]   Antonucci, R.: 1993, ARAA 31, 473

[BaG94]   Balbus, S.A., Gammie, C.F., Hawley, J.F.: 1994, Fluctuations, dissipation and turbulence in accretion disks, preprint

[BaH91]   Balbus, S.A., Hawley, J.F.: 1991, ApJ 376, 214

[Bar89]   Barthel, P.D.: 1989, ApJ 336, 606

[BaV92]   Basri, G., Geoffrey, W.M., Valenti, J.A.: 1992, ApJ 390, 622.

[BDC88]   Bastian, T.S., Dulk, G.A., Chanmugan, G.: 1988, ApJ 324, 431

[Bec94]   Beckwith, S.V.W.: 1994, in Theory of Accretion Disks – 2, eds. W.J. Duschl, J. Frank, F. Meyer, E. Meyer-Hofmeister, W.M. Tscharnuter, Kluwer Acad. Publ., Dordrecht, p. 1

[Ber94]   Berger, M.A.: 1994, in Fragmented Energy Release in Sun and Stars, ed. G.H.J. van den Oord, Space Sci. Rev., 68, 3

[BGI93]   Beskin, V.S., Gurevich, A.V., Istomin, Ya.N.: 1993, Physics of the Pulsar Magnetosphere, Cambridge Univ. Press

[Bis93]   Biskamp, D.: 1993, Nonlinear Magnetohydrodynamics, Cambridge Univ. Press, Ch. 6

[BiW89]   Biskamp, D., Welter, H.: 1989, Solar Phys. 120, 49

[Bla76]   Blandford, R.D.: 1976, MNRAS 176, 465

[Bla91]   Blandford, R.D.: 1991, in Particle Acceleration near Accreting Compact Objects, eds. J. van Paradijs, M. van der Klis, A. Achterberg, KNAW Verh., Afd. Natuurkunde, Eerste Reeks, deel 35, p. 9

[Bla92]   Blandford, R.D.: 1992, in Physics of Active Galactic Nuclei, eds. W.J. Duschl, S.J. Wagner, Springer-Verlag, Berlin, p. 3

[BlP82]   Blandford, R.D., Payne, D.G.: 1982, MNRAS 199, 883

[BlZ77]   Blandford, R.D., Znajek, R.L.: 1977, MNRAS 179, 433

[BrR74]   Brown, G.L., Roshko, A.: 1974, J. Fluid Mech. 64, 4775

[BuL93]   Burgarella, D., Livio, M., O'Dea, C.: 1993, Astrophysical Jets, Cambridge Univ. Press

[Bur88]   Burm,H., Kuperus, M.: 1988, A&A 192, 165

[Cam90]   Camenzind, M.: 1990, Rev. Mod. Astron. 3, ed. G. Klare, Springer-Verlag, Heidelberg, p. 234

[CC93]   Cameron, A.C., Campbell, C.G.: 1993, A&A 274, 309

[CDS94]   Chakrabarti, S.K., D'Silva, S.: 1994, ApJ 424, 138

[Ch93]   Chakrabarti, S.K., Wiita, P.J.: 1993, ApJ 411, 602

[Cha61]   Chandrasekhar, S.: 1961, Hydrodynamic and Hydromagnetic Stability, Oxford Univ. Press, §66, Chap. IX (and references therein)

[ChK86]   Choudhuri, A.R., Königl, A.: 1986, ApJ 310, 96

[Cor85]   Coroniti, F.V.: 1985, in: Unstable Current Systems and Plasma Instabilities in Astrophysics, IAU Symp. No. 107, eds. M.R. Kundu, G.D. Holman, Reidel, p. 453

[DeJ94]   De Jager, O.C.: 1994, ApJ Suppl Series 90, 775

[deV93]   De Vries, M.: 1993, Magnetic Relaxation in Active Galactic Nuclei, Ph.D. Thesis, Utrecht University, Chs. 2, 3

[deVK92] De Vries, M., Kuijpers, J.: 1992, A&A 266, 77

[Dix89]   Dixon, A.M., Berger, M.A., Browning, P.K., Priest, E.R.: 1989, A&A 225, 156

[Dra83]   Draine, B.T.: 1983, ApJ 270, 519

[Dub94]   Dubois, M.A., Nakach, R., Sabot, R., Hennequin, P.: 1994, in Fragmented Energy Release in Sun and Stars, ed. G.H.J. van den Oord, Space Sci. Rev., 68, 81

[DuK93]   Dubrulle, B., Knobloch, E.: 1993, A&A 274, 667

[DuV92]   Dubrulle, B., Valdettaro, L.: 1992 A&A 263, 387

[EaL75]   Eardley, D.M., Lightman, A.P.: 1975, ApJ 200, 187

[FiR93]   Field, G.B., Rogers, R.D.: 1993, ApJ 403, 94

[Fou90]   Foukal, P.V.: Solar Astrophysics, John Wiley and Sons Inc., New York, Ch. 9, 10

[FrK92]   Frank, J., King, A.R., Raine, D.J. : 1992, Accretion Power in Astrophysics, Cambridge Univ. Press, Ch. 5.8

[Gah88]   Gahm, G.F.: 1988, in Formation and Evolution of Low Mass Stars, eds. A.K. Dupree and M.T.V.T. Lago, NATO ASI, Kluwer Acad. Publ., Dordrecht, The Netherlands, p. 295

[GaR79]   Galeev, A.A., Rosner, R., Vaiana, G.S.: 1979, ApJ 229, 318

[GeA92]   Geertsema, G.T., Achterberg, A.: 1992, A&A 255, 427

[GL79]   Ghosh, P., Lamb, F.K.: 1979, ApJ 234, 296

[Gul94]   Gullbring, E.: 1994, Polar accretion and light variability on T Tauri stars, A&A in press

[Haa93]   Haardt, F., Maraschi, L.: 1993, ApJ 413, 507

[Har94]   Hartmann, L.: 1994, in Theory of Accretion Disks – 2, eds. W.J. Duschl,
          J. Frank, F. Meyer, E. Meyer-Hofmeister, W.M. Tscharnuter, Kluwer Acad.
          Publ., Dordrecht, p. 19
[HaN87]   Hartmann, L.W., Noyes, R.W.: 1987, ARAA, 25, 271
[Has85]   Hasegawa, A.: 1985, Advances in Physics 34, 1
[Haw92]   Hawley, J.F., Balbus, S.A.: 1992, ApJ 400, 595
[Hei87]   Heiles, C.: 1987, in Physical Processes in Interstellar Clouds, eds. G.E. Morfill
          and M. Scholer, NATO ASI Series C, Vol. 210, p. 429
[Hey91]   Heyvaerts, J.: 1991, in Structure and Emission Properties of Accretion Disks,
          IAU Coll. No 129, eds. C. Bertout, S. Collin-Souffrin, J.P. Lasota and J. Tran
          Tranh Van, Editions Frontiéres, 91192 Gif sur Yvette, France, p. 109
[Hey92]   Heyvaerts, J.: 1992, in Physics of Active Galactic Nuclei, eds. W.J. Duschl,
          S.J. Wagner, Springer-Verlag, Berlin, p. 445
[HN89]   Heyvaerts, J., Norman, C.A.: 1989, ApJ 347, 1055
[HnP89]   Heyvaerts, J., Priest, E.R.: 1989, A&A 216, 230
[Hop93]   Hopfinger, E.J., van Heijst, G.J.F.: 1993, Annual Review of Fluid Mechanics
          25, 241
[Hor94]   Horne, K.D.: 1994, in Theory of Accretion Disks – 2, eds. W.J. Duschl, J.
          Frank, F. Meyer, E. Meyer-Hofmeister, W.M. Tscharnuter, Kluwer Acad.
          Publ., Dordrecht, p. 77
[Horn94] Horne, K.D.: 1994, personal communication
[KCa92]   Khanna, R., Camenzind, M.: 1992, A&A 263, 401
[Kui89]   Kuijpers, J.: 1989, Solar Phys. 121, 163
[Kui90]   Kuijpers, J.: 1990, in Active Close Binaries, ed. C. Ibanoğlu, NATO ASI,
          Kuşadasi, Turkey, Sept. 89, Kluwer Acad. Publ., Dordrecht, p. 761.
[KuK94]   Kuijpers, J., and Kuperus, M.: 1994, Astron. Astrophys. 286, 491.
[Kum94]   Kumar, S., Coleman, C.S., Kley, W.: 1994, MNRAS, 266, 379
[Kup90]   Kuperus, M.: 1990, Computer Physics Reports 12, 275
[KupI85] Kuperus, M., Ionson, J.A.: 1985, A&A 148, 309
[KZd94]   Kusunose, M., Zdziarski, A.A.: 1994, ApJ 422, 737
[LeR92]   Lemmens, A.F.P., Rutten, R.G.M., Zwaan, C., 1992, A&A 257, 671
[LiP77]   Liang, E.P., Price, R.H.: 1977, ApJ 218, 247
[LiS91]   Livio, M., Spruit, H.C.: 1991, A&A 252, 189
[Lo76]   Lovelace, R.V.E.: 1976, Nature 262, 649
[LoB91]   Lovelace, R.V.E., Berk, H.L., Contopoulos, J.: 1991, ApJ 379, 696
[LoW87]   Lovelace, R.V.E., Wang, J.C.L., Sulkanen, M.E.: 1987, ApJ 315, 504
[LuP93]   Lubow, S.H., Pringle, J.E.: 1993, ApJ 409, 360
[Lyn69]   Lynden-Bell, D.: 1969, Nature 223, 690.
[LyP74]   Lynden-Bell, D., Pringle, J.E.: 1974, MNRAS 168, 603
[Mar93]   Marcus, P.S.: 1993, ARAA 31, 523
[Mas88]   Mason, K.O., Rosen, S.R., Hellier, C. 1988, Adv. Space Res., 8, No. 2, p. 293
[Maz93]   Mazets, E.I., Bykov, A.M.: 1993, Astronomy Letters 19, 184
[McW84]   McWilliams, J.C.: 1984, J. Fluid Mech. 146, 21
[Mes61]   Mestel, L.: 1961, MNRAS 122, 473
[Mes68]   Mestel, L.: 1968, MNRAS 138, 359.
[Mic84]   Michel, F.C.: 1984, ApJ 279, 807
[Mic91]   Michel, F.C.: 1991, Theory of Neutron Star Magnetospheres, Univ. of
          Chicago Press
[MiL94]   Mikić, Z., Linker, J.A.: 1994, ApJ 430, 898

[Min94] Mineshige, S., Ouchi, N.B., Nishimori, H.: 1994, Publ. Astron. Soc. Japan 46, 1

[Mor94] Morse, J.A., Hartigan, P., Heathcote, S., Raymond, J.C., Cecil, G.: 1994, ApJ 425, 738

[Mun88] Mundt, R.: 1988, in Formation and Evolution of Low Mass Stars, eds. A.K. Dupree and M.T.V.T. Lago, NATO ASI, Kluwer Acad. Publ., Dordrecht, The Netherlands, p.257

[Mus93] Mushotzky, R.F., Done, C., Pounds, K.A.: 1993, ARAA, 31, 717

[Nel88] Nelson, R.F., Spencer, R.E.: 1988, MNRAS 234, 1105

[NeN92] Newman, W.I., Newman, A.L., Lovelace, R.V.E.: 1992, ApJ 392, 622

[NoH83] Norman, C.A., Heyvaerts, J.: 1983, A&A 124, L1

[Oor77] Oort, J.H.: 1977, ARAA 15, 295

[Par94] Park, S.J., Vishniac, E.T.: 1994, ApJ 426, 131

[Park79] Parker, E.N. 1979, Cosmical Magnetic Fields, Clarendon Press, Oxford, UK

[PoM88] Pounds, K.A., McHardy, I.M.: 1988, in Physics of Neutron Stars and Black Holes, ed. Y.Tanaka, Tokyo, Japan, p. 285

[Poz83] Pozdnyakov, L.A., Sobol, I.M., Sunyaev, R.A.: 1983, Sov. Sci. Rev. E 2, 189

[Pri81] Pringle, J.E.: 1981, ARA&A 19, 137

[Pri92] Pringle, J.E.: 1992, Rev. Mod. Astron. 5, ed. G. Klare, Springer-Verlag, Heidelberg, p. 97

[PuO94] Pudritz, R.E., Ouyed, R.: 1994, in Theory of Accretion Disks – 2, eds. W.J. Duschl, J. Frank, F. Meyer, E. Meyer-Hofmeister, W.M. Tscharnuter, Kluwer Acad. Publ., Dordrecht, p. 35

[RFa93] Ross, R.R., Fabian, A.C.: 1993, MNRAS 261, 74

[Rou94] Roumeliotis, G., Sturrock, P.A., Antiochos, S.K.: 1994, ApJ 423, 847

[Roz94] Różyczka, M., Bodenheimer, P., Bell, K.R.: 1994, ApJ 423, 736

[RSp93] Rozyczka, M., Spruit, H.C.: 1993, ApJ 417, 677

[Sch62] Schatzman, E.: 1962, A&A 25, 121

[Sch90] Schnack, D.D., Mikić, Z., Barnes, D.C.: 1990, Computer Phys. Comm. 59, 21

[Sci94] Schramkowski, G.P.: 1994, The Dynamics of Intermittent and Diffuse Magnetic Fields in Accretion Disks, PhD Thesis, Utrecht Univ., Ch. 1

[Scj94] Schramkowski, G.P.: 1994, Dynamics of Slender Flux Tubes in Accretion Disks II, A&A, submitted

[Sck94] Schramkowski, G.P., Achterberg, A.: 1994, in Fragmented Energy Release in Sun and Stars, ed. G.H.J. van den Oord, Space Sci. Rev., 68, 331

[SeM94] Serabyn, E., Morris, M.: 1994, ApJ 424, L91

[SoF73] Shakura, N.I., Sunyaev, R.A.: 1973, A&A 24, 337

[SFW86] Sofue, Y., Fujimoto, M., Wielebinski, R.: 1986, ARA&A 24, 459

[StS86] Strom, S.E., Strom, K.M.: 1986, in IAU Symp. 115, Star Forming Regions, eds. M. Peimbert, J. Jugaku, Reidel Publ Cy., Dordrecht, p. 255

[Sve94] Svensson, R.: 1994, The Nonthermal Pair Model for the X- and Gamma-Ray Spectra from AGN, ApJ Suppl Series, in press

[SvZ94] Svensson, R., Zdziarski, A.A.: 1994, Black hole accretion disks with coronae, submitted to ApJ

[Tag90] Tagger, M., Henriksen, R., Pellat, R., Sygnet, J.F.: 1990, ApJ 353, 654

[Tay74] Taylor, J.B.: 1974, Phys. Rev. Lett. 33, 1139

[Tay86] Taylor, J.B.: 1986, Rev. Mod. Phys. 58, 741

[ThP88] Thorne, K.S., Price, R.H., Macdonald, D.: 1988, Black Holes: The Membrane Paradigm, Yale

[Tor93]   Torkelsson, U.: 1993, A&A 274, 675
[Tou92]   Tout, C.A., Pringle, J.E.: 1992, MNRAS 259, 604
[USh85]   Uchida, Y., Shibata, K.: 1985, PASJ 37, 515
[vaB94]   Van Ballegooijen, A.A.: 1994, in Fragmented Energy Release in Sun and
          Stars, ed. G.H.J. van den Oord, Space Sci. Rev., 68, 299
[vaK89]   Van der Klis, M.: 1989, ARA&A 27, 517.
[vaO94]   Van Oss, R.F.: 1994, in Fragmented Energy Release in Sun and Stars, ed.
          G.H.J. van den Oord, Space Sci. Rev., 68, 309
[vBO94]   Van Oss, R.F., Belyanin, A.A.: 1994, A&A submitted
[vdO]   Van Oss, R.F., van den Oord, G.H.J., Kuperus, M.: 1993, A&A 270, 275
[VeP94]   Vekhstein, G.E., Priest, E.R., Steele, C.D.C.: 1994, ApJ Suppl Series 92, 111
[Ver87]   Vermeulen, R.C., Schilizzi, R.T., Icke, V., Fejes, I., Spencer, R.E.: 1987, Na-
          ture 328, 309
[Ver93]   Vermeulen, R.C. and 7 coauthors: 1993, A&A 270, 177
[Vla94]   Vlahos, L., Paschos, P., Kluiving, R., Georgoulis, M.: 1994, A magnetic dipole
          percolation model for the evolution of active solar regions, A&A submitted
[VOK93]   Volwerk, M., van Oss, R.F., Kuijpers, J.: 1993, A&A 270, 265
[VoW48]   Von Weizsäcker, C.F.: 1948, Z. f. Naturforsch. 3a, 524
[Wa93]   Wardle, M., Königl, A.: 1993, ApJ 410, 218
[WeD67]   Weber, E.J., Davis, L.: 1967, ApJ 148, 217
[WHO93]   Welsh, W.F., Horne, K.D., Oke, J.B.: 1993, ApJ 406, 229
[Wie93]   Wierenga, M.H., de Bruijn, A.G., Jansen, D., Brouw, W.N., Katgert, P.:
          1993, A&A 268, 215
[Yec92]   Yecko, P.: 1992, preprint.
[Yus88]   Yusef-Zadeh, F., Morris, M.: 1988, ApJ 329, 729
[Zah90]   Zahn, J-P.: 1990, in Structure and Emission Properties of Accretion Disks,
          eds. C. Bertout, S. Collin, J-P. Lasota, J. Tran Thanh Van, Editions
          Frontières, p. 87
[Zir88]   Zirin, H.: 1988, Astrophysics of the Sun, Cambridge Univ. Press

Non-Linear Data Analysis and Statistical Techniques in Solar Radio Astronomy

Jürgen Kurths, Udo Schwarz and Annette Witt

Arbeitsgruppe Nichtlineare Dynamik der Max-Planck-Gesellschaft an der Universität Potsdam**, Am Neuen Palais, D-14415 Potsdam, PSF 601553, Germany

Abstract. We have discussed some tools from nonlinear dynamics which may help to analyze transient phenomena, such as solar bursts. The structure function known from turbulence theory is an appropriate method to find out some scaling behavior of fluctuations in time. More generally, the wavelet analysis, which is some generalization of the power spectrum, exhibits information on the location as well as the size of hidden characteristic features. Applying both techniques to microwave bursts, we have found some scaling properties that refer to the existence of hierarchic time structures. This is in good accordance with the electric circuit model for describing the flare-particle energization process.

1 Introduction

Astronomy and mathematical statistics have been closely related for a long time. Many fundamental contributions to statistics came from astronomers or from challenges in astronomy, such as Kepler's principle of model building or Gauss' least squares for celestial mechanics. Modern observational astronomy characterized by an enormous amount of data has yielded a lot of new insights in fine structures as well as in (mostly weak) large scale phenomena. The main challenges to statistics has, however, not changed during the last centuries. Most important problems are:

– What are significant signals in noisy observations?
– Which are the structural properties underlying these signals?
– How can a theoretical hypothesis be validated by measurements?

There is now a large variety of statistical techniques to treat these questions, mostly devoted to linear structures (cf. Feigelson & Babu [FB1]). Of special interest for solar radio astronomy are methods for finding hidden periods, for filtering out (Gaussian) colored noise in time series (cf. Priestley [Pr1]) or for image processing.

The paradigm of deterministic chaos developed in the 80ies has substantially extended the area of structural properties which should be taken into account,

---

** E-Mail: juergen@agnld.uni-potsdam.de

especially to nonlinear deterministic dynamics (cf. Ott [Ot1]). The application of chaos theory has become popular in many disciplines and has also led to a somewhat better understanding of fine structure in flares, especially of coronal pulsations (Kurths & Herzel [KH1], Kurths, Benz & Aschwanden [KB1]). However, the nowadays wide-spread approach that is based on finding a low dimensionality seems to be successful only for rather coherent phenomena, such as coronal pulsations (Kurths & Schwarz [KS1], Isliker & Benz [IB1]).

Therefore, we discuss here further techniques of nonlinear dynamics which look more appropriate to the analysis of solar flare data. Firstly, some problems of pre-processing observations are considered (Section 2). We study, in particular, the question how to filter out some parts in the data (e.g. some slowly varying components) to make them stationary and we present tests for stationarity. Because several kinds of non-stationary phenomena typically occur two different approaches to describe such a dynamics are presented. Firstly, fluctuations in a broad range of time are uniformly characterized by means of the scaling exponent of the structure function, which is closely related to a $1/f$-like behavior in the power spectrum (Section 3). Secondly, we introduce the wavelet analysis which is suitable to find out rather coherent substructures in inhomogeneous or transient processes (Section 4). All these techniques are applied to observations of microwave bursts. Some discussion of our experiences is presented in Section 5.

2 Problems with Pre-Processing

Transient phenomena, such as flares or outbursts, are of special interest in astronomy. It is in general unavoidable that the corresponding observations include several components and not only that one under study. Measurements of solar radio emission often consist of parts caused by the quiet sun, some continuum (slowly varying), bursts and instrumental noise. To analyze only one of these components, one usually tries to filter out the other ones. Simplest manipulations, which are, nevertheless, often efficient, are subtraction of the mean value, linear detrending or modification of the variance. As is well-known, there is a bunch of more sophisticated filters (Elliot [El1]).

It is important to note that all these procedures have some side effects. They can modify the signal of the interesting component significantly, or - more probably - can destroy possible relationships between different parts, or different time scales, such as between the continuum component and some fine structures in bursts. In order to avoid artifacts from filtering, we, therefore, highly recommend to analyze different kinds of filtered data and to compare the results obtained.

Another important point in data analysis is that most well-known techniques assume rather strong properties of the observations. Some of them are very difficult to check, e.g. Gaussian distribution of the noise or ergodicity. **Stationarity** is a further condition often assumed. It means that statistical quantities of a process are independent of absolute time; they are at most a function of rela-

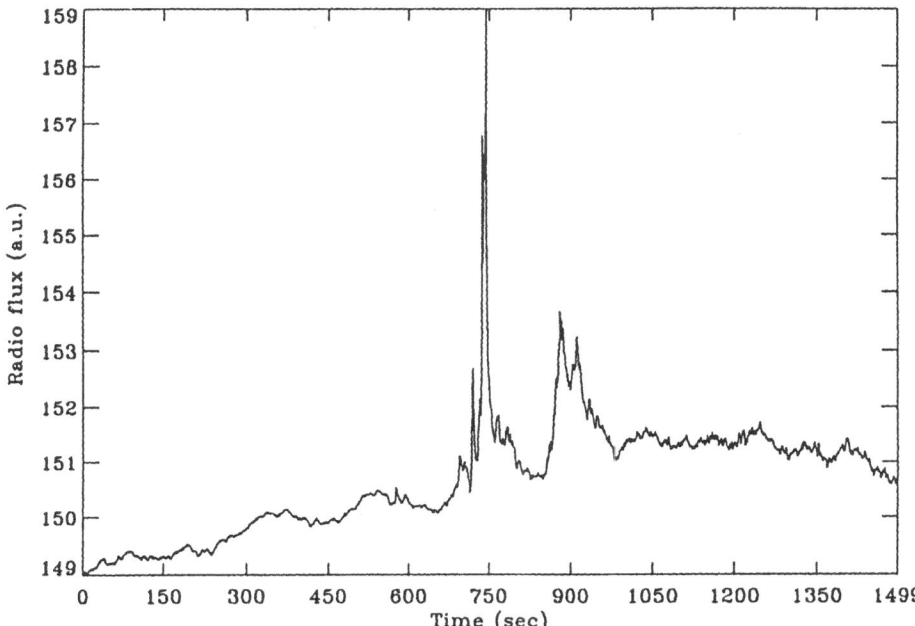

Fig. 1. Radio flux time profile of a burst at 37 GHz observed on March 23, 1991, 11:58 UT, at the Metsähovi Radio Research Station near Helsinki

tive times. If all statistical quantities satisfy this property the process is called stationary in the strong sense. Isliker & Kurths [IK1] have recently proposed a method to identify phases in time series whose invariant measures are independent of time, and are therefore apt for considering as strongly stationary.

This strong stationarity is a rather hard condition and many methods of data analysis suppose another type of stationarity – weak stationarity – which claims only the time independence of the first and second moments: mean value, standard deviation and autocorrelation function, which is equivalent to a power spectrum independent of time.
Actually, we have also developed a method which finds out weakly stationary phases in observations, i.e. these phases where at least the power spectrum is time invariant (Witt & Kurths [WK1]). The main idea of this method is to test whether the power spectra depend on time. The test goes as follows: We, firstly, subdivide the observational interval into different parts. Each of these subseries is transformed into Fourier components which present a sample of the normalized spectral distribution function (Priestley [Pr1]) for each. If the series is stationary, all the samples represent the same distribution which can be tested with the χ^2 test or the Kruskal-Wallis test. This method is performed hierarchically, i.e. we firstly compare the spectrum obtained from the entire interval with those of the subintervals. If the Null hypothesis – weak stationarity

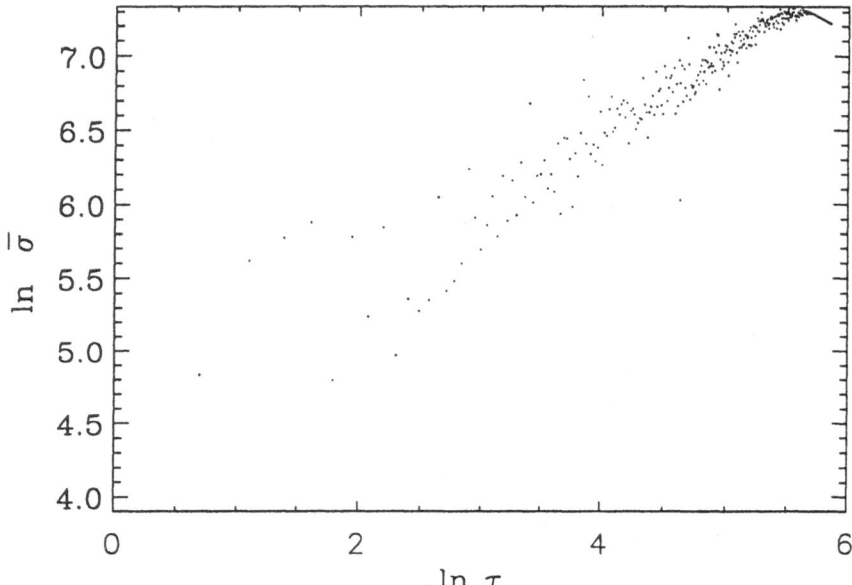

Fig. 2. Dependence of the logarithm of the mean variance $\ln \bar{\sigma}_X$ on time $\ln \tau$ (τ in units of 0.5 sec) obtained from the radio flux of the event presented in Fig. 1

of the whole record – has to be rejected the analogous procedure is repeated for some part of the record which is then compared with subintervals of this part etc.

Next, we analyze microwave bursts observed at 37 GHz at the Metsähovi Radio Research Station near Helsinki (Tab. 1). The data length is about 1500 sec with a time resolution up to 0.5 sec. (Krüger *et al.* [KK1]). The burst flux time profile (Fig. 1) display various time structures. To remove the slowly varying background flux, we use a high-pass filter basing on locally fitted linear polynomials. The filter length is 90 points (45 sec).
Applying the test described above, we find that all burst events under study show a strongly non-stationary behavior. The power spectra calculated from different time intervals have different shapes, i.e. they depend on time essentially. Next, we try to quantify this non-stationarity by analyzing the dependence of the mean standard deviation of the radio flux $\bar{\sigma}_X(\tau)$ on its duration τ for T different time intervals

$$\bar{\sigma}_X(\tau) = \frac{1}{T} \sum_{t=1}^{T} \sigma_X(t, \tau) \tag{1}$$

with the variance of the radio flux on the time interval of length τ at time t

$$\sigma_X^2(t, \tau) = \frac{1}{\tau} \sum_{i=1}^{\tau} (x(t + i\ \Delta t) - \bar{x}(t, \tau))^2 \tag{2}$$

and the mean value of the radio flux on the time interval of length τ at time t

$$\bar{x}(t, \tau) = \frac{1}{\tau} \sum_{i=1}^{\tau} x(t + i\, \Delta t)\,. \tag{3}$$

We find first hints to a power-law scaling in time (Fig. 2)

$$\bar{\sigma}_X(\tau) \propto \tau^a\,, \tag{4}$$

but this is not a reliable result. Therefore we introduce in the next section another description, the structure function.

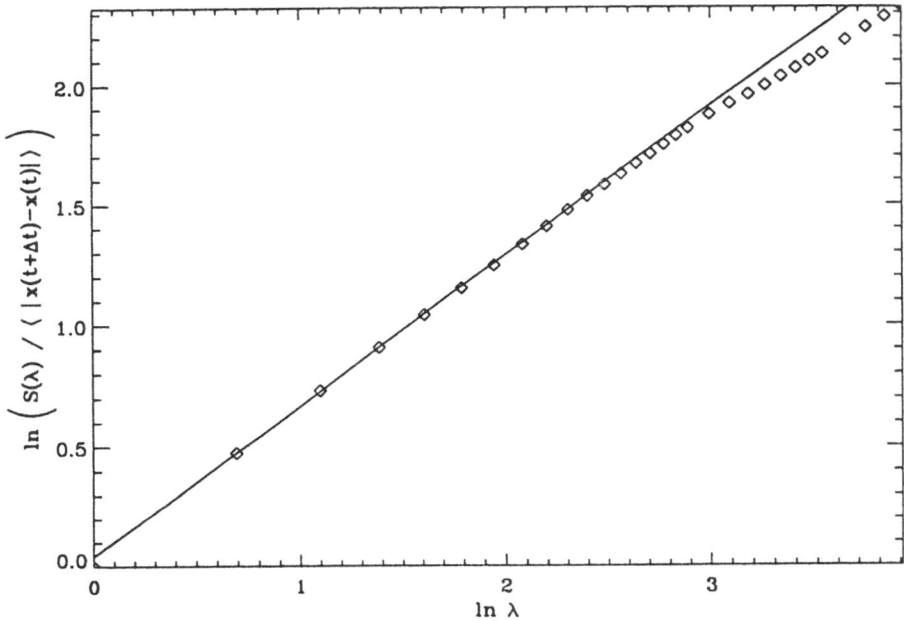

Fig. 3. Double logarithmic plot of the structure function $S(\lambda)$ calculated from the radio flux of the event presented in Fig. 1

3 Broad-Range Time Scales in Solar Observations

Many observations of nature exhibit variations on both a short- and a long-term time scale. This raises the question whether or not this behavior can be described in a uniform manner. The concept of self-affinity presented in the following seems to be a promising approach to characterize a uniform broad-range scaling in time, although the underlying process can be non-stationary (Feder [Fe1]).

It is basing on the structure function

$$S(\lambda) = \langle |x(t + \lambda \Delta t) - x(t)| \rangle_t \qquad (5)$$

where $\langle . \rangle_t$ is the average over time t, Δt denotes the sampling rate and $x(t)$ the time series observed. If a process is **self-affine** then this structure function fulfills for λ the following power-law scaling

$$S(\lambda) = \lambda^H \langle |x(t + \Delta t) - x(t)| \rangle_t . \qquad (6)$$

H is then the characteristic scaling exponent. Self-affinity means that a scaling exponent H does exist independent of the shift λ in time. The fractional Brownian motion (fBm) that generalizes the classical Brownian motion or random walk is a typical self-affine process (Feder [Fe1]). Such fBm is characterized by a power-law scaling in time (eq. 6) as well as in frequency range

$$P(f) \propto f^{-\alpha} . \qquad (7)$$

Both scaling exponents are simply related by

$$H = \frac{\alpha - 1}{2}. \qquad (8)$$

It is important to note that this process is not stationary because the standard deviation σ depends on the length of the time interval in which it is calculated and also scales as power-law Eq. (4). Therefore, fractional Brownian motion cannot be characterized by a fractal dimension (this means, as usually, the box dimension (Feder [Fe1])). The main difference to self-similar processes is that fBm exhibits a different scaling in space and time. The concept of fractal dimensions requires, however, that both scalings are identical.

For $H = 0.5$, we have the classical Brownian motion, i.e. the increments are not correlated. In the case $H > 0.5$, there is a positive persistence. Therefore, an increasing trend in the past implies an increasing trend in the future. This way, processes with $H > 0.5$ are characterized by long-range correlations. Because

Table 1. Spectral indices and scaling exponents of the analyzed microwave bursts

Date	Starting Time UT	Duration (min)	Spectral index α	Scaling exponent H
March 23, 1991	11:58	58.51	2.25	0.62
March 23, 1991	12:58	48.79	2.35	0.68
March 24, 1991	10:00	63.16	2.51	0.76
March 24, 1991	11:09	68.63	2.02	0.51

the microwave bursts analyzed in Section 2 are far from stationary we calculate the structure function $S(\lambda)$ (eq. 5) for records of four such bursts. Surprisingly, we find for each event a well-expressed power-law scaling (cf. Fig. 3 and Tab. 1) in a rather broad region of λ. There is a longer λ-range if the unfiltered data are used, meaning that this uniform scaling is valid also for longer time scales of the bursts. It comes out that only one of the four events is compatible with the usual Brownian motion. The other ones are characterized by scaling exponents larger than 0.5 which is typical for fractional Brownian motion with rather long-range correlations (Table 1).

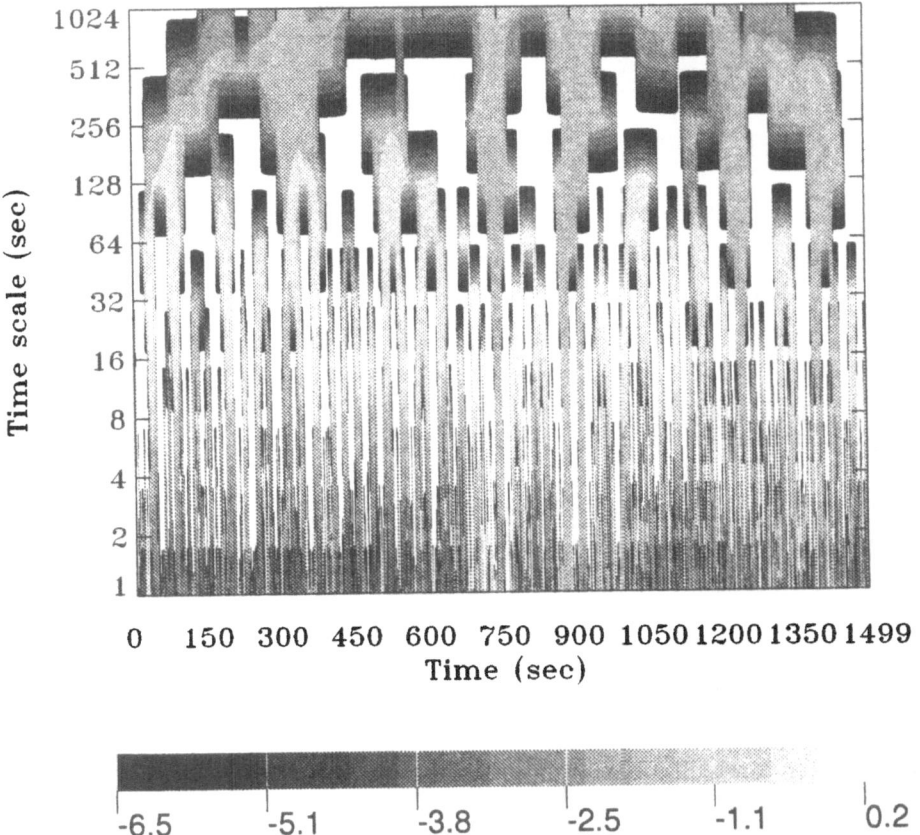

Fig. 4. Scalogram of the high-pass filtered event presented in Fig. 1 without noise reduction. We have plotted the logarithm of the positive wavelet coefficients. The white background represents wavelet coefficients with very small amplitudes

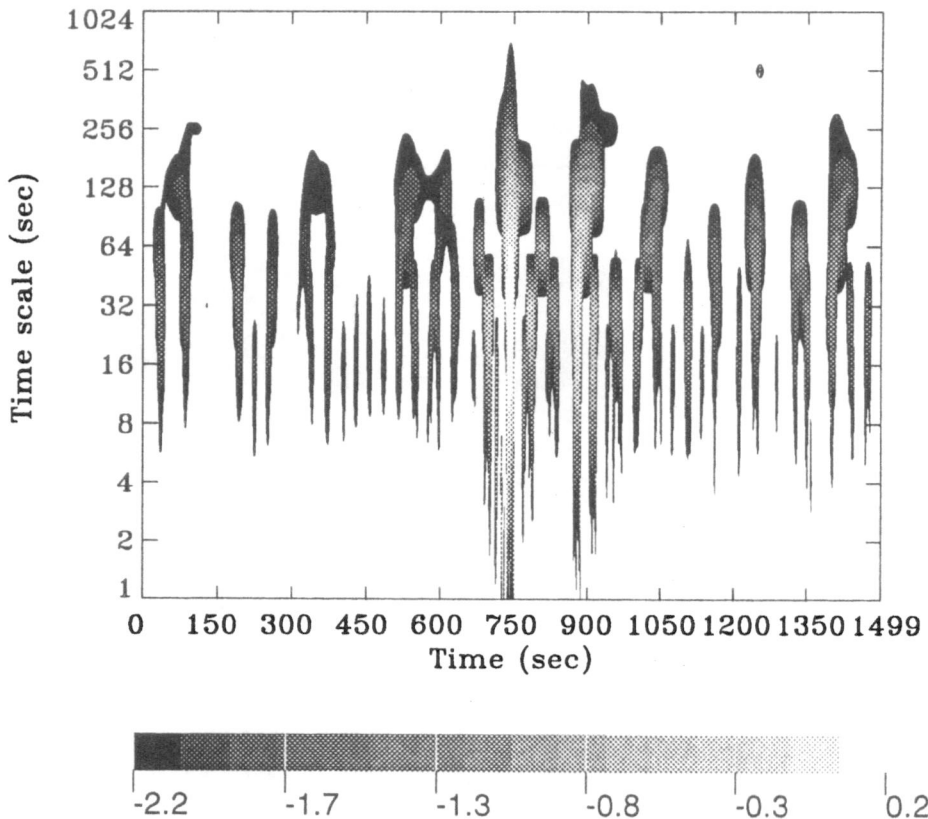

Fig. 5. De-noised scalogram of the high-pass filtered event presented in Fig. 1

4 Wavelet Analysis

In the general case of non-stationarity, we have to expect an inhomogeneous scaling behavior in space and/or in time. Hence, another way to study transient phenomena must be to look for detailed information on both the location and the size of the characteristic features. Because traditional concepts, such as power spectrum analysis, are basing on a global description of the system, transient processes challenge to apply other approaches. Wavelets introduced in the following are a proper tool to analyze such phenomena. It should be mentioned that the well-known windowed Fourier transform is another tool to study local behavior, but it is a more coarse-grained one (cf. Daubechies [Da1]).

We first recall a few basics about the most common global technique, the power spectrum, which is basing on the **Fourier transform**. It is an efficient tool at giving some dominant frequencies (or characteristic size). This transform

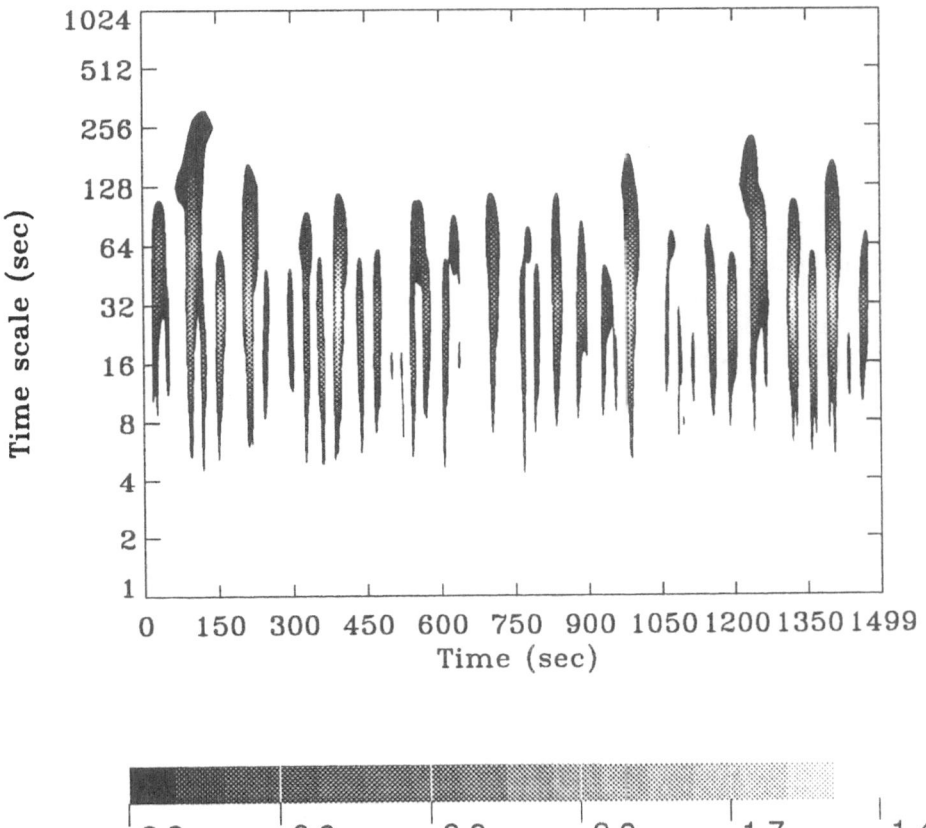

Fig. 6. De-noised scalogram of a fBm realization with $H = 0.625$. The data pre-processing of the realization was the same as in the case of the data

is a projection on an orthogonal basis consisting of sinusoids. Hence, there exists a unique decomposition and reconstruction formula for a given function x_t, but there is no simple relationship between the local behavior of x_t and the Fourier coefficients. This information is so deeply buried in the phases of the coefficients that it is very difficult to retrieve.

In generalization, the notion of **wavelet analysis** addresses to both unknown periodicities and non-stationary structure. The wavelet analysis is basing on time-limited elements, the wavelets. This way, one has the possibility of dealing with non-stationary time series, where some coherent structures are changing over time. The wavelet transform of $x(t)$ is the decomposition into a basis of functions $w_{a,b}(t)$ with

$$w_{a,b}(t) = |a|^{-1/2} w(a^{-1}(t-b)) \,, \tag{9}$$

all derived from a unique function $w(t)$, called the "mother wavelet", by trans-
lation b and scaling a. Some functions have been recommended as wavelets, e.g.
Daubechies's wavelets and Gabor-Malvar wavelets. For our purpose, the simple triangle-like wavelets (cf. Mallet [Ma1], Meyer & Ryan [MR1], Vigouroux &
Delache [VD1]) are appropriate, because their shape fits the time profile of the
radio flux quite well.
We use a bi-orthogonal wavelet basis consisting of the collection $w_{j,k}(t)$, $j \in N$,
$k \in Z$ (set of relative integers), together with $q(t-k)$, $k \in Z$, where $q(t)$ is a
smooth function with a rapid decay at infinity. On this basis, $x(t)$ can uniquely
be written as

$$x(t) = \sum_{k \in Z} c_k q(t-k) + \sum_{j \in N} \sum_{k \in Z} d_{j,k} w_{j,k}(t) \,. \tag{10}$$

Obviously, the two functions $q(t)$ and $w(t)$ cannot be chosen independently.
Among the many possibilities for the two functions $q(t)$ and $w(t)$, following
Bendjoya *et al.* [BP1] we choose

$$q(t) = \begin{cases} 1 - |t| & \text{for} \quad |t| \le 1, \\ 0 & \text{otherwise} \end{cases} \tag{11}$$

and

$$w(t) = \begin{cases} \frac{1}{4}|t| - \frac{1}{2} & \text{for} \quad 1 < |t| \le 2, \\ \frac{1}{2} - \frac{3}{4}|t| & \text{for} \quad |t| \le 1, \\ 0 & \text{otherwise.} \end{cases} \tag{12}$$

The function $q(t)$ is a smoothing function that allows us to cut the doubly infi-
nite sum of Eq. (10). The $q(t-k)$'s account for the large scale fluctuations, the
$w_{j,k}$'s for the small ones.
One main advantages with wavelets is that the coefficients can be calculated
recursively. The first trivial step is to take the function $q(x)$ with the same resolution as the sampled signal. We only have to compute the c_k^0's (the superscript
refers to the number of the step in the iterative process). ¿From the formula for
$q(t)$, it is easy to see that the c_k^0's are simply the sampled values of $x(t)$.
Then we double the resolution. We choose the normalizing coefficients such that
we replace $q(t)$ by $q(t/2)/2$. There is only one level of fine fluctuations to add to
the smoothed part to recover the signal, and it is defined by the set of $d_{1,k} = d_k^1$.
We can now repeat the same procedure to the smoothed part of level 1. This
will give us two new sets of coefficients c_k^2 and d_k^2.
With our choice of $q(t)$ and $w(t)$, the formulae to obtain the c_k^j's and the d_k^j's
are very simple:

$$c_k^j = \frac{1}{2} c_k^{j-1} + \frac{1}{4} \left(c_{k-2^{j-1}}^{j-1} + c_{k+2^{j-1}}^{j-1} \right), \quad j \ge 1, \tag{13}$$

and

$$d_k^j = c_k^{j-1} - c_k^j. \tag{14}$$

Hence, it is clear that the c_k^j's are obtained by applying a low–pass filter to the c_k^{j-1}'s. On the other hand, the d_k^j's being obtained by a difference between two levels of c_k^j's are in fact the result of a band–pass filter applied to the signal.

The decomposition described above is best portrayed in a $2^j \Delta t - k\Delta t$ - plot, called **scalogram**, that shows the scaling behavior of the radio flux $x(t)$ in dependence on the time location $k\Delta t$, where Δt denotes the sampling rate. Different intensities of the coefficients (amplitudes) are displayed by different colors. Using the functions given by (11) and (12) we observe that the positive wavelet coefficients d_k^j reflect the burst-like behavior of the radio flux quite well.

In the Fig. 4 - 6 we have plotted the logarithm of the positive wavelet coefficients $\log d_k^j$ (for $d_k^j > 0$). Such a representation that gives three parameters, here: scaling, time and amplitudes, is sometimes called $2\frac{1}{2}$ - dimensional. Summing up the wavelet coefficients over the time location (index k), we get a picture similar to a power spectrum.

The computational effort of the recursive procedure we used is similar to that for calculating the fast Fourier transform. Another advantage of these wavelets is that they enable a rather simple method to reduce some noise. We follow the method proposed by Donoho [Do1] for reconstructing an unknown function from noisy data. The reconstruction is defined in the wavelet domain by translating all the estimated wavelet coefficients towards 0 by a specially-chosen threshold.

Next, we apply the wavelet analysis to the microwave burst events listed in Table 1. The wavelet transform calculated from the high-pass filtered data gives some indications that a rather broad range of time scales is involved (Fig. 4). This hierarchy becomes more clear if we consider the scalogram after some noise reduction (Fig. 5). The characteristic scales range from five seconds to two minutes.

To get some more insight into the nature of process underlying these bursts, we also calculate the scalograms for typical formal models, such as white noise, linear colored noise (autoregressive processes) and fractional Brownian motion (cf. Section 3). If one compare the scalogram plots of the different models by eye it comes out that the scalogram of an fBm with $H = 0.62$ after noise reduction (Fig. 6) is most similar to that of the burst structure. This fits well into the results we obtain from the variance and the structure function (Section 2 and 3). For a more detailed analysis, including the comparison of the pre- and post-burst phase by means of wavelets, we refer to Schwarz *et al.* [SK1].

5 Discussion and Conclusions

Observations in solar radio astronomy are often related to transient phenomena, such as bursts. Therefore, the classical correlation and spectral analysis as well as the rather new concept of fractal dimensions mostly fail. We have discussed in this contribution other techniques from nonlinear dynamics which look more appropriate to analyze such data.

In order to check whether or not we can apply the methods mentioned above, some statistical tests for stationarity have been introduced. Using them, it comes indeed out that there are no stationary intervals in the solar microwave bursts analyzed here. Nevertheless, we have found other structural behavior: these events are characterized by a power-law scaling in time which is well expressed in terms of the structure function.

Such a dynamics can be well described by fractional Brownian motion which exhibits a homogeneous scaling in time as well as in space, but both are different. This uniform scaling found from the bursts in a rather broad range in time refers to the existence of hierarchic time structures and means especially that also small scale fluctuations are due to the burst. It is, however, important to note that such fBm can be only a tool for a description of the dynamics, but not for an explanation of the underlying physical mechanism.

Therefore, our findings have to be compared with physical models for microwave bursts. Such a hierarchy in time has only been predicted by means of the electronic circuit model for describing the flare-particle energization process (Krüger et al. [KK1]).

A more detailed insight into these data is possible using the wavelet analysis which exhibits information on the location as well as the size of characteristic features. It is a generalization of the well-known power spectrum. The results obtained from this analysis are in good accordance with those from the scaling of the variance and the structure function. Another advantage of the wavelet analysis is that one can easily filter out some components. ¿From such filtered burst data, our analysis has yielded some indications for a rather small change of the scaling behavior during the bursts (cf. Schwarz et al. [SK1] for a detailed analysis).

To summarize: nonlinear dynamics also offers techniques to describe structural properties of transient phenomena usually observed in solar radio bursts. Although there are still many open problems with these tools, they can be helpful for a deeper physical understanding of processes in the solar corona. Our results obtained from this kind of data analysis call for more quantitative models in order to compare the properties of both and to come to a better plasma diagnostics. Our findings must, of course, be confirmed as typical by further observations.

Acknowledgements

We thank A. Krüger for helpful discussions and S. Urpo for submitting the data. The authors are greatly indebted to Stefan Schmidt for assistance in preparing the programs used in this paper. Finally, we thank the referee for his helpful criticism.

References

[BP1] Bendjoya, Ph., Petit, J.-M., Spahn, F.: 1993 Wavelet analysis of the Voyager data on planetary rings. I. Description of the method, ICARUS **105**, 385-399

[Da1] Daubechies, I.: Ten Lectures on Wavelets, Society for Industrial and Applied Mathematics, Philadelphia 1992
[Do1] Donoho, D. L.: Wavelet shrinkage and W.V.D.: A 10-minute tour, in: Meyer, Y., Roques, S. (eds.): Progress in Wavelet Analysis and Applications. Proceedings of the International Conference "Wavelets and Applications", Toulouse, France, June 1992, Editions Frontieres 1993, pp. 109-129 or anonymous-ftp: playfair.stanford.edu, Preprint: /pub/reports/toulouse.tex
[El1] Elliott, D. F. (ed.): Handbook of Digital Signal Processing. Engineering Applications, Academic Press, San Diego 1987
[Fe1] Feder, J.: Fractals, Plenum Press, New York 1988
[FB1] Feigelson, E. D., Babu, G. J. (Eds.): Statistical Challenges in Modern Astronomy, Springer-Verlag, New York 1992
[IB1] Isliker, H., Benz, A. O.: Non-linear properties of the dynamics of bursts and flares in the solar and stellar coronae. Astron. Astrophys. 285 (1994) 663-674
[IK1] Isliker, H., Kurths, J.: A test for stationarity: Finding parts in time series apt for correlation dimension estimates, Int. J. Bifurcation and Chaos 3 (1993) 1573-1579
[KK1] Krüger, A., Kliem, B., Hildebrandt, J., Zaitsev, V. V.: Microwave burst timescales and solar flare acceleration processes, Astrophys. J. Suppl. Series 90 (1994) 683-688
[KB1] Kurths, J., Benz, A. O., Aschwanden, M.: The attractor dimension of solar decimetric radio pulsations, Astron. Astrophys. 248 (1991) 270-276
[KH1] Kurths, J., Herzel, H.: An attractor in a solar time series, Physica D 25 (1987) 165-172
[KS1] Kurths, J., Schwarz, U.: Chaos theory and radio emission, Space Science Reviews 68 (1994) 171-184
[Ma1] Mallet, S.G.: A Theory for multiresolution signal decomposition: The wavelet representation, IEEE Transactions on Pattern Analysis and Machine Intelligence 11 (1989) 674-693
[MR1] Meyer, Y., Ryan, R. D.: Wavelets. Algorithms and Applications, Society for Industrial and Applied Mathematics, Philadelphia 1993
[Ot1] Ott, E.: Chaos in Dynamical Systems, Cambridge University Press 1993
[Pr1] Priestley, M. B.: Spectral Analysis and Time Series, Academic Press, London and New York 1981
[Sc1] Scargle, J. D.: Wavelet methods in astronomical time series analysis, in: Lessi, O. (ed.) 1993 International Conference on Applications of Time Series Analysis in Astronomy and Meteorology, Padua 1993
[SK1] Schwarz, U., Kurths, J., Krüger, A., Urpo, S.: Wavelet analysis of solar radio bursts, Sol. Phys. 1994 (submitted)
[VD1] Vigouroux, A., Delache, Ph.: Fourier versus wavelet analysis of solar diameter variability, Astron. Astrophys. 278 607-616
[WK1] Witt, A., Kurths, J.: Problems with stationarity 1994 (in preparation)

Interplanetary Scintillation Imaging of Disturbances in the Solar Wind

A. Hewish and G. Woan

Mullard Radio Astronomy Observatory, Cavendish Laboratory,
Madingley Road, Cambridge CB3 0HE, UK.

Abstract. Small scale turbulence in the solar wind causes radio sources to scintillate and it has been found that this effect depends upon the mean plasma density along the line of sight. Routine daily observations of interplanetary scintillation (IPS) on about 1000 sources can be used to construct whole-sky maps showing the two-dimensional projection of large scale features of enhanced, or reduced density such as corotating streams or compression zones following interplanetary shocks. This ground-based method provides a global picture of major interplanetary disturbances and gives information over a wide range of ecliptic latitude which complements the detailed sampling provided by individual spacecraft. Continuous monitoring of the interplanetary medium has been carried out at Cambridge since March 1990. The method will be described and some typical disturbances observed during this period will be shown. The results provide new insights into the nature of coronal magnetic energy releases responsible for the most powerful interplanetary shocks.

1 Introduction

In the study of energetic transients which leave the outer corona there is a large observational gap between about 5 solar radii, where white light coronagraphs provide the last glimpse of departing coronal mass ejections (CMEs), and about 1 AU where interplanetary shocks are recorded by spacecraft or indirectly by their interaction with the magnetosphere and their generation of magnetic storms and Forbush decreases of the flux of galactic cosmic rays. Some data for limited periods in the region 0.3 to 1 AU has been provided by the satellites Helios 1 and 2. One technique which can partially fill this gap is the use of interplanetary scintillation (IPS) to make images of large scale perturbations of plasma density in the solar wind. Such a global view provides unique information on the location and nature of large scale transients which complements the much more detailed local sampling provided by spacecraft [1]. Knowledge of the location, for example, gives information on the position of the source at the Sun, whilst whole-sky viewing enables long-lived disturbances to be monitored for many days in situations where solar rotation would carry disturbances away from any one spacecraft.

In the context of this meeting on the release of coronal magnetic energy our new data raise some difficult questions for the generally accepted theory that the most powerful interplanetary shocks are driven by the CME process. We have found that strong shocks are frequently followed by prolonged outflows of unusually high-speed wind which demands a more continuous supply of energy than could be obtained from the destabilisation of coronal magnetic fields, which gives only short-lived inputs. Until quite recently it was widely believed that the strongest interplanetary shocks were caused by solar flares, but the evidence from CMEs now shows that flares are merely peripheral events, often associated with the eruption of CMEs but not causally effective [2]. Disentangling cause and effect in solar activity, where a number of different phenomena are associated in time and occur in adjacent regions of the corona and the chromosphere, is evidently not straightforward and demands some caution.

At present, the only known source of prolonged inputs of excess kinetic energy into the interplanetary medium are coronal holes, which generate high-speed streams lasting from less than one rotation up to many months. It has been suggested that these streams are driven by the deposition of Alfvén waves at high level in the corona where the flow exceeds sonic (or Alfvénic) speeds. Energy supply at lower levels would increase the density, rather than the velocity, of the solar wind [3]. Our IPS images have shown a strong and statistically significant correlation between the source locations of powerful interplanetary shocks and coronal holes. This has lead us to suggest that coronal holes themselves could be the cause of the shocks [4]. In this case the associated CMEs, like the solar flares, would also be peripheral events. Alternatively there might be some interaction between a CME, and an adjacent coronal hole, such that the sustained high-speed outflow following the shock was 'turned on' by modification of the geometry of the coronal hole associated with the erupting CME [5].

In March 1990 a new series of IPS observations was initiated at Cambridge to obtain more extensive data and to explore the possibility of predicting the arrival of strong shocks at the Earth. The observations were made with the 3.6–hectare array, which operates at 81.5 MHz and covers the range $-80°$ to $+70°$ in declination. Daily measurements were carried out on about 1000 sources which were observed for about 2–10 minutes at meridian transit. In this review we briefly describe the IPS-imaging technique and present some of the recent results.

2 The imaging method

The imaging technique is shown schematically in Figure 1. Disturbances crossing the lines of sight to a large grid of extragalactic radio sources are seen as patterns of enhanced or reduced scintillation over the celestial sphere. Provided that the disturbances have simple shapes their three-dimensional structures may be estimated by comparison with theoretical patterns computed for corresponding models.

To obtain useful images it is, of course, necessary to remove systematic effects,

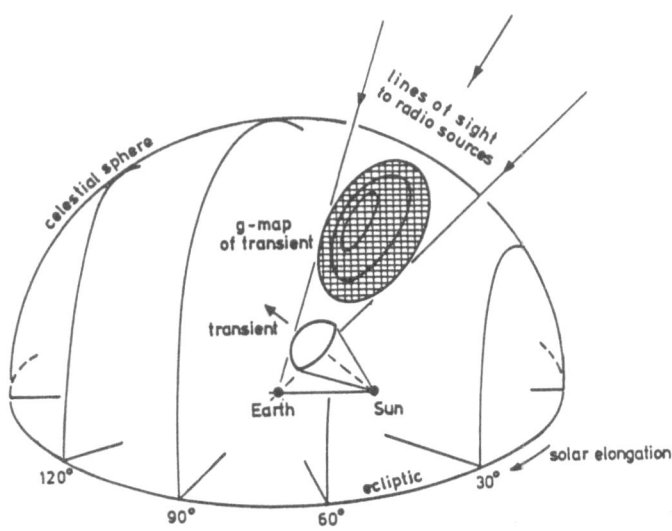

Fig. 1. Schematic diagram of the IPS-imaging method.

the most important of which is the strong variation of scintillation with radial distance from the Sun which reflects the inverse square law dependence of the average plasma density. Account must also be taken of the angular sizes of the individual radio sources which are usually large enough to cause substantial blurring of the speckle pattern. Both effects can be allowed for by observations for about one year to derive the average dependence of scintillation upon solar elongation for each source separately. At 81.5 MHz, scintillation peaks at an elongation near 30°. This marks the onset of strong scattering where an ideal point source observed with negligible bandwidth would show 100% scintillation. The reduction at smaller elongations results from the combined action of source-blurring and a finite bandwidth. At higher frequencies the maximum moves closer to the Sun; at 327 MHz, for example, it occurs near an elongation of 10°.

If ΔS^2 is the mean square scintillating flux density of a source observed on a given day, and $\overline{\Delta S^2}$ is the long term average value, we define the perturbation (g) by $g^2 = \Delta S^2 / \overline{\Delta S^2}$. When g-values for a large grid of sources have been obtained, g-maps in the form of individual values, contours or grey-scale pixels may be presented.

Quantitative models may be fitted to g-maps to obtain estimates of the shapes and densities of observed disturbances. The weak scattering theory is used, which assumes a linear relation between the scattering power and density of the small scale irregularities, and a suitable model of the solar plasma must be adopted. If r is radial distance from the Sun, and z the distance from the

Earth along a line of sight, the scintillation flux density is given by

$$\Delta S^2 = \int_0^\infty f(r, \theta, \phi) \, \beta(r) \, F(z, \theta_s) \, dz \, . \tag{1}$$

In this expression $f(r, \theta, \phi)$ is a numerical factor defining the perturbation of density in the assumed model, $\beta(r)$ is the average scattering power of the plasma irregularities, and $F(z, \theta_s)$, where θ_s is the angular size of a radio source, is a function defining both source-blurring and the Fresnel filter. The latter takes account of near-field effects which control the growth of intensity variations in the diffraction pattern from a phase-modulated wavefront.

To investigate the region of the heliosphere within which disturbances should be detectable consider how different elements along the line of sight contribute to the integrated scintillation. Let $H(z)$ be a weighting function such that

$$\Delta S^2 \propto \int_0^\infty H(z) \, dz \, . \tag{2}$$

For a spherically symmetrical model of the heliosphere with no disturbances $H(z)$ behaves as shown in Figure 2. Here the systematic dependence on elongation has been removed and $H(z)$ has been normalised to constant scintillation for any line of sight. Broadly speaking, disturbances travelling radially will be most easily detected when they are directed within about ±45° of the Sun-Earth line, and they become undetectable at angles > 70°. Disturbances are also undetectable within an elongation of 30 at 81.5 MHz because of strong scattering. Thus the zone of detection extends from about 0.5 to 1.5 AU in a cone of width ±70° having its apex at the Sun.

It is interesting to compare the IPS detection zone with that of white-light coronagraphs. The weighting function for Thomson scattering is dependent upon the plasma density, but since coronagraphs can only study regions within a few solar radii the most important elements of the line of sight occur near the solar limb. Transients seen in white light are therefore most easily detected when they travel at around 90° to the Sun-Earth line. For this reason it is not easy to observe the same disturbances by both techniques. Since a close association has been found between coronal mass ejections seen in white light and interplanetary shocks, and transients seen on g-maps are also associated with interplanetary shocks, some fraction of these events should be detectable by both methods. Attempts to find well-correlated events have so far been inconclusive.

Quantitative estimates of plasma density demands some method of calibration as IPS responds only to a small fraction of the total density, and it is not clear that this fraction is independent of varying conditions in the solar wind. The small-scale irregularities might, for example, constitute a larger fraction of the mean density in post-shock flows. Extensive calibrations were made by Tappin [6] who selected about thirty periods during 1978–81 when the line of sight appeared to be embedded in regions of roughly constant density which were also accessible to spacecraft. These occasions included high-speed streams, corotating interaction regions, compressions associated with shocks and post-shock outflows. The relation $g^2 = N/9$, where N cm^{-3} is the plasma density calculated by averaging hourly spacecraft values over 24 hours, was found to give the best fit. This holds over the range $2 < N < 50$ and confirms that scintillation is well correlated with density under widely differing conditions. Our more recent data during 1990–94 have been compared with IMP-8 data and have confirmed this correlation.

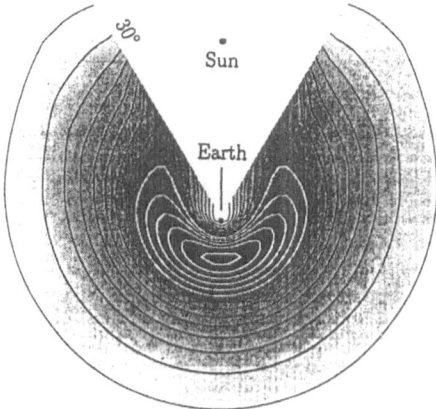

Fig. 2. The IPS weighting function $H(z)$ describing the relative contributions of points along a line of sight to the total value of g^2.

3 Observations of corotating streams

The use of scintillation images (g-maps) to monitor the long term behaviour of a corotating disturbance which would pass over a single spacecraft in about one day is well exemplified in Figure 3. In this sequence of images made during January, 1993, enhanced scintillation corresponding to increased plasma density is plotted darker on the grey scale. The coordinates are right ascension and declination displayed using the equatorial Hammer-Aitoff projection, with the Sun's position indicated on the ecliptic. Enhanced scintillation is first seen in the east. As the high density stream advances towards the Earth, through corotation with the Sun, the zone of enhanced scintillation progressively fills a larger area of sky. After engulfing the Earth the stream is seen in the west until fading from view. The approximate shape of the stream projected onto the ecliptic plane is sketched in Figure 3. Theoretical modelling shows that sequences of images like this are in good agreement with simple corotating structures [7]. The quite rapid disappearance of enhanced scintillation after the stream has passed over the Earth is due to low density in the high speed wind which typically follows such high-density streams. The spiral geometry explains why low density along

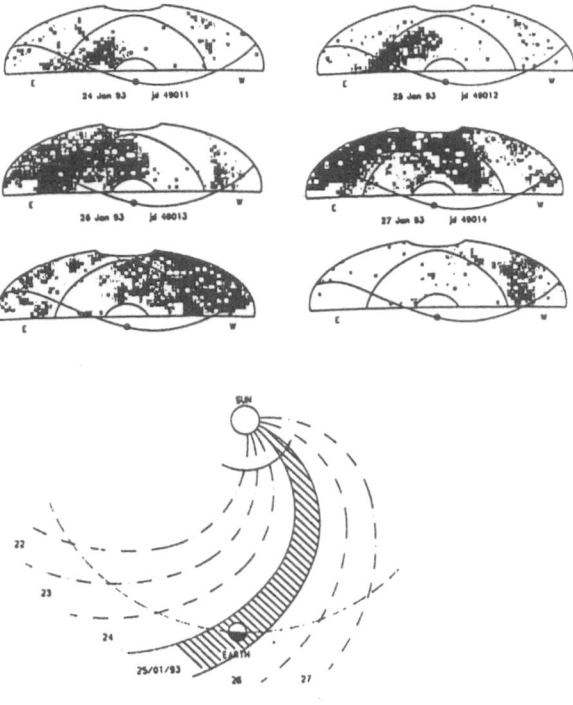

Fig. 3. Consecutive g-maps showing enhanced density (dark) in a long-lived corotating stream. The ecliptic plane and lines of constant solar elongation (30° and 90°) are shown.

the dominant part of the line of sight effectively removes enhanced scintillation in the west under these conditions.

Even during periods of high solar activity there is usually an underlying pattern of corotating streams which is readily seen when g-maps are displayed in a compressed synoptic format. Each daily map is reduced to a strip showing g as a function of solar elongation only, plotted separately for the eastern and western halves of the sky. This analysis removes information about the solar latitude dependence but allows long temporal sequences to be inspected for corotating features which are distinguished by having a characteristic signature.

The compressed synoptic format is illustrated in Figure 4 which shows data for several solar rotations. To emphasise high speed streams of low density we plot reduced scintillation ($g < 1$) as darker on the grey scale. Dark bands indicating reduced plasma density in the Earth's environment, which take several days to pass through our detection zone and then repeat at 27 days intervals, are clearly seen. Also seen is a 'herring-bone' pattern of thinner light bands which correspond to corotating high density outflows as illustrated in Figure 3. The coronal hole which generated the long-lived high speed stream (denoted by arrows) is clearly seen on He 10830 Å images of the solar disk as indicated in Figure 4.

4 Interplanetary shocks

Short-lived transients, which travel radially outwards from the Sun, pass through the scintillation detection zone in two or three days and appear as white strips in Figure 4 which do not show the slanting 'herring-bone' pattern. Exceptionally fast transients may be seen on only one g-map. A rather slow transient of this type which took four days to traverse the detection zone is shown in Figure 5. A very strong shock travelling at more than 1000 km s^{-1} is illustrated in Figure 6. This was seen on one g-map only so its speed could not be estimated from scintillation imaging, but the same shock passed the spacecraft Ulysses at 2.5 AU just over one day later so the speed of the transit could be obtained. The best-fit model to the g-map indicated that the high density plasma immediately behind this shock took the form of a spherical shell about 0.15 AU in radial thickness, with an overall angular extent of ±35° and direction of travel about 20° above the plane of the ecliptic. This was one of the most powerful shocks observed during the Ulysses mission.

Scintillation maps such as Figure 5 and Figure 6 demonstrate the use of this technique for obtaining global information about interplanetary shocks and many similar images were obtained in routine monitoring carried out almost continuously during 1978–79, 1980–81 and 1990–94. As noted earlier, the evidence from 1978–79 showed that coronal holes were usually located near the estimated solar sources of interplanetary shocks [8]. This correlation was highly significant and could not be explained by chance association. The same association is seen in Figure 6 where the source location given by the best-fit model lies close to a coronal hole. This outstanding shock was also followed by a sustained high-speed solar wind stream, exemplifying the behaviour that has been found to be typical of the global picture provided by IPS. Following the March 1991 shock at Ulysses the solar wind speed rose to about 1000 km s^{-1} and remained above 800 km s^{-1} for three days. The IPS-images were severely contaminated by intense radio emission from the Sun until 28 March, but thereafter showed a corotating low density stream for several days which was probably the same stream passing the Earth somewhat later.

Returning to the question of the sustained coronal magnetic energy release involved in the most powerful interplanetary shocks, it has been suggested that CMEs could create open-field regions which would allow the escape of high speed streams [9]. This might account for the observed sustained outflows but would require new coronal holes to be formed at CME sites. There is, however, no observational support for the sudden appearance of coronal holes large enough to produce the observed high speed streams. The simplest assumption is that these streams originate in the existing coronal holes that lie near the locations

Compressed synoptic maps

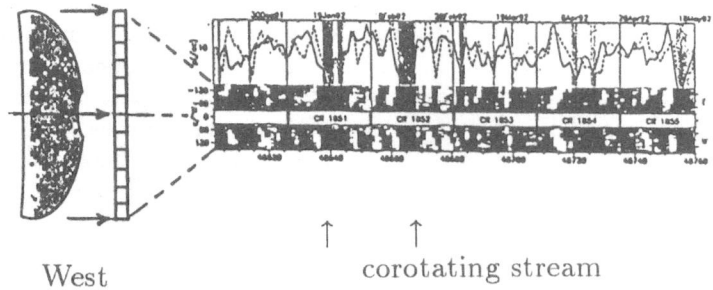

West   corotating stream

↓ coronal hole source

Fig. 4. Compressed synoptic maps for nearly six solar rotations. Enhance scintillation is shown in logarithmic grey-scale from $g < 0.77$ (black) to $g > 1.3$ (white). The dotted and solid lines show derived densities to the east and west of the Sun. Arrows indicate a long- lived high-speed stream due to a prominent coronal hole.

of the shock sources as determined from IPS-imaging. Whether the streams are caused by enhanced energy generation within the hole and any associated CME is merely peripheral activity, or whether the CME changes the coronal hole geometry and is thus a necessary cause of the outflow, are currently unanswered questions.

Further evidence suggesting that CMEs alone may not be the drivers of strong interplanetary shocks lies in the anomalous lack of strong shocks during solar

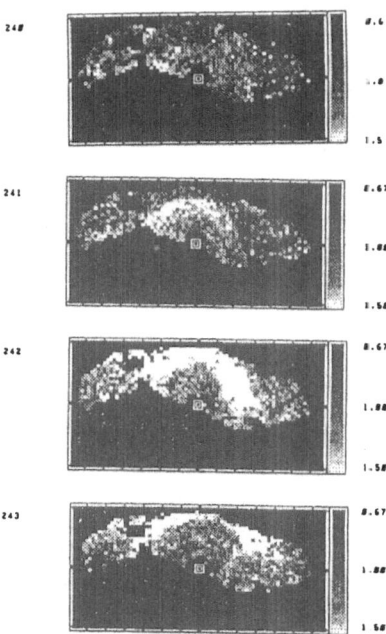

Fig. 5. *g*-maps for 26–30 September 1980 showing a major transient leaving the Sun above the ecliptic plane. Day numbers are shown on the left and *g*-values are indicated by the grey scale on the right. Ecliptic (Mollweide) coordinates are used.

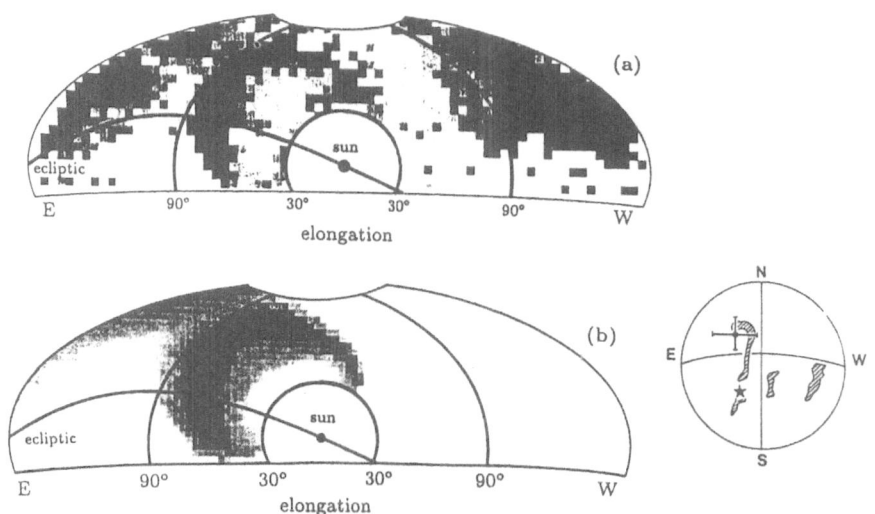

Fig. 6. (a) *g*-map of a strong interplanetary shock to the east of the Sun observed near elongation 90° on 23 March 1991. (b) The best-fit simulation is a model density enhancement of radial thickness 0.15 AU and angular width ±35°. The solar image shows the location of coronal holes and the estimated source of the shock above the solar equator.

maximum in 1980. At this time the occurrence of CMEs rose to a maximum, yet the number of strong shocks registered by spacecraft, and revealed less directly by geomagnetic activity, proton events and Forbush decreases fell to a value more appropriate to years near solar minimum. It is notable that the number of low latitude coronal holes also fell to a low value in 1980, which suggests that the presence of low latitude coronal holes is necessary for the generation of strong interplanetary shocks.

References

1. Behannon K.W., Burlaga L.F., and Hewish A., *J. Geophys. Res.*, **96**, 21213, 1991.
2. Kahler S.W., *Annv. Rev. Astron. Astrophys.*, **30**, 113, 1992.
3. Leer E., Holzer T.E., *J. Geophys. Res.*, **85**, 4681, 1980.
4. Hewish A., Tappin S.J. and Gapper G.R., *Nature*, **314**, 137, 1985.
5. Bravo S. and Perez-Enriquez R., *Rev. Mexicana. Astron. Astrofis.*, **28**, 17, 1994.
6. Tappin S.J., *Planet. Space Sci.* **34**, 93, 1986.
7. Tappin S.J., Hewish A. and Gapper G.R., *Planet. Space Sci.*, **32**, 1273, 1984.
8. Hewish A. and Bravo S., *Solar Phys.*, **106**, 185, 1986.
9. Borrini G., Gosling J.T., Barnes S.J. and Feldman W.C., *J. Geophys. Res.*, **87**, 4365, 1982.

Theory and Observations of Coronal Shock Waves

Gottfried Mann

Astrophysikalisches Institut Potsdam,
Observatorium für solare Radioastronomie,
D-14552 Tremsdorf, Germany

Abstract. Coronal shock waves are generated by flares and/or coronal mass ejections. They manifest themselves in solar type II radio bursts. The observational features of these radio bursts and their relationship to coronal mass ejections and interplanetary shocks are briefly reviewed. Finally, the various theoretical models of solar type II radio bursts are presented and compared with the observations. All these models base on the production of highly energetic electrons by coronal shock waves since nonthermal radio radiation is generally assumed as being generated by suprathermal electrons. These high energy electrons excite Langmuir waves or upper hybrid waves, which convert into radio waves.

1 Introduction

In the solar corona shock waves are generated due to flares and/or coronal mass ejections (CME's). They manifest themselves in solar type II radio bursts (Wild et al. 1959; Smerd et al. 1962) (cf. also Bougeret (1985) as a review). An special example of such a type II burst is presented in Fig. 1. Some of these coronal shock waves are able to enter into the interplanetary space where they are appearing as interplanetary shocks. These shocks may emit radio waves in terms of interplanetary type II radio bursts below 1 MHz.

In dynamic radio spectra solar type II radio bursts appear as stripes of enhanced radio emission slowly drifting from high to low frequencies. Generally, the spectral features of solar type II bursts consists of two components, the "backbone" and the "herringbones". The "backbone" is the aforementioned emission stripe with a slow drift of roughly -0.1 MHz/s. The "herringbones" are rapidly drifting emission stripes shooting up from the "backbone" to high and low frequencies. They resemble type III radio bursts and are interpreted as the radio signature of electron beams acceleratd at the associated coronal shock wave. All these features are seen in the example shown in Fig. 1.

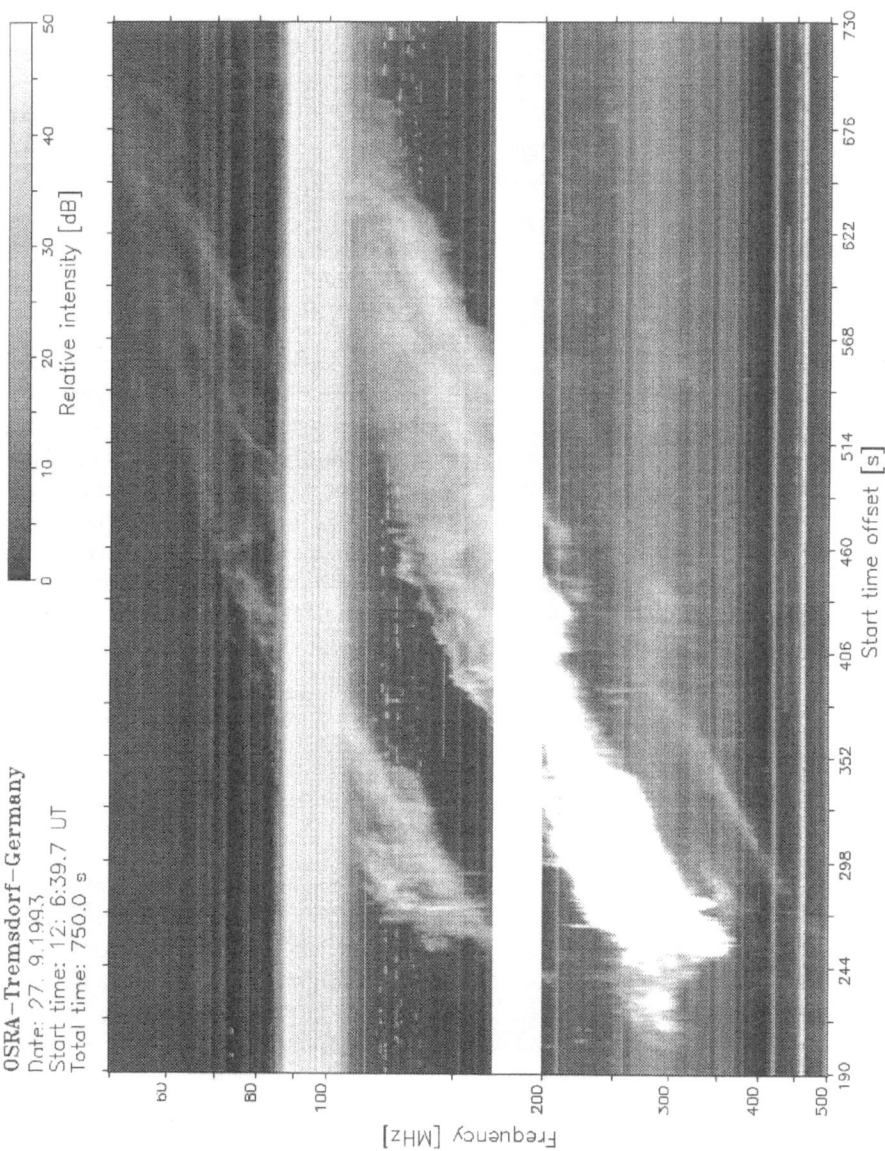

Fig. 1. Solar type II radio burst recorded by the new radiospectrograph of the Astrophysikalisches Institut Potsdam in a frequency range between 40 and 800 MHz on September 27, 1993. The starting time is given in the top panel. Note that the second harmonic is seen in this special case.

2 Observations

2.1 Properties of the Backbone

The backbones of solar type II radio bursts show typical drift rates with a mean value of $\langle D_f \rangle \approx -0.1\,\text{MHz/s}$ (Nelson and Melrose 1985; Urbarz 1990). Generally, there is a relationship between the drift rate D_f and the starting frequency f_s of solar type II radio bursts (cf. Smerd et al. 1975) as indicated in Fig. 2. Figure 2 results from a statistical analysis of 25 solar type II radio bursts recorded by the new radiospectrograph of the Astrophysikalisches Institut Potsdam in Tremsdorf (Mann et al. 1992). This instrument covers a frequency range between 40 and 800 MHz with a sweep rate of 10 Hz. The same sample provides a mean value and a standard deviation of $\langle \Delta f/f \rangle = 0.32 \pm 0.08$ for the relative instantaneous bandwidth $\Delta f/f$ of the backbone (Mann et al. 1994c). The corresponding distribution is presented in Fig. 3.

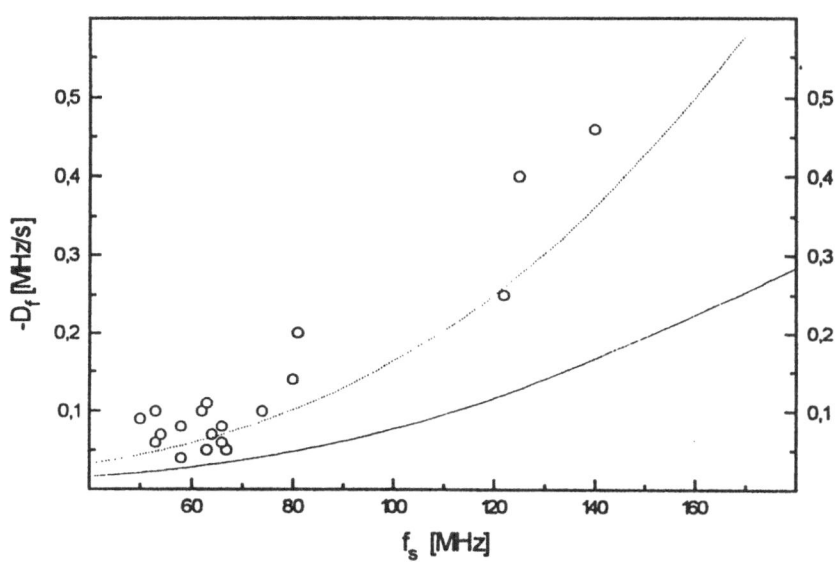

Fig. 2. Drift rate D_f versus starting frequency f_s derived from a sample of 25 solar type II radio bursts. The dotted and straight line represents the theoretically computed curve according to Eq. (1) by means of a two- and fourfold Newkirk model (cf. Section 2.4), respectively.

Most of the solar type II radio bursts show a fundamental-harmonic structure (cf. Fig. 1). The ratio between the frequencies at the low frequency edges of

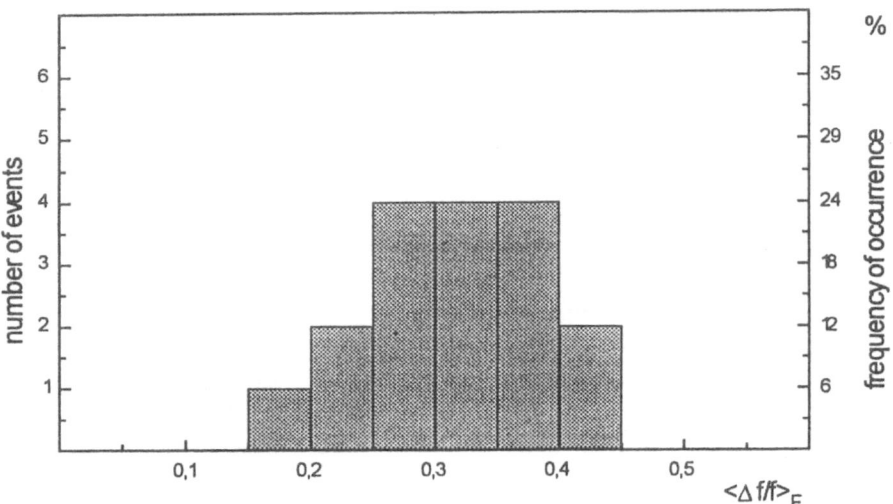

Fig. 3. Histogram of the distribution of the relative instantaneous bandwidth $\Delta f/f$ for a sample of 17 solar type II radio bursts

the fundamental band f_F and the harmonic band f_H has a mean value and a standard deviation of $\langle f_H/f_F\rangle = 1.97 \pm 0.10$ (Nelson and Melrose 1985; Mann et al. 1994c). The corresponding distribution is shown in Fig. 4. It also results from the aforementioned sample of 25 solar type II radio bursts. In rare cases the second harmonic band is also visible (cf. Fig. 1) (Roberts 1959; Kliem et al. 1992). The fundamental-harmonic structure is also observed in interplanetary type II radio bursts (Lengyel-Frey et al. 1985).

Sometimes, both the fundamental and harmonic band of the backbone split into two bands with a frequency separation of $\Delta f/f \approx 0.1$. This phenomenon is called band splitting (McLean 1967; Wild and Smerd 1972; Smerd et al. 1975; Nelson and Robinson 1975). Furthermore, the backbone shows a multiple lane structure in some cases, i.e., the backbone consists of few emission stripes with slightly different drift rates (Nelson and Melrose 1985). But the individual lanes (emission stripes) do not show a fundamental-harmonic structure as in the case of band splitting.

The radio radiation of the backbone is weakly polarized in the ordinary mode (Roberts 1959) and has brightness temperatures of typically 10^{12} K (Nelson and Melrose 1985).

2.2 Properties of Herringbones

20 % of all type II bursts are accompanied by herringbones (Roberts 1959). They are drifting from the backbone to both high and low frequencies. Their drift rates are similar to those of type III bursts and correspond to an exiter velocity of

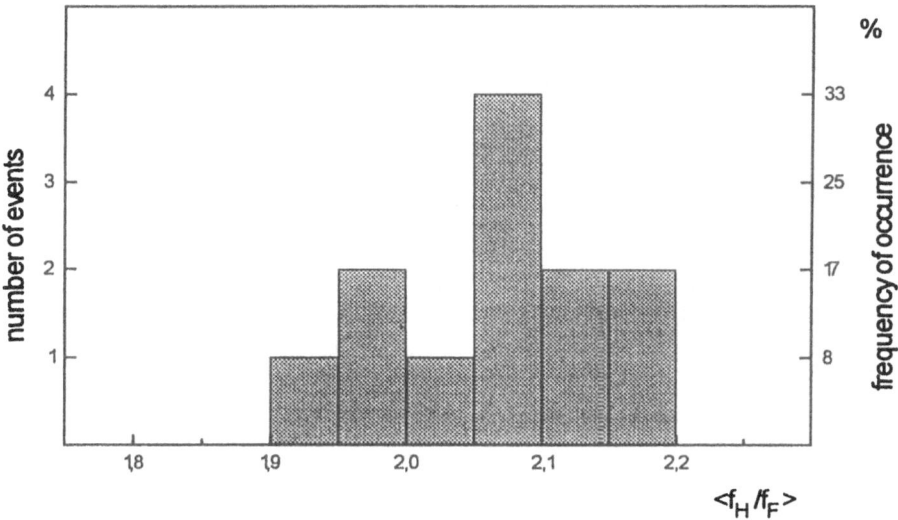

Fig. 4. Histogram of the distribution of the ratio between the harmonic and fundamental emission frequency (f_H/f_F) for a sample of 13 solar type II radio bursts.

$0.1 - 0.4$ times the velocity of light (Cairns and Robinson 1987). The herringbones also show a fundamental harmonic structure as the backbone. The radio radiation of herringbones is always polarized in the same sense as the corresponding backbone. But it shows a higher degree of circular polarization, namly $\approx 70\,\%$ (Stewart 1966). The brightness temperatures of the herringbones are slightly higher than that of the backbone, namely $\approx 10^{13}\,\mathrm{K}$ (Stewart and Magun 1980).

The radio measurements of solar type II bursts demonstrate that the backbone and the herringbones are generated by different mechanisms (Benz and Thejappa 1988). Although the herringbones and the type III bursts have similar drift rates, their spectral features are quite different (Cairns and Robinson 1987). While type III busts show a fan-like shape in dynamic radio spectra, i. e., its duration is monotonically increasing towards lower frequencies, a herringbone resembles a thorn, i. e., it has a longer duration nearer its starting frequency. Thus, the emission of the herringbones is different from type III burst generation (Cairns and Robinson 1987).

2.3 Further Observational Results

Coronal shock waves can only be observed by remote sensing techniques, e.g., radioastronomical methods as aforementioned. On the other hand interplanetary shock waves and Earth's bow shock can be studied by extraterrestrial in-situ measurements as special examples of collisionless shocks in space plasmas. If one assumes that the basic physical processes are essentially the same in all collisionless shocks, one can learn from the in-situ measurements of interplanetary shocks and Earth's bow shock for a better physical understanding of shock waves in the corona.

Filbert and Kellog (1979), Treumann et al. (1986), and Cairns (1986) investigated the radio radiation at Earth's bow shock and demonstrated that the radio emission is generated in the upstream region of Earth's bow shock.

Lengyel-Frey (1992) studied 20 interplanetary type II radio bursts observed by the radiospectrograph aboard the ISEE III satellite. She argued that the radio radiation is coming from the downstream region.

Hoang et al. (1992) reported on plasma wave in-situ measurements at a single interplanetary shock by the ULYSSES satellite. They deduced from the plasma wave spectra, that the fundamental and harmonic radiation is emitted from the downstream and upstream region, respectively. But this would lead to a ratio of f_H/f_F, which is different from $2 : 1$ due to the density jump across the shock wave. Even the observations show a ratio of f_H/f_F very close to 2 (cf. Fig. 4). From this point of view, this result indicates, that the fundamental and harmonic radiation have the same source region.

Thus, the question of the source region of type II radio burst radiation is still open.

2.4 Interpretation of the Observational Results

Solar type II burst observations provide a relationship between the drift rate D_f and the starting frequency f_s (cf. Fig. 2). Generally, it is assumed that the radio emission is generated near the local electron plasma frequency, i.e., $f_{obs} \approx \omega_{pe}/2\pi$ with $\omega_{pe} = (4\pi e^2 N/m_e)^{1/2}$ (e, elementary charge; N, particle number density; m_e, electron mass). Then, the definition of the drift rate, i.e., $D_f = \mathrm{d} f_{obs}/\mathrm{d} t$, provides a relationship between the drift rate and the velocity V of the radio source due to the density inhomogeneity of the solar corona.

$$D_f = \frac{f_{obs}}{2} \cdot \frac{1}{N} \cdot \frac{\mathrm{d} N}{\mathrm{d} s} V \tag{1}$$

Here, $\mathrm{d} N/\mathrm{d} s$ denotes the density gradient along the progation path of the radio source. The Newkirk (1961) model

$$N(R) = \alpha \cdot N_0 \cdot 10^{4.32 \cdot R_S/R} \tag{2}$$

($N_0 = 4.2 \cdot 10^4 \, \mathrm{cm}^{-3}$; $R_S = 6.958 \cdot 10^5 \, \mathrm{km}$, radius of Sun) is chosen as a special density model of the corona. Dulk and McLean (1978) presented an expression describing the magnetic field behaviour in the corona:

$$B(R) = \frac{1}{2} \cdot \left(\frac{R}{R_S} - 1 \right)^{-1.5} \quad \text{in G} \tag{3}$$

The behaviour of the density and the magnetic field in the corona is presented in Fig. 5 according to the fourfold Newkirk model ($\alpha = 4$; cf. Eq. (2)). Of course, the solar corona is highly structured with respect to the density, the temperature, and the magnetic field. Thus, Eqn. (2) and (3) should only be regarded as a rough approximation. Choosing the local Alfvén velocity v_A as the source velocity, i.e., $V = v_A = B/(4\pi m_p N)^{1/2}$ (m_p, proton mass), a relationship between the drift rate and the frequency can be derived by means of Eqn. (1), (2), and (3). In Fig. 3 the dotted and full line represents this relationship for a radio source travelling with the local Alfvén velocity in the case of a two- and fourfold Newkirk model (cf. Eq. (2)), respectively. The inspection of Fig. 2 shows that the exciter velocity of solar type II radio bursts well exceeds the Alfvén velocity (cf. also Smerd et al. (1975)). This demonstrates the justification of the basic assumption that coronal shock waves generate solar type II radio bursts.

3 Association with Coronal Mass Ejections

For a long time a close relationship between solar type II radio bursts and coronal mass ejections (CME's) was generally supposed. The comprehensive investigation of the observation of CME's and solar type II bursts (Gosling et al. 1976; MacQueen 1980; Wagner 1982; Sheeley et al. 1984; Robinson et al. 1984) demonstrated that CME's with a velocity > 400 km/s are associated with both solar and interplanetary type II bursts. Note that 41 % of all CME's have a velocity > 400 km/s. On the other hand 30 % of all solar type II radio bursts are not related to both CME's and interplanetary shocks. Furthermore, not all interplanetary shocks are accompanied by solar type II radio bursts.

These observational results can be interpreted in the following way:

1. CME's with a velocity > 400 km/s generate piston driven shocks in the corona and the interplanetary space (i.e., a bow shock is formed ahead the CME). Such piston driven shocks are able to generate solar and interplanetary type II bursts.
2. Impulsive flares cause blast shock waves, which are damped during their propagtion through the corona. Thus they are not able to reach the interplanetary space. In this case solar type II bursts were observed without association of interplanetary shocks.

Recently, Gopalswamy and Kundu (1992) investigated 12 events of solar type II radio bursts and CME's with the aim to compare between the piston-driven and blast wave model. Positional informations simultaneously obtained by the SKYLAB, SOLWIND and SMM-C/P instruments in white light and by the Culgoora and Clark Lake Radioheliograph are available in all cases. The analysis showed that 8 solar type II radio bursts had a source location well below the CME leading edge. Basing on other observational characteristics the blast wave model is favoured for the remaining 4 cases. Thus, Gopalswamy and Kundu (1992) concluded that solar type II radio bursts are generated by flare blast waves according to the available data of CME's and solar type II radio bursts.

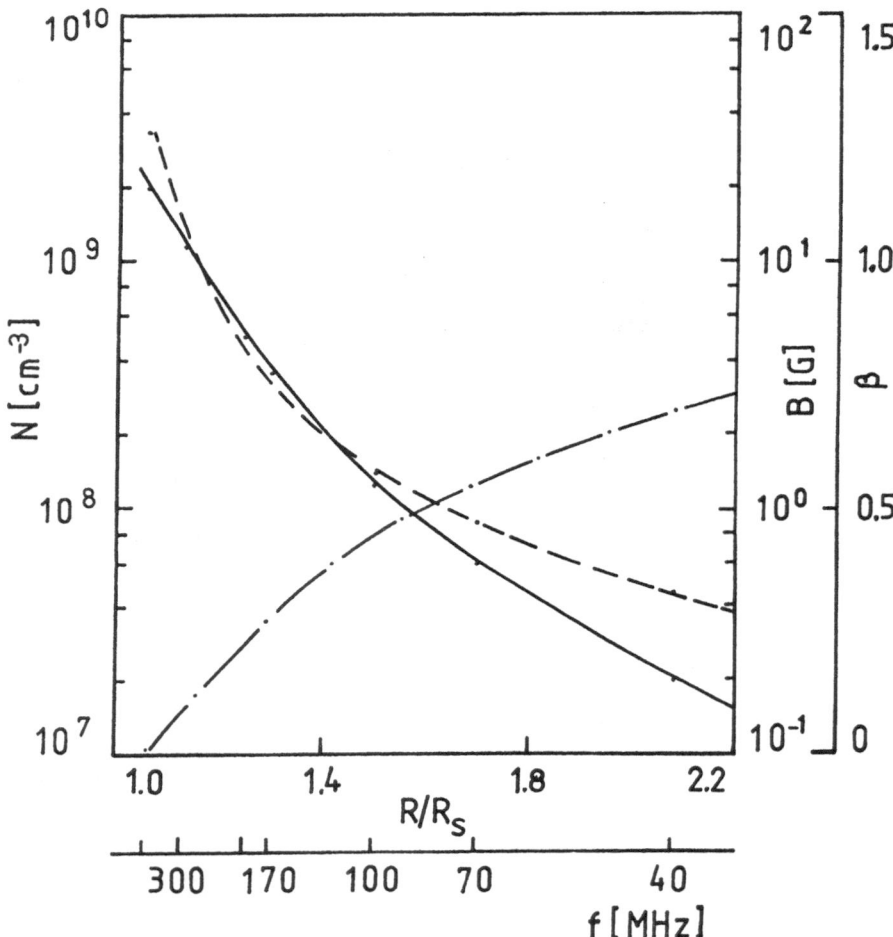

Fig. 5. Radial behaviour of the particle number density N, the magnetic field $\mathbf{B}$, and the plasma β in the solar corona. The curves and the relatioship between height and emission frequency are derived by means of a fourfold Newkirk model (cf. Eqn. (2) and (3)). Here, the radio emission is assumed to occur near the local electron plasma frequency. The plasma beta is calculated by $\beta = 8\pi N k_{\mathrm{B}} T / B^2$ (k_{B}, Boltzmann's constant) with a coronal temperature of $2 \cdot 10^6$ K.

4 Theory

Generally, the solar type II radio emission is assumed to be generated by supra-thermal and/or high energy electrons. Such electrons are able to excite Langmuir waves and upper hybrid waves. These high frequency electrostatic plasma waves convert into escaping radio waves by scattering at thermal ion density fluctuations and/or by coalescence with low frequency plasma waves. This mechanism is responsible for the fundamental radiation, while the coalescence of two high frequency electrostic waves produces the harmonic radiation. Thus, all type II radio burst theories must basically explain the production of suprathermal and high energetic electrons.

Ginzburg and Zheleznyakov (1958) were the first to present a solar type II burst model. It based on the close analogy of type III and type II radio emission, especially on their fundamental-harmonic structure. Therefore, they concluded that the same emission mechansim is responsible for the type III and type II radio emission and, consequently, electron streams are the source of this mechanism. These electron streams are accelerated at the shock front and, subsequently, propagating in the up- and downstream region, where they excite Langmuir waves. These Langmuir waves convert into radio waves by the aforementioned manner.

Furthermore, there are models in which the radio emission takes place within the shock front. At nearly perpendicular shocks, i.e., shocks with an angle $\theta_{n_s,B_{up}}$ between the shock normal $\mathbf{n}_s$ and the upstream magnetic field $\mathbf{B}_{up}$ close to 90°, an electric current is induced within the shock front according to Maxwell's equations. This current is mainly carried by the electrons due to their great mobility. Then, the current $\mathbf{j}$ is represented by $\mathbf{j} = -eN\mathbf{v}_d$ (e, elementary charge; N, particle number density; $\mathbf{v}_d$, electron drift velocity). If this induced current becomes so strong, that the corresponding drift velocity v_d of the electrons exceeds the thermal electron velocity $v_{th,e}$, i.e., $v_d > v_{th,e}$, the Buneman instability begins to act and excites Langmuir waves necessary for radio wave emission. The thermal electron velocity $v_{th,e}$ has a value of $5500\,\mathrm{km/s}$ in the solar corona with a temperature of $T = 2 \cdot 10^6\,\mathrm{K}$. This model was mainly worked out by Zheleznyakov (1965), Zaitsev (1966), (1977), and Stepanov (1970).

In 1983 Holman and Pesses (1983) suggested that the high energy electrons needed for type II radio burst emission are generated by shock drift acceleration. But the electrons become only suprathermal if the angle $\theta_{n_s,B_{up}}$ between the shock normal and the upstream magnetic field exceeds 80° ($\theta_{n_s,B_{up}} > 80°$), i.e., in the case of a nearly perpendicular shock geometry. On the other hand the production of herringbone structures requires a final electron speed of roughly $10^5\,\mathrm{km/s}$. Under coronal circumstances such velocities can only be reached by a single encounter with the shock if the initial electron velocity is greater than $20 \cdot 10^3\,\mathrm{km/s}$. Thus, more than one shock encounter of the electrons is needed for the generation of herringbone structures. Furthermore, Holman and Pesses (1983) interpreted the appearance of a backbone only, a backbone with herringbones, herringbones without a backbone, and a backbone with bandsplitting in terms of shock drift acceleration with different geometries between the shock normal, the coronal density gradient, and the direction of the upstream magnetic field.

Recently, Benz and Thejappa (1988) suggested that solar type II radio bursts are generated by super-critical, quasi-perpendicular shocks. In-situ measurements at Earth's bow shock showed that a significant amount of the solar wind energy is converted into high energetic particles appearing in the upstream region. Electron beams are especially observed on interplanetary field lines newly connected with the tangential point of Earth's bow shock, i.e, at an quasi-perpendicular shock geometry. This area forms the electron foreshock. The incident electrons are accelerated by shock drift acceleration and reflected back into the upstream region, since the magnetic field compression at the front of a fast magnetosonic shock acts as a mirror. Finally, a shifted loss-cone distribution of electrons is established in front of an quasi-perpendicular shock (Leroy and Mangeney 1984), (Wu 1984). Such a distribution is unstable and able to excite upper hybrid waves, which convert into escaping radio waves by coalescence with ion-acoustic and/or lower hybrid waves. Ion-acoustic and lower hybrid waves are also generated by a shifted loss-cone distribution of protons. Note that the same mechansim also produce a shifted loss-cone distribution for the protons. This emission mechanism was suggested by Benz and Thejappa (1988) and is only responsible for the backbone radiation.

5 Electron Acceleration at Quasi-Parallel Shocks in the Corona

The solar type II burst theories mentioned in the previous section mainly base on the assumption that the high energy electrons needed for the radio emission are accelerated at nearly perpendicular shocks in the corona. Obviously, one would expect that the shock wave associated with the type II radio burst is radially moving through the corona and, thus, quasi-parallel to the magnetic field, especially in the upper corona, i.e., below 100 MHz (cf. Fig. 5). Nevertheless, a substantial deviation from the mainly radial movement of coronal shock waves may occur. When a large-scale shock wave is moving through the corona there are probably always regions where the shock geometry is nearly perpendicular, in particular at its flanks. But this would mean that the type II burst source is consisting of spatially different subsources. On the other hand, the duration of solar type II radio bursts of typically 8 minutes (cf. Fig. 1) and their drift velocity of roughly 800 km/s indicate that the associated shock wave travels $4 \cdot 10^5$ km through the corona. This leads to the question whether a shock wave sees a nearly perpendicular shock geometry with respect to the ambient magnetic field over such a long distance.

Recently, Mann and Lühr (1994a) suggested a mechanism of electron acceleration at strong magnetohydrodynamic turbulence in front of super-critical, quasi-parallel shocks in the solar corona.

As already mentioned Earth's bow shock is the most observed collisionless shock in space plasmas. Because of its curvature it has regions of a quasi-parallel and a quasi-perpendiclular shock geometry. Figure 6 shows the behaviour of the magnitude of the magnetic field during a quasi-parallel crossing of Earth's bow shock by the AMPTE/IRM satellite.

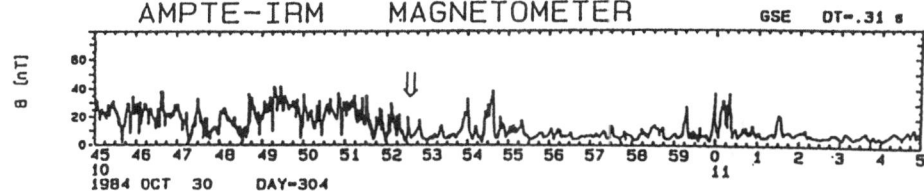

Fig. 6. Behaviour of the magnitude of the magnetic field recorded by the magnetometer aboard the AMPTE/IRM satellite during a crossing of the quasi-parallel region of Earth's bow shock. The shock transition appeared at 10:52:30 UT according to the plasma data as indicated by the arrow.

Recently, so-called SLAMS (Short Large Amplitude Magnetic Field Structures) have been observed as a common feature in the vicinity of the quasi-parallel region of Earth's bow shock (Thomsen et al. 1990), (Schwartz et al. 1992). In Fig. 6 SLAMS were appearing in the upstream region at 10:54:50, 10:59:15, and 11:01:30 UT, for example. According to Schwartz and Burgess (1991) a supercritical, quasi-parallel shock transition should be regarded as a patchwork of SLAMS. Although SLAMS are propagating quasi-parallel to the upstream magnetic field the magnetic field is locally swung into a quasi-perpendicular magnetic field geometry within SLAMS (Mann and Lühr 1992). Thus, electrons can be accelerated to suprathermal energies at SLAMS by means of shock drift acceleration. Assuming that SLAMS also appear at supercritical, quasi-parallel shocks in the solar corona the aforementioned mechanism may provide the suprathermal energy electrons necessary for type II radio burst radiation from quasi-parallel coronal shocks.

5.1 Properties of SLAMS

During the time interval 10:20 – 11:05 UT on October 30, 1984 the AMPTE/IRM satellite experienced several quasi-parallel shock crossings of Earth's bow shock and observed SLAMS in the upstream and downstream region (Schwartz et al. 1992). Mann et al. (1994b) investigated 18 SLAMS appearing in the upstream region with the aim of studying the local deformation of the magnetic field with respect to the background magnetic field, the propagation direction, and the shock normal. This statistical analysis provided the following mean values and standard deviations:

1. enhancement of the magnetic field magnitude:
   $(B_{\max} - B_0)/B_0 = 2.63 \pm 1.11$
2. angle between the propagation direction $\mathbf{n}_1$ of the SLAMS and the background magnetic field $\mathbf{B}_0$:
   $\theta_{\mathbf{n}_1,\mathbf{B}_0} = 33.9° \pm 17.2°$

3. angle between the proapagtion direction $\mathbf{n}_1$ of the SLAMS and the maximum magnetic field within the SLAMS $\mathbf{B}_{max}$:
$$\theta_{n_1, B_{max}} = 80.0° \pm 9.6°$$
4. angle between the shock normal $\mathbf{n}_s$ and the maximum magnetic field within the SLAMS $\mathbf{B}_{max}$:
$$\theta_{n_s, B_{max}} = 76.3° \pm 9.9°$$

Here, B_{max} denotes the magnitude of the maximum magnetic field $\mathbf{B}_{max}$. The distribution of the angle $\theta_{n_s, B_{max}}$ is presented in Fig. 7. Figure 7 shows that the magnetic field within SLAMS is locally rotated into a quasi-perpendicular geometry although SLAMS are propagating quasi-parallel with respect to the unperturbed upstream magnetic field (Mann and Lühr 1992). SLAMS have a typical spatial width of 1000 km, i.e., approximately 10 ion inertial length, and a velocity of 3.6 times the Alfvén velocity in the plasma rest frame (Mann et al. 1994b).

5.2 Shock Drift Acceleration

SLAMS represent moving structures of large magnetic field compressions, i.e., moving particle mirrors, in the vincinity of a supercritical, quasi-parallel shock. Thus, particles can be accelerated at these structures by shock drift acceleration. In this case the particles gain energy by their reflection at SLAMS under the conseration of their magnetic moment. Following the approach given by Sonnerup (1969), Paschmann et al. (1980), Schwartz et al. (1983), and Krauss-Varban (1989) the calculations are conveniently performed in the the de Hoffmann-Teller frame. In this special frame the component of the particle velocity parallel to the magnetic field changes its sign, i.e., $V_{r,\parallel}^{HT} = -V_{i,\parallel}^{HT}$, while the magnitude of the particle velocity perpendicular to the magnetic field stays unchanged, i.e., $\left(V_{r,\perp}^{HT}\right)^2 = \left(V_{i,\perp}^{HT}\right)^2$, during the reflection process. The transformation back into the laboratory frame provides

$$V_{r,\parallel} = 2v_s \cdot \sec(\theta) - V_{i,\parallel} \tag{4}$$

v_s represents the velocity of the SLAMS, at which the particle is reflected. θ is the angle between the propagation direction of the SLAMS and the magnetic field within the SLAMS. $V_{i,\parallel}$ and $V_{r,\parallel}$ denote the component of the velocity parallel to the magnetic field of the incident and reflected particle, respectively.

5.3 Discussion

As already mentioned SLAMS should be considered as a common feature in the vicinity of a supercritical, quasi-parallel shock (Schwartz and Burgess 1991). Assuming that the physical processes of collisionless shocks are essentially the same in space plasmas (Benz and Thejappa 1988) it is justified to suppose that SLAMS also appear at supercritical, quasi-parallel shocks in the corona. Thus, these SLAMS can accelerate electrons to suprathermal velocities.

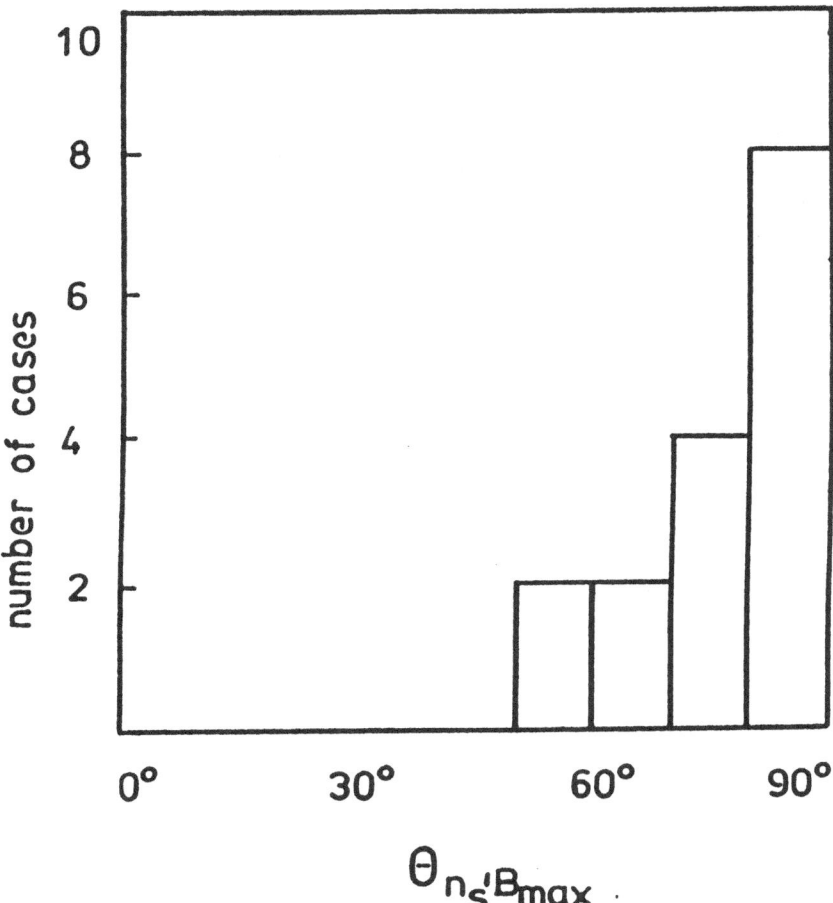

Fig. 7. Histogram of the distribution of the angle $\theta_{n_s, B_{max}}$ between the shock normal and the maximum magnetic field within the SLAMS

In order to estimate the velocity gain of an electron during an encounter with a SLAMS at coronal shocks the plasma parameters occuring at the 70 MHz level are employed. There, the following plasma parameters are found: particle number density $7 \cdot 10^7 \, cm^{-3}$, temperature $2 \cdot 10^6 \, K$, and magnetic field strength $1 \, G$ (Krüger 1979). These values result in a plasma beta $\beta = 0.4$, an Alfvén velocity $v_A = 260 \, km/s$, an ion sound speed $c_s = 170 \, km/s$, an ion inertial length $c/\omega_{pi} = 3000 \, cm$ (c, velocity of light; ω_{pi}, proton plasma frequency), a thermal speed of the electrons $v_{th,e} = 5500 \, km/s$, and an electron gyroradius $r_L = v_{th,e}/\omega_{ce} = 35 \, cm$ (ω_{ce}, electron gyrofrequency). Under these circumstances the condition for conservation of the magnetic moment of electrons, i.e., $r_L \|\nabla \mathbf{B}\|/\|\mathbf{B}\| \ll 1$ or $r_L \ll 10 \cdot c/\omega_{pi}$, is well fulfilled within SLAMS. Thus, an electron makes

860 gyrations during the penetration into a SLAMS. During this movement and, subsequently, at the reflection process the electron experiences an quasi-perpendicular magnetic field geometry, i.e., $\theta_{n_s,B_{max}}$ is about $80°$. Thus, the shock drift acceleration can efficiently act for electrons at SLAMS. Adopting for the initial particle velocity $V_{i,\parallel} = -2^{1/2} v_{th,e} = -7800\,km/s$, the velocity of SLAMS $v_s = 800\,km/s$, and the angle $\theta = 80°$, the Eq. (4) provides a velocity parallel to the magnetic field of the reflected electron $V_{r,\parallel} = 17,000\,km/s > 3\,v_{th,e}$, which is well suprathermal (Mann and Lühr 1994a).

Which velocity distribution of electrons are formed in front of SLAMS? A Maxwellian distribution of the form

$$f_i = C \cdot \exp\left\{\frac{-\left(V_{i,\parallel}^2 + V_{i,\perp}^2\right)}{2v_{th,e}^2}\right\} \tag{5}$$

is assumed as the initial distribution of electrons in the upstream region. Here, C and $v_{th,e}$ denote a normalization constant and the thermal electron velocity. In the de Hoffmann-Teller frame the electric field is vanishing. Thus, the electrons are adiabatically reflected, i.e., they form a loss-cone distribution after the reflection. Thus, electrons, which fulfill the conditions $V_{i,\parallel}^{HT} \le 0$ and $V_{i,\perp}^{HT}/V_{i,\parallel}^{HT} \ge \tan(\alpha_{lc})$ with $\alpha_{lc} = \arcsin(B_{up}/B_{max})^{1/2}$ are only reflected. After tranformation back into the laboratory frame a shifted loss-cone distribution of the form

$$f_r = C \cdot \Theta\left(V_{r,\parallel} - v_s \sec\theta\right) \cdot \Theta\left(V_{r,\perp} - \left[V_{r,\parallel} - v_s \sec\theta\right] \cdot \tan\alpha_{lc}\right)$$

$$\times \exp\left\{\frac{-\left[\left(-V_{r,\parallel} + 2v_s \sec\theta\right)^2 + V_{r,\perp}^2\right]}{2v_{th,e}^2}\right\} \tag{6}$$

is established for the reflected particles in the upstream region (cf. Leroy and Mangeney (1984), Wu (1984), and Krauss-Varban et al. (1989)). Here, Θ represents the step-function. $V_{r,\parallel}$ and $V_{r,\perp}$ denote the velocity components parallel and perpendicular to the magnetic field of the reflected electrons, respectively. The transition from an isotropic initial Maxwell distribution to a shifted loss-cone distribution during the reflection of electrons is shown in Fig. 8. Thus, the electron velocity distribution in the vicinity of SLAMS, i.e., in the upstream region of a supercritical, quasi-parallel shock, is a superposition of an isotropic Maxwellian distribution (cf. Eq. (5)) and a shifted loss-cone distribution (cf. Eq. (6)). Such anisotropic velocity distribution of electrons is unstable and gives rise to excite upper hybrid waves (Benz and Thejappa 1988), which can convert into escaping radio waves by the aforementioned manner (cf. Sec. 4). Thus, the mechanism originally suggested for the backbone radiation at quasi-perpendicular shocks by Benz and Thejappa (1988) (cf. also Sec. 4) is also able to act locally in the vicinity of SLAMS, i.e., super-critical, quasi-parallel shocks in the corona can also generate solar type II radio bursts.

According to Schwarz and Burgess (1991) a quasi-parallel shock transition should be regarded as a patchwork of SLAMS. This picture was confirmed by the

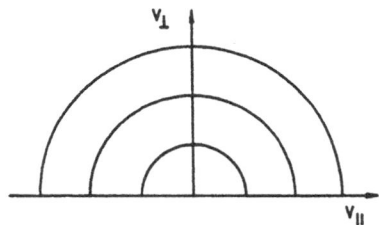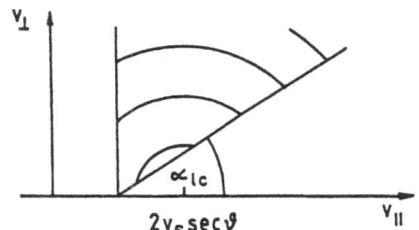

Fig. 8. Illustration of a shifted loss-cone velocity distribution (right) (cf. Eq. (6)) in comparision of an isotropic Maxwellian distribution (left) (cf. Eq. (5)). $v_\parallel$ and $v_\perp$ denote the velocity components parallel and perpendicular to the magnetic field, respectively.

numerical particle simulations of Scholer (1993). It is well known, that a quasi-perpendicular shock exhibits a clearly localized shock transition. Therefore, a solar type II radio burst with homogeneous emission features with respect to the backbone and the herringbones should be expected as being generated by an quasi-perpendicular shock in the corona. In contradiction to this, a solar type II radio burst associated with an quasi-parallel shock should show more irregular (or patchy) emission features in its dynamic radio spectrum.

6 Final Remarks

In the solar corona shock waves generated by flares and/or coronal mass ejections are a common phenomenon observed by radioastronomical methods in form of solar type II radio bursts. Their frequency of occurrence is obviously depending on the cycle of solar activity. The study of the origin of solar type II radio bursts is very important for understanding the active processes in the solar corona. In the last three decades solar type II radio bursts were investigated by different manners, i.e., observational and theoretical. Nevertheless, many questions are still open. Without the aim of completeness, I would like to mention some of these problems:

1. What is the origin, i.e., blast wave and/or piston driven, of the shock waves related to solar type II radio bursts?
2. Which global structure show coronal shock waves accompanied by solar type II radio bursts?
3. What kind of collisionless shock, i.e., supercritical, subcritical, quasi-parallel or quasi-perpendicular shock, is responsible for the type II burst radiation?
4. What is the relationship between solar type II radio bursts, coronal mass ejections, and interplanetary shocks?
5. What is the radio emission mechanism for the generation of the fundamental and harmonic radiation in solar type II radio bursts?

6. Where is the source of the fundamental and harmonic type II burst radiation located with respect to the shock front, i.e., in the upstream or downstream region?

7. As already mentioned the radiation mechanism of the backbone and herring-bone radiation is different. What mechanism is responsible for the backbone and the herringbones?

8. What is the physical reason of the difference of the herringbone radiation and solar type III radio bursts although both features are regarded as being generated by high energy electron beams?

Some of these problems are discussed in this paper and have partly been solved by many authors in the last decades. But a complete understanding of the solar type II burst phenomena is a task in the future.

Finally, I would like to emphasize that just a comprehensive study of phenom-ena at collisionless shocks in space plasmas, i.e., observational investigations by in-situ measurements and remote sensing techniques as well as theoretical stud-ies, is expected to lead to a solution of the aforementioned problems, as already done by several authors and, for instance, in Sec. 5.

References

Benz, A. O., Thejappa, G.: Radio emission of coronal shock waves. Astron. Astrophys. **202** (1988) 267-274

Bougeret, J.-L.: "Observations of Shock Formation and Evolution in the Solar Atmo-sphere". in Collisionless Shocks in the Heliosphere: Reviews of Current Research, ed. by B. S. Tsurutani and R. G. Stone, Geophys. Monogr. Ser. Vol. **35** (1985), Washington DC, pp. 13-32

Cairns, I. H.: New waves at multiples of the plasma frequency upstream of the Earth's bow shock. J. Geophys. Res. **91** (1986) 2975-2988

Cairns, I. H., Robinson, R. D.: Herringbone bursts associated with type II solar radio emission. Solar Phys. **111** (1987) 365-383

Dulk, G. A., McLean, D. J.: Coronal magnetic fields. Solar Phys. **57** (1979) 279-291

Filbert, P. C., Kellog, P. J.: Electrostatic noise at the plasma frequency beyond the Earth's bow shock. J. Geophys. Res. **84** (1979) 1369-1381

Ginzburg, V. L., Zheleznyakov, V. V.: On the possible mechanisms of sporadic solar radio emission. Sov. Astron.-AJ. **3** (1958) 235-247

Gopalswamy, N., Kundu, M. R.: "Are coronal shocks piston driven". in Particle Ac-celeration in Cosmic Plasmas, ed. by G. P. Zank and T. K. Gaisser, American Institute of Physics, (1992), New York, pp 257-260

Gosling, J. T., Hildner, E., MacQueen, R. M., Munroe, R. H., Poland, A. I., Ross, C. L.: The speeds of coronal mass ejection events. Solar Phys. **48** (1976) 389-397

Hoang, S., Pantellini, F., Harvey, C. C., Lacombe, C., Mangeney, A., Meyer-Vernet, N., Perche, C., Steinberg, J.-L., Lengyel-Frey, D., MacDowell, R. J., Stone, R. G., Forsyth, R. J.: "Interplanetary fast shock diagnosis with the radio receiver on ULYSSES". in Solar Wind Seven, ed. E. Marsch and R. Schwenn, Pergamon Press (1992), Oxford, pp. 465-468

Holman, G. D., Pesses, M. E.: Solar type II radio emission and shock drift acceleration of electrons. Astrophys. J. **267** (1983) 837-843

Kliem, B., Krüger, A., Treumann, R. A.: Third plasma harmonic radiation in type II bursts, Solar Phys. **140** (1992) 149-160

Krauss-Varban, D.: Fast Fermi and gradient drift acceleration of electrons at nearly perpendicular collisionless shocks. J. Geophys. Res. **94** (1989) 15,367-15,372

Krauss-Varban, D.,Burgess, D., Wu, C. S.: Electron acceleration at nearly perpendicular collisionless shocks 1. One-dimensional simulations without electron scale fluctuations. J. Geophys. Res. **94** (1989) 15,089-15,098

Krüger, A.: "Introduction to Solar Radioastronomy", D. Reidel Publ. Co., (1979), Dordrecht

Lengyel-Frey, D.: Location of the radio emitting regions of interplanetary shocks. J. Geophys. Res. **97** (1992) 1609-1617

Lengyel-Frey, D., Stone, R. G., Bougeret, J.-L.: Fundamental and harmonic emission in interplanetary type II radio bursts. Astron. Astrophys. **151** (1985) 215-223

Leroy, M. M., Mangeney, A.: A theory of energetization of solar wind electrons by Earth's bow shock. Ann. Geophys. **2** (1984) 449-456

MacQueen, R. M.: Coronal transients: a summary. Phil. Trans. R. Soc. London, Ser. A **297** (1980) 605-643

Mann, G., Auraß, H., Voigt, W., Paschke, J.: Preliminary observations of solar type II radio bursts with the new radiospectrograph in Tremsdorf. ESA-Journal **SP-348** (1992) 129-132

Mann G., Claßen, T., Auraß, H.: Characteristics of coronal shock waves and solar type II radio bursts. Astron. Astrophys. (1994c) in press

Mann, G., Lühr, H.: On short large amplitude magnetic structures in the vicinity of the quasi-parallel region of the Earth's bow shock. ESA-Journal **SP-346** (1992) 91-93

Mann, G., Lühr, H.: Electron acceleration at quasi-parallel shocks in the solar corona and its signature in solar type II radio bursts. Astrophys. J. suppl. series **90** (1994a) 577-581

Mann, G., Lühr, H., Baumjohann, W.: Statistical analysis of short large amplitude magnetic structures at a quasi-parallel shock. J. Geophys. Res. **99** (1994b) 13,315-13,323

McLean, D. J.: Band splitting in type II solar radio bursts. Proc. Astron. Soc. Aust. **1** (1967) 47-49

Nelson, G. S., Melrose, D.: "Type II bursts", in Solar Radiophysics, ed. by D. J. McLean and N. R. Labrum, Cambridge University Press, (1985), Cambridge, pp 333-360

Nelson, G. J., Robinson, R. D.: Multi-frequency heliograph observations of type II bursts. Proc. Astron. Soc. Aust. **2** (1975) 3670-378

Newkirk, G.: The solar corona in the active regions and the thermal origin of the slowly varying component of the solar radiation. Astrophys. J. **133** (1961) 983-992

Paschmann, G., Sckopke, N., Asbridge, J. R., Bame, S. J., Gosling, J. T.: Energetization of solar wind ions by reflection from the Earth's bow shock. J. Geophys. Res. **85** (1980) 4689-4693

Roberts, J. A.: Solar radio bursts of spectral type II. Aust. J. Phys. **12** (1959) 327-334

Robinson, R. D., Stewart, R. T., Cane, H. V.: Properties of metre-wavelength solar bursts associated with interplanetary type II emission. Solar Phys. **91** (1984) 159-168

Scholer, M.: Upstream waves, shocklets, short large-amplitude magnetic structures and the cyclic behaviour of oblique, quasi-parallel collisionless shocks. J. Geophys. Res. **98** (1993) 47-57

Schwartz, S. J., Burgess, D.: Quasi-parallel shocks: a patchwork of three-dimensional structures. Geophys. Res. Lett. **18** (1991) 373-376

Schwartz, S. J., Burgess, D., Wilkenson, W. P., Kessel, R. L., Dunlop, M., Lühr, H.: Observations of short large amplitude magnetic structures at a quasi-parallel shock. J. Geophys. Res. **97** (1992) 4209-4232

Schwartz, S. J., Thomsen, M. F., Gosling, J. T.: Ions upstream of the Earth's bow shock: A theoretical comparison of alternative source populations. J. Geophys. Res. **88** (1983) 2039-2047

Sheeley, N. R., Stewart, R. T., Robinson, R. D., Howard, R. A., Koomen, M. J., Michels, D. J.: Associations between coronal mass ejections and metric type II bursts. Astrophys. J. **279** (1984) 839-853

Smerd, S. F., Sheridan, K. V., Stewart, R. T.: Split-band structure in type II radio bursts from the Sun. Astrophys. Lett. **16** (1975) 23-28

Smerd, S. F., Wild, J. P., Sheridan, K. V.: On the relative position and origin of harmonics in the spectra of solar radio bursts of spectral types II and III. Aust. J. Phys. **15** (1962) 180-189

Sonnerup, B. U. Ö.: Acceleration of particles reflected at a shock front. J. Geophys. Res. **74** (1969) 1301-1304

Stepanov, A. V.: On the intensity of radio emission from the front of a collisionless shock wave. Radiophys. Quantum electron. **13** (1970) 1034-1042

Stewart, R. T.: The polarization of herringbone features in solar radio bursts of spectral type II. Aust. J. Phys. **19** (1966) 209-204

Stewart, R. T. and Magun, A.: Radio evidence for electron acceleration by transverse shock waves in herringbone type II solar bursts. Proc. Astron. Soc. Aust. **4** (1980) 53-57

Thomsen, M. F., Gosling, J. T., Bame S. J., Russell, C. T.: Magnetic pulsations at the quasi-parallel shock. J. Geophys. Res. **95** (1990) 957-968

Treumann, R. A., Bauer, O. H., LaBelle, J., Haerendel, G., Christiansen, P. J., Darbyshire, A. G., Norris, A. J., Wolliscroft, L. J. c., Anderson, R. R., Gurnett, D. A., Holtzworth, R. W., Koons, H. C., Roeder, J.: Electron plasma waves in the solar wind: AMPTE/IRM and UKS observations. Adv. Space Res. **6** (1986) 93-97

Urbarz, H.: Type II solar radio bursts recorded at Weissenau 1967 - 1987. report **UAG-98** (National Geophsical Data Center) (1990) Boulder

Wagner, W. J.: SERF studies of mass motions arising in flares. Adv. Space Res. **2**, no. 11 (1982) 203-206

Wild, J. P., Sheridan, K. V., Trent, G. H.: "The tranverse motions of the source of solar radio bursts", in Proc. IAU/URSI Symp. Paris Symposium on Radio Astronomy, ed. by R. N. Bracewell, Stanford Univ. Press (1959), p. 176-182

Wild, J. P., Smerd, S. F.: Radio bursts from the solar corona. Annu. Rev. Astron. Astrophys. **10** (1972) 159-173

Wu, C. S.: A fast Fermi process: energetic electrons accelerated by a nearly perpendicular bow shock. J. Geophys. Res. **89** (1984) 8857-8862

Zheleznyakov, V. V.: On the origin of solar radio bursts in the meter-wavelength range. Sov. Astron.-AJ. **9** (1965) 191-203

Zaitsev, V. V.: A theory for type II bursts of solar radio emission. Sov. Astron.-AJ. **9** (1966) 572-583

Zaitsev, V. V: Theory of type II and type III radio bursts. Radiophys. Quantum Electron. **20** (1977) 135-147

Numerical Simulations of Shock Electron Acceleration in Solar Physics

Bertrand Lembege

C.E.T.P. / C.N.E.T.
38-40 rue du general Leclerc
92131 Issy-les-Moulineaux (France)

Abstract. The connection between solar radiobursts of type II and shocks has been largely recognized since many years; however, the details of radiobursts generation are not been fully understood yet. A two-step mechanism is often invoked including (i) the formation of electrons streams by shock waves and (ii) the excitation of radio emission by these streams. In the present report, we will focuss mainly on step (i) through results of numerical simulations mainly based on full particle electrostatic and electromagnetic codes. Such codes are appropriate to follow the overall dynamics of the shock itself and to identify the different sources of acceleration and heating of particles in a self-consistent way.

1 Introduction

Collisionless shocks are very common in solar physics. It is well known that such shocks are generated by coronal mass ejections (CME) in the solar corona. Moreover, it is generally that the solar radiobursts of spectral type II are generated by the passage outward through the solar corona of fast-mode MHD shock waves, which originate in relatively intense solar flares. At any given height in the solar corona, these shock waves excite radio emission of the appropriate plasma frequency (Wild and Smerd, 1972; Mc Lean and Nelson, 1977; Mc Lean, 1980). Estimates of the velocities of the shocks, made from the frequency drift-rate of the type II bursts, in conjunction with appropriate models of the electron densities, generally give velocities of about thousands kilometers per second (Maxwell and Thompson, 1962; Maxwell and Dryer, 1981). Despite the fact that the connection between solar radiobursts of type II and shocks has become evident, the details of radiobursts generation are not clear yet. Various mechanisms for the bursts generation have been proposed which can be summarized as follows:

a) The burst generation by collisionless shock waves moving across the magnetic field is connected to the Buneman instability developing at the shock front (Pikelner and Gintzburg, 1963; Zaitsev, 1977). This instability is due to the relative motions of ions and electrons. However, the excited plasma waves have low frequencies. The rising of the plasma wave frequency needs subsequent induced scattering on nonthermal electrons that leads to the broadening of the spectrum up to $2\omega_{pe}$ and to its isotropisation. So the transformation of plasma waves into electromagnetic ones gives use to the radio emission in a wide frequency band $\omega_{pe} < \omega < 2\omega_{pe}$, where ω_{pe} is the electron plasma frequency.

b) Type II burst emission can be associated with acceleration of electrons by lower-hybrid waves followed by nonlinear conversion. The lower-hybrid waves are assumed to be generated by a current-driven instability in the shock front (Lampe and Papadopoulos, 1977).

However, the proposed mechanisms are not general in the sense that they do not describe adequately an analogy between type III and type II bursts often observed (herringbone structure is a typical example). Type III radiobursts are produced by fast electron streams propagating along the magnetic field lines (Ginzburg and Zheleznyakov, 1958). From the analogy between the components of type III and type II bursts, one can assume that types II are also generated by streams of accelerated electrons (Mc Lean and Nelson, 1977). On the same, it is important to note that intense streams of electrons and ions are very often observed near the Earth's bow shock (Eastman et al., 1981; Anderson et al., 1981) .

Then, it seems natural to connect the radio emission of the coronal shock with the fluxes of electrons as the most general and effective way of direct generation of plasma waves (Langmuir oscillations). The key questions are those of the effective electron acceleration and of the transformation of electron energy into electromagnetic radiation. Let us remind that the radio emission of the type II bursts may be very strong (brightness temperature $T_b \approx 10^{11}$ K) as noticed by Nelson and Robinson (1975).

In summary, one distinguishs two different steps in the radiation mechanisms: (i) one step based on the formation of particle streams (ions and/or electrons) by a shock wave, and (ii) a second step focussed on the radio emission triggered by such streams. In the present report, we will focuss mainly on the first step (i). Such an analysis requires a detailed study on the microphysical phenomena which develop through the shock itself and are responsible of various processes for particle acceleration (and in extenso of radio emissions). In such a case, MHD numerical approach is not sufficient, and particle numerical simulations represent a very powerful tool for analyzing the various mechanisms of energy dissipation via a self-consistent approach. The different steps of the energy conversion (including both ions and electrons) taking place through a collisionless shock will be shortly reviewed. Numerical results are presented with a particular emphasis on the electron acceleration mechanisms and on the corresponding numerical technics under common use. The paper is arranged as follows: Section 2 summarizes the main characteristics of a collisionless shock; the main sources of particle acceleration through the shock will be identified in Section 3 (so-called macroscopic processes) and in Section 4 (microscopic processes). Conclusions will be presented in Section 5; a particular emphasis will be given on the necessity for choosing and adapting carefully the numerical code as function of the microphysical effects one intends to simulate.

2 Main characteristics of collisionless shocks

Our understanding of the dynamics of collisionless shocks has been greatly improved thanks to succesfull space missions (in particular the ISEE bi-satellite mission which has collected an impressive amount of detailed local data on the Earth's bow shock), and to a very strong effort involved simultaneously in numerical simulations and motivated initially by these space missions. Presently, we will summarize the main characteristics of the shock by using a few snapshots in order to keep a tutorial approach of the intricated shock structures. The sketch of Figure 1 represents the profile of the magnetic field when crossing the terrestrial collisionless shock. It results from experimental data gathered on board of space mission (ISEE); V_{sw} indicates the direction of the solar wind, and the different profiles are represented versus the shock angle θ measured between the normal to the shock front and the upstream magnetic field ($\theta = (\mathbf{n}, \mathbf{B}_o)$).

2.1. One main characteristic is the collisionless nature of the shock since the associated scalelength of the field jump is much less than the mean free path of particles in the solar wind; this means that any energy exchange will take place through wave-particle interactions.

2.2. Main difficulties in the analysis appear due to the fact that the shape of the shock, its dynamics and all related macroscopic and microscopic phenomena change drastically versus a large number of plasma parameters and of geometrical factors. Two parameters are of particular importance: the Mach regime (which leads to so-called subcritical and supercritical shocks) and the propagation angle θ (so called quasi-perpendicular and quasi-parallel shocks). The first point, related to the critical Alfven Mach number MA^*, is associated with the fact that the sources and the nature of the energy dissipation changes according to the shock regime: Viscosity is dominant for $MA > MA^*$, (super-critical shocks) and becomes weak for $MA < MA^*$ (sub-critical shocks). For the shocks of concern herein, the fast magnetosonic Mach number is the relevant number for fast shocks. The last point is illustrated in Figure 1: quasi-perpendicular (or Q-$\perp$) shocks, defined for $45° < \theta < 90°$, have a transition region characterized by a clear and well pronounced jump in the magnetic field (and associated plasma momenta) over a narrow scalelength, while quasi-parallel (or Q-$\parallel$) shocks, defined for $0° < \theta < 45°$, have a very turbulent transition region more difficult to identify and spreading over a very large scalelength.

2.3. Foreshock upstream of the front: This pattern is due to the presence of field-aligned beams emitted from the front and flowing upstream in the direction opposite to the solar wind. Indeed, particles carried by the solar wind suffer some intricated interactions with the fluctuating fields when approaching the shock front (Sections 3 and 4). Some of these reach locations in the front where they are reflected, can easily escape from the front and stream back freely into the incoming solar wind. This leakage of electrons takes place from the tangential point between the upstream B field and the shock front. Energetic supra-thermal ions can also escape from the shock, but since they move more slowly than

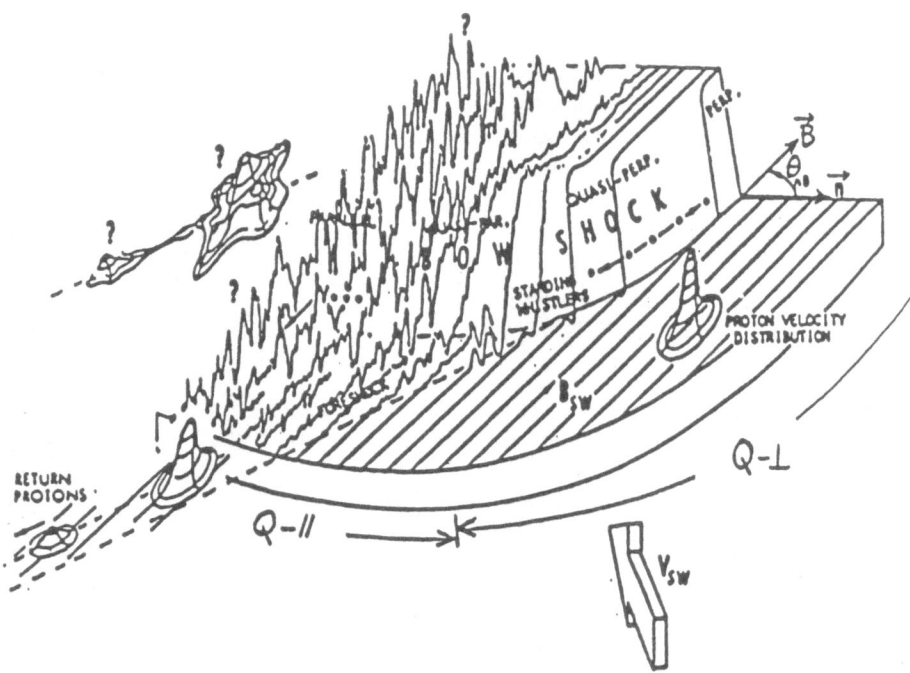

Fig. 1. Sketch of a collisionless shock as manifested in the Earths curved bow shock. Unshocked interplanetary direction B_{sw} (also named B_o in the simulations referred later in this report) is indicated in the foreground field "plateform". Field magnitude is plotted vertically. The superimposed 3D sketches represent solar wind proton thermal properties as distribution in velocity space; V_{sw} is the direction of the solar wind flow. The foreshock defines the region when particles are accelerated back from the shock into the solar wind (from Greenstadt and Fredricks, 1979).

electrons, the ion foreshock will be located behind the electron foreshock. Particle beams are formed and interact with the incoming solar wind plasma; instabilities are excited and are convected back towards the shock.

In terms of snapshots, collisionless shocks can be considered as:

a) An antenna: a precursor (whistler electromagnetic mode) is often evidenced in oblique shocks and seems to play a dominant role as the angle θ varies progressively from quasi-perpendicular to quasi-parallel directions (Pantellini, 1992).

b) An energy converter, which converts the kinetic energy of the incoming particles of the solar wind (upstream) into particles thermal energy (downstream). This conversion takes place via various processes (Sections 3 and 4) and over different time and space scales for ions and electrons.

c) A source of energetic particles as evidenced both in the upstream region (streaming of particles in the foreshock region) and in the downstream region (see b) above).

d) A source of entropy jump: the conversion from kinetic energy (upstream) to thermal energy (downstream) takes place by irreversible processes. The entropy jump results from strong changes in the distribution functions both for ions and electrons, which lead to the triggering of microinstabilities as discussed in Section 4.

An appropriate description of collisionless shocks, where waves-particles interactions play an important role, require the inclusion of kinetic effects. Since the overall dynamics is mainly controlled by ions, numerical codes where ions, at least, are described as particles are necessary. "Hybrid" codes, where ions are included as individual particles and electrons as massless fluid, and "particles" codes (including both ions and electrons as individual particles) are quite satisfactory for this aim .

3 Macroscopic wave-particle interactions through a collisionless shock

A collisionless (magnetosonic) shock can be considered both as a magnetic and an electric barrier. Electrons and ions react differently to this double barrier, because of their different masses. Although we focuss herein on the electrons dynamics, it is necessary to summarize the behavior of ions which play an important role in the overall shock dynamics.

3.1 Ion dynamics

Incoming (upstream) ions suffer different acceleration mechanisms according to the angle θ:

(i) the shock drift mechanism (or more simply, drift mechanism) accelerates a particle as the particles nearly helical orbit intersects the shock discontinuity during the process of reflection or transmission at a shock encounter (Amstrong et al., 1985). This drift mechanism is most efficient for quasi-perpendicular shocks ($45° < \theta < 90°$), (Decker and Vlahos, 1985, 1986). Energy gains due to drift are fast ($<$ a few tens of gyroperiods) and can be fairly large (≈ 10 and more times the initial energy) but in the absence of a return mechanism (e.g. pitch angle scattering for instance), particles will escape the shock and never return: these contribute to the ion-foreshock.

More precisely, ions divide into two populations when crossing the shock front. A large fraction succeeds to pass through the front and suffer some heating: these correspond to the so-called directly transmitted ions. Other ions (smaller fraction) have not enough energy and are reflected against the barrier. During their reflection, they suffer a combined effect of both the electric and magnetic fields at the front, and gain an important acceleration ($\mathbf{E} \times \mathbf{B}$ drift) which guides them into a very coherent motion. Then, they gain enough energy to overcome the barrier and succeed to reach the downstream region. The coherent motion forces the reflected ions to describe a ring in the velocity space. Let us remind that the density of reflected ions is high (very low) in supercritical (subcritical) regime, and increases when the velocity (i.e. the regime) of the shock increases.

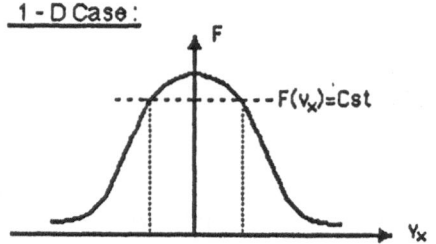

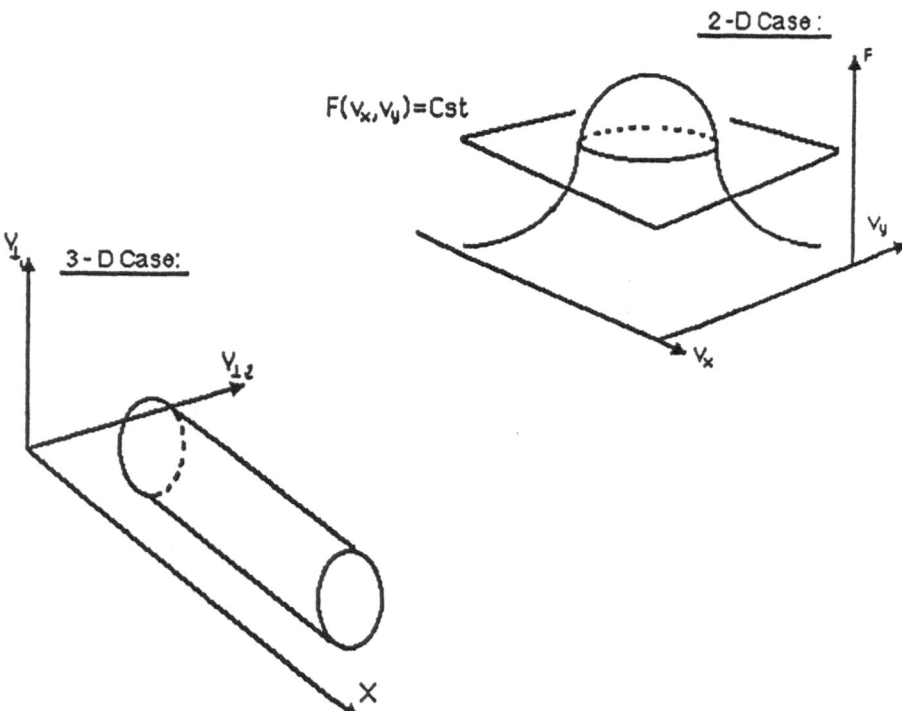

Fig. 2a For a 1-D distribution function a cut at certain value of $F_i(v)$ = constant allows to access to 2 points; for a 2-D distribution function, a cut at certain value of $F_i(v_x, v_y)$ = constant defines a circle. Following this iso-value versus the distance x through an unperturbed plasma leads to a tube (3-D case).

An overall picture of the ions dynamics is illustrated on Figure 2 issued from the post-processing of numerical data obtained with a 2-D full particle electromagnetic code (Lembege and Savoini, 1992; Savoini and Lembege, 1994). Figure 2a (3-D case) represents the reference frame used in the simulation box: x-axis is the main direction of the shock propagation, and y is the direction parallel to the shock front (planar shock). Figure 2b represents the progressive deformation of an iso-tube of the ion distribution function F_i from the upstream region (right hand side) through the shock front and to the downstream region (left hand side): ions are followed within a tube defined by $F_i(v_{\perp 1}, v_{\perp 2}) = $ constant versus the real axis x; $v_{\perp 1}$ and $v_{\perp 2}$ are the velocities components of the particles perpendicular to the local magnetic field. Indeed, the magnetic field changes its amplitude and its direction with respect to the upstream (unperturbed) field when crossing the shock. The upstream iso-tube becomes splitted into two parts at the shock front: one part is characterized by an enlarged tube with a cycloidal deformation. This tube corresponds to directly transmitted ions with an increased heating; the cycloidal deformation corresponds to the enlarged cyclotron ion motion in the increased magnetic field. The second part is given by the ion ring described by reflected ions and evidenced in the figure; a recent video-movie has clearly shown the progressive destruction (by relaxation) of the ion ring when penetrating further into the downstream region (Section 4).

(ii) The first order Fermi or diffusive shock acceleration mechanism is a statistical process in which particles undergo spatial diffusion along field lines and are accelerated as they scatter back and forth across the shock, thereby being compressed between scattering centers fixed in the converging upstream and dowstream flows (Drury, 1983). Energy gains per diffusive cycle (consisting of diffusion from upstream to dowstream and back again), are the largest at turbulent shocks for quasi-parallel ($0° < \theta < 45°$) or parallel shocks ($\theta = 0°$), (Tsurutani and Lin; 1985).

3.2 Electron dynamics

In contrast with ions, electrons have a completely different answer to the double barrier formed by the shock. For the purpose of clarity, let us consider separately the role of the magnetic and electric fields.

(i) Magnetic field

For almost perpendicular shocks, electrons have no particular dynamics; they cannot stream freely along the B field and suffer only an adiabatic compression in a first approach (Feldman and al., 1983a, 1983b). However, in a second and refined approach, this adiabaticity can be broken in various cases: (i) when microinstabilities at the shock front are strong enough, or when the Mach regime of the shock is high (Tokar et al., 1986). In summary, this means that adiabaticity is not the only mechanism for perpendicular heating.

(ii) Electric field

Over the ion scale, electrons are strongly magnetized and suffer important drifts ($E \times B$ drifts due to the gradients of magnetic field, of density, of temperature etc...), while ions can be considered as unmagnetized. This difference is quite

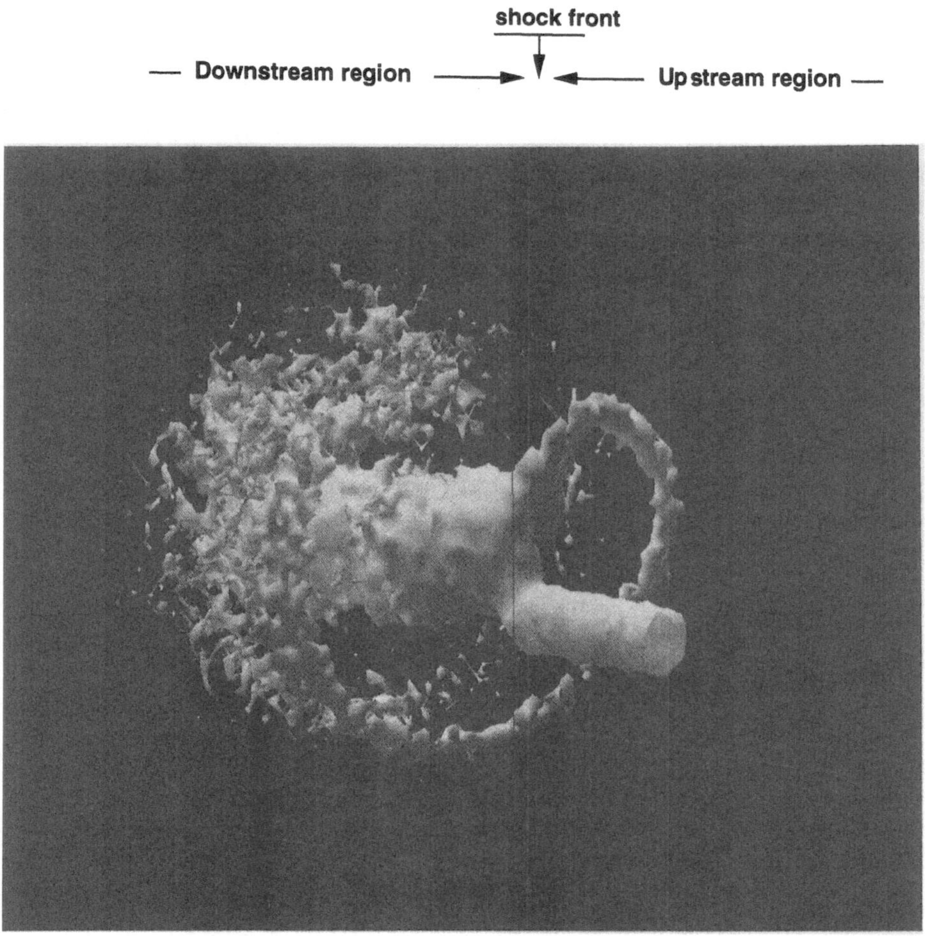

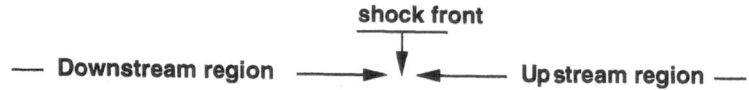

Fig. 2b Iso-tube of ion distribution function deduced from 2-D full particle electromagnetic simulations $F_i(v_{\perp 1}, v_{\perp 2})$ = constant versus x, as explained in case 3-D of Figure 2a. Upstream and downstream regions of the shock are on the right and left hand sides of the Figure, respectively. In this Figure, data of the simulations have been y-averaged; the x-axis is the main direction of propagation of the shock wave (from B. Lembege and P. Martel, 1992).

appropriate to feed self-consistently an electrostatic field due to space charge effects. Then, electrons may suffer an important acceleration by the macroscopic electric field which builds up at the shock front, as clearly evidenced by Forslund and Shonk (1970) with 1-D full particle electrostatic simulations. Because of the large trapping potential associated with the shock, electrons have a marked increase in pressure behind the shock with secondary partial trapping regions asociated with large amplitude oscillations downstream. The importance of this electric field on the acceleration and heating of electrons has been confirmed in recent 1-D and 2-D full particle, electromagnetic simulations by Savoini and Lembege (1994).

A theoretical approach has been performed by Balikin et al. (1989, 1993), Balikin et Gedalin (1994), and Krasnosel'skikh et al. (1994) on the role of the electric field inhomogeneity. As mentioned above, electrons are usually considered as magnetized. However, they can become unmagnetized (breakdown of adiabaticity) in an electric field (perpendicular to B_c field), whose spatial inhomogeneity is strong enough so that its typical scalelength remains much larger than the electron Larmor radius. In such conditions, an electron can be efficiently accelerated by $E \times B$ drift the across the B field; this leads to an effective transfert of cross-field potential into downstream electron gyration energy.

An extensive study has allowed to follow the changes in the electrons dynamics versus the angle θ, with both 1-D (Lembege and Dawson, 1987, 1989a) and 2-D full particle electromagnetic code (Savoini and Lembege, 1994). The amplitude of the electric field and its scalelength at the shock front increases, as the direction of propagation varies from perpendicular to oblique. This change is due to that ions and electrons behave differently not only in direction perpendicular but also parallel to B. Partial reflection of both particle species take place in both directions and leads to some particle accumulation; in particular, a parallel electric field builds up. Under this field effect, energetic electrons can be formed parallel to B and lead to the formation of a nonthermal electron tail or to a flat-top distribution as illustrated in Figure 3.

In summary, most of the ram energy is transferred to ions when $\theta = 90°$ (Section 3). As θ decreases, the ram energy is progressively transferred partially to ions and to electrons; while ions continue to suffer strong acceleration and heating in the plane perpendicular to the B field, electrons develop a similar dynamics but mainly along the direction parallel to B. Electron beams are formed by these interactions and important deformations result in the electron distribution function. Recent numerical simulations (Liewer et al., 1991 for 1-D simulations; Savoini et Lembege, 1994 for 2-D simulations) have shown that electrons suffer a pre-heating in the whistler precursor; this heating is strongly dependent on the ratio $v_{\phi\parallel}/v_{the}$, where $v_{phi\parallel}$ and v_{the} are the parallel phase velocity of the whistlers and the electron thermal velocity, respectively.

However, interactions of electrons with macroscopic fields are not the only source of energetic particles. Other mechanisms take place and are summarized in Section 4.

4 Microscopic wave-particles interaction through a collisionless shock

The previous section has emphasized the interactions of particles with macroscopic E and B fields, in particular around the shock front which plays the role of an energy converter. All these interactions lead to strong changes in the

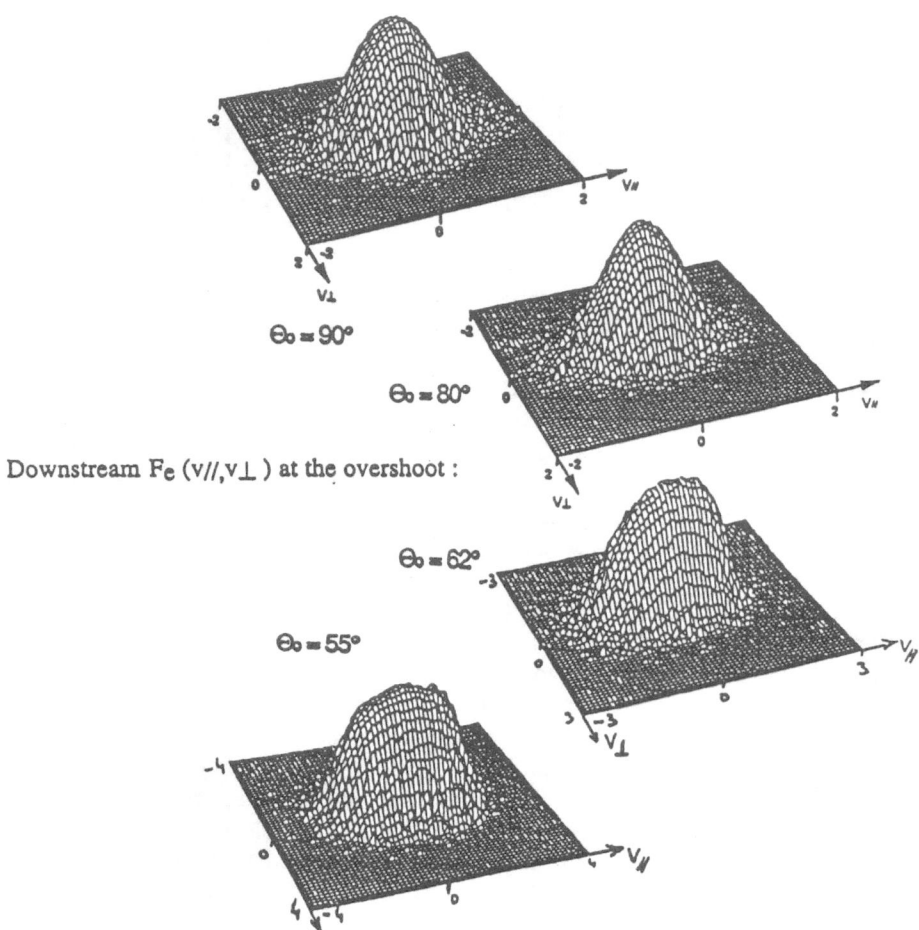

Downstream F_e $(v//, v\perp)$ at the overshoot :

Fig. 3 Variation of the electron distribution function $F_e(v_\perp, v_\parallel)$ measured at the over-shoot of the shock (first maximum of the magnetic field profile located immediately behind the front) for different directions θ of the shock normal with respect to the unperturbed upstream field; $v_\perp$ and $v_\parallel$ are the velocities components of particles perpendicular and parallel to the local B field, respectively . Results are issued from 2-D full particle electromagnetic simulations (from Savoini, 1991).

particle distribution functions which, in return, are sources of microturbulence (fluctuating fields) and affect the particle dynamics (kinetic effects).

Numerical simulations represent a very powerful tool to identify the different dissipation mechanisms which can take place through a shock front. Two kinds of energy dissipation are usually invoked: resistivity and viscosity. MHD approach allows only a phenomenological description of these effects via fixed constants introduced in MHD codes. However, kinetic analysis based on the use of hybrid and/or particle codes allow (i) to include these dissipation effects in a partially and/or totally self-consistent way and (ii) to determine the source-mechanisms which are at their origin. Thanks to this kinetic approach, resistivity and viscosity can be determined in a collisionless plasma in a more precise way. For the sake of clarity, we will distinguish two kinds of free energy sources which result from the energy conversion through a shock front; these are intrinsically linked and are illustrated on Figure 4.

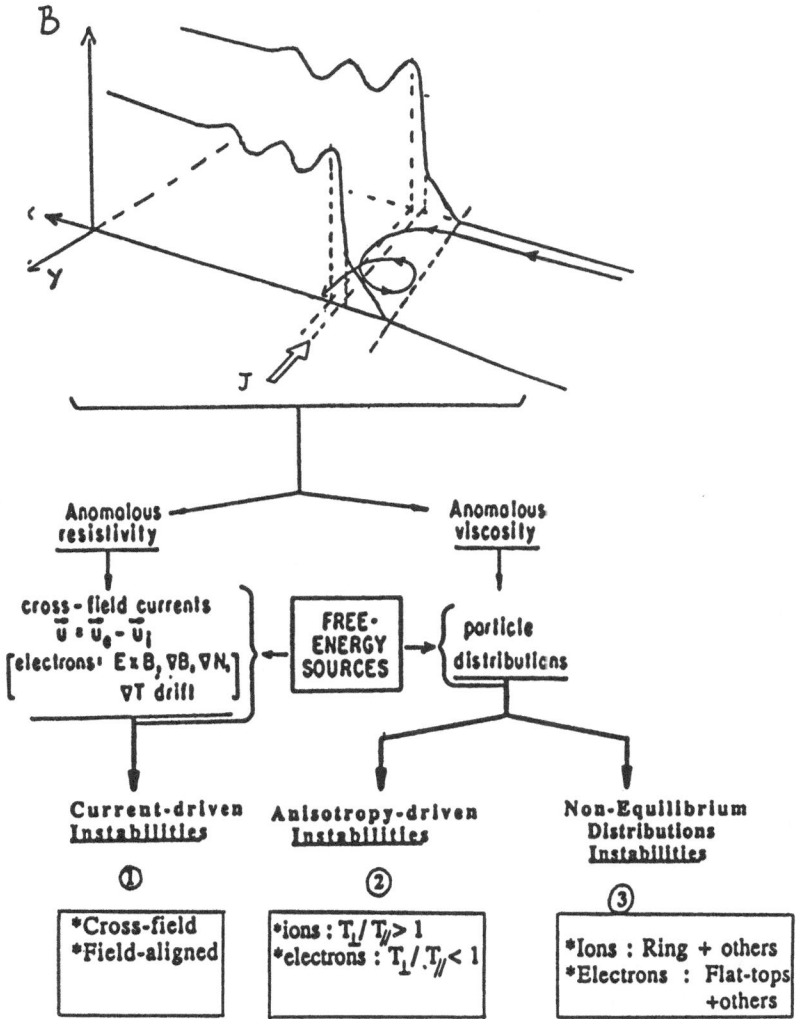

Fig. 4 Identification of the various free-energy sources created through the front of a collisionless shock and which are at the origin of three different groups of instabilities. The sketch on the top of the figure shows the space profile of the magnetic field with a typical trajectory of a reflected ion; J is the diamagnetic current supporting the magnetic field jump at the front.

4.1 Diamagnetic current

At the shock front, large jumps both in the electric and magnetic fields (transverse components) are supported by an important diamagnetic current mainly carried by electrons. This current can be large enough to be above a certain threshold where some instabilities are triggered. A certain number of candidates, field-aligned and/or cross-field current-driven instabilities, are excited within certain spectra and vary acccording to their respective threshold; these waves can play a more or less important role according to their growth rate, time for triggering , etc... In other words, the energy of the current (which plays the role of the energy source) is converted into wave energy. In return, these waves interact with electrons which gave birth to them. A diffusion of particles result both in the velocity space (equivalent to particles heating), and in the real space (equivalent to a decrease of the local current). This decrease is equivalent to a resistance which builds up in the current circuit of the shock front. Anomalous resistivity raises up in a self consistent way as a consequence of wave-particle interactions (instead of binary collisions for a neutral gas shock).

Two-dimensional full particle electromagnetic numerical simulations are quite appropriate to study such effects and have shown that, for conditions of the terrestrial shock, resistive dissipation leads to a poor electron heating and does not affect the overall dynamics of the shock (Lembege and Savoini, 1992). The numerical procedure under use is summarized in Figure 5. The reference set includes again the x-axis as the direction of the shock propagation, and y is the direction parallel to the shock front (which is planar). Two B_o-configurations can be considered for simulating a perpendicular shock (which is the simplest case):

a) In the first case, the upstream magnetostatic field B_o is along z; then, the main current is aligned along the y-axis and lies within the simulation plane (x,y). As a consequence, current-driven instabilities are fully included and this configuration allows to study their impact on the shock dynamics. This procedure is valid provided that characteristics of the simulation (size of the box and time length of the run) are correctly adapted to involve the instability features (wavelength and growth rate) estimated from a separate study (dispersion relation). This configuration corresponds to the so-called resistive case.

b) In the second case, B_o is inside the simulation plane (along y), and the main current and associated current-driven instabilities are outside this plane. All resistive effects based on these instabilities are excluded in such simulations (so called non-resistive case).

For the resistive case, Figure 5 shows clearly the presence of instabilities which propagate along the shock front (y-axis) and which are at the origin of the front rippling; for the plasma conditions of concern, these have been identified as lower-hybrid frequency waves. In contrast, the shock does not exhibit any rippling in the non-resistive case. Except this change in the orientation of the static upstream field, conditions for plasma and for the simulation run are identical in both cases. The comparison between both results allow to identify the

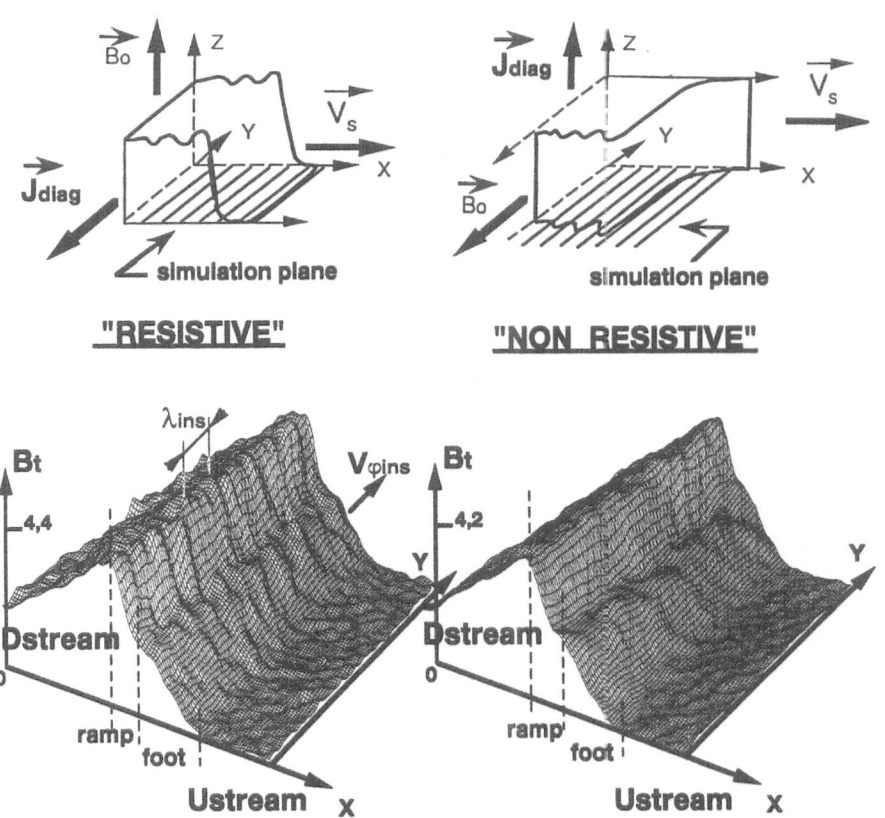

Fig. 5 Results of 2-D full particle electromagnetic simulations for a perpendicular shock wave. The top part shows the two types of the upstream B-field configuration leading to the so-called "resistive" and "non-resistive" cases; the lower part shows an enlarged view of the corresponding self-consistent magnetic field around the shock front (from Lembege and Savoini, 1992)

sources of resistive effects in a self-consistent way and to estimate quantitatively their impact on the shock dynamics.

Similar applications to shocks in solar physics can be performed in the future with such codes. On the other hand, some efforts have been also invested to analyze in detail such instabilities and to estimate their efficiency for creating energetic electrons. Basically, two groups of simulations have been performed, devoted respectively to cross-field currents and field-current instabilities:

(i) Cross-field current instabilities have been studied in two frequency ranges: one, devoted to low frequencies (lower-hybrid range), has been analyzed with a 2-D full particle electromagnetic code with an initial electron-ion drift (Winske and al., 1985). Instabilities grow up to a level high enough for nonlinear effects to become important (wave saturation); as an example, for high β plasma conditions, instabilities propagate at angles approaching $\theta_{\text{inst}} = 0°$ and suffer a saturation by electrons which get a parallel acceleration. The other study is devoted to the high-frequency range (ion-acoustic range), and has been analyzed by Dum et al. (1974) with a 2-D full particle electromagnetic code but applying a constant current J_{ext}, i.e. runs were not self-consistent.

More generally, low-energy electrons that have gyroradii comparable to or smaller than the shock thickness can interact with the electrostatic turbulence (lower-hybrid waves) driven by the cross-field driven currents in the shock transition. As a result of wave-particle interactions, these electrons will diffuse in velocity space and can gain hundreds of keV in energy at quasi-perpendicular shocks (Lampe and Papadopoulos, 1977; Tanaka and Papadopoulos, 1983)

(ii) Field-current instabilities have been also analyzed in the ion-acoustic regime (Thomsen et al., 1983) and in the whistler regime (Tokar et al., 1984); important electron diffusion takes place. Most of these studies are helpful to know quantitatively the characteristics of such instabilities and their contribution in the formation of energetic electrons; however, these need to be reinserted and performed, if possible, within the full framework of a collisionless shock. This type of study is one of the challenges for extended numerical simulations in the near future.

4.2 Changes in the particle distribution functions

In this case, the free energy sources are provided by the non-Maxwellian character of the distribution functions that particles acquire when crossing the shock front (Section 3). First, particle distributions become strongly anisotropic in temperature ($T_\perp/T_\parallel$ >1 for ions, and $T_\perp/T_\parallel$ < 1 for electrons) at the front, and may excite some electromagnetic instabilities when relaxing progressively with further penetration into the downstream region; distributions return back to an isotropic state. Second, non-equilibrium distributions are formed at the front region and are appropriate to excite some instabilities, too (Figure 4). It is out of the scope of this report to list or to analyze all instabilities expected in a collisionless shock; the reader is invited to refer to previous reviews on this topics (Wu et al., 1984; Akimoto et al., 1985; Papadopoulos, 1985; Lembege, 1990). As an illustration, the ion ring (reflected ions) is progressively destroyed when penetrating further in the downstream region, as shown in Figure 2b.

A very important effort has been already invested to identify the possible candidates for instabilities. Again, self-consistent analyses of such instabilities within the frame of collisionless shocks have been rather limited. The reason is mainly due to computer ressources since these instabilities are very different in wavelength, in frequency, in growth rate, etc. and are mixed in the downstream region; then, these require completely different time and space scales for sampling.

4.3 Discussion

Various numerical attemps have been made in solar physics, to explain electron energization and the formation of electrons beams. These approaches have been based either on the direct interactions of particles with $\mathbf{E}$ and $\mathbf{B}$ fields (as in Section 3), or on the presence of non-equilibrium ion distribution functions (as in Section 4.2). Three typical examples can be given; the two last cases (ii) try to account for the simultaneous energization of electrons and ions:

(i) The first case is related to reflected ions which can be considered locally, in the foot region immediately upstream of the front, as ion beams (Papadopoulos, 1981; Vaisberg et al., 1983; Krasnoselskikh et al., 1985). When interacting with background plasma, this beam drives low-frequency waves $\omega_{ci} < \omega < \omega_{ce}$ propagating almost perpendicular to $\mathbf{B}_o$. These waves are absorbed by magnetized electrons due to the resonance $\omega = k_z v_z$. A non-Maxwellian tail results which corresponds to electrons accelerated along the $\mathbf{B}$ field. In return, these hot electrons interact with the background plasma and drive high-frequency Langmuir oscillations with $\omega = \omega_{pe}$ up to the high level $W_L/n_o T_e \approx 10^{-5} - 10^{-4}$. The conversion of plasma waves into electromagnetic waves is caused by the induced scattering of plasma waves on ions ($\omega \approx \omega_{pe}$) or by merging of two Langmuir waves ($\omega \approx 2\omega_{pe}$).

(ii) The second case tries to account for the simultaneous energization of electrons and ions. Indeed, let us remind that before the launch of the Solar Maximum Mission (SMM) spacecraft in 1980, it was believed that electrons with energies up to ≈ 100 keV were produced in a first phase process and relativistic protons and electrons were generated in a second process, which begins several minutes after the first process and lasts for several tens of minutes. Hence, the acceleration of protons and electrons to relativistic energies was thought to be a rather slow process taking more than several minutes. However, after the launch of the SMM and Hinotori (Tanaka, 1987) spacecrafts, it has been shown that protons and electrons are simultaneously accelerated within 1s up to ≈ 20 GeV for protons and up to ≈ 100 MeV for electrons. The hard X-ray and γ-ray quasi-periodic bursts in the 1980 June 7 solar flare showed the first evidence of simultaneous acceleration of protons and electrons process (Forrest and Chupp, 1983; Kane and al., 1983, 1986). Two different approaches have been proposed:

- the first approach is based on the direct interaction of particles with macroscopic $\mathbf{E}$ and $\mathbf{B}$ fields, by using 1-D full particle electromagnetic simulations (Ohsawa and Sakai, 1988). A strong non-stochastic acceleration occurs in a rather strong magnetic field such that $\omega_{ce} > \omega_{pe}$. The time needed for the proton acceleration is of the order of the ion-cyclotron period and is quite short (much less than 1 s for solar plasma). The electron acceleration time is shorter than the ion-cyclotron period. These shock mechanisms could explain three important features of particle acceleration observed in the impulsive phase of solar flares:

first, the acceleration occurs simultaneously to protons and electrons; second, the acceleration is rapid (within 1s); and third, protons and electrons are accelerated to relativistic energies. A similar numerical approach has been also given by Lembege and Dawson (1989b, 1989c).

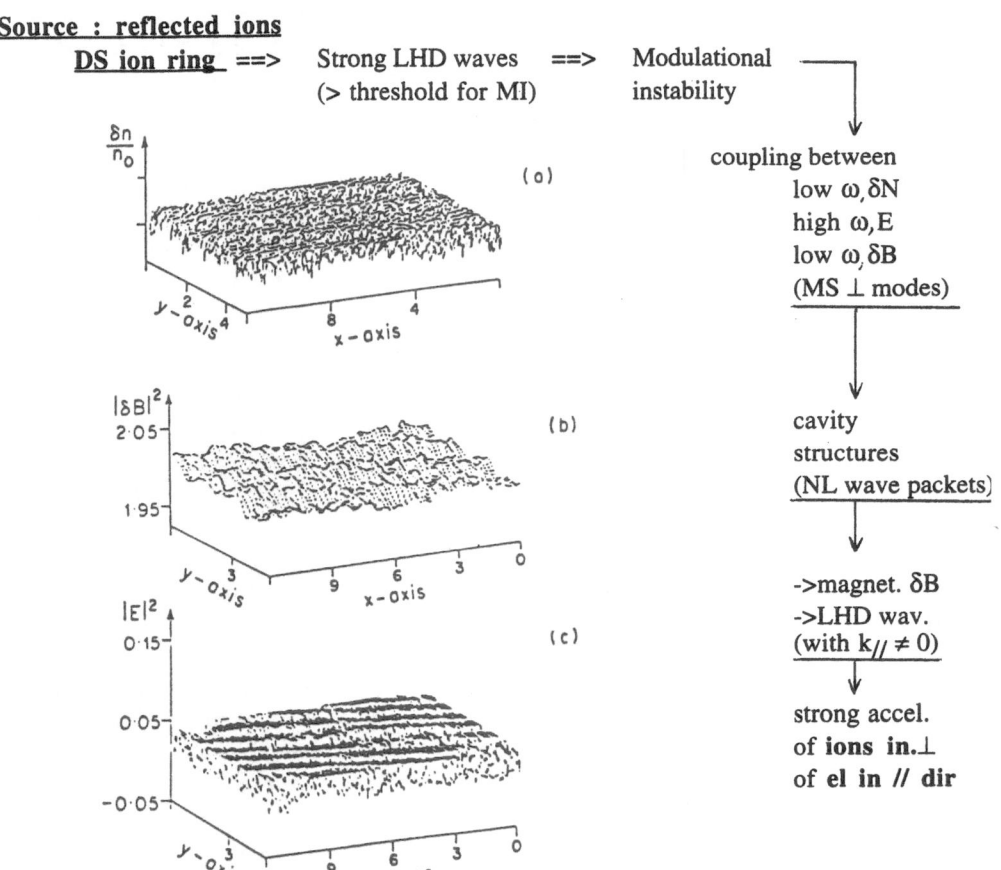

Fig. 6 Two-dimensional plots versus x and y of the particle density fluctuations, of the magnetic field and electric field energy; results issued from 2-D full particle electromagnetic simulations (from Bingham and al, 1991).

- the second approach is related again to reflected ions, as shown by Bingham et al. (1991) who have used 2-D full particle simulations (Figure 6). The input ingredients of the simulation consist of an ion ring relaxing in a background plasma, as that evidenced in the shocked downstream region; this differs from the above case (i) using the beam feature of reflected ions evidenced in the foot region. Strong lower-hybrid frequency waves are excited as ion ring relaxes (Mikailovski, 1974). Their level is strong enough to be above the threshold of themodulational instability; cavity structures occur as a result of wave collapse. This leads to a coupling between low-frequency density fluctuations δN, high-frequency electric field fluctuations δE and low-frequency magnetic field fluctuations δB. The modulational instability not only results in the creation of magnetic structures slowly varying with time but also in the appearance of waves with $k_\parallel \neq 0$ in the lower-hybrid wave spectrum. Cavity correspond to wave packets within which the wave energy is concentrated. Then, two different types of interactions result from simulations: high-frequency electric fields lead to efficient electrons acceleration along the B field, i.e. electron beams result, and low-frequency magnetosonic waves lead to efficient acceleration of ions perpendicular to B. While magnetic fluctuations are responsible for ions acceleration, waves from lower-hybrid spectrum propagating obliquely to B_o with $k_\parallel \approx (1/3v_{the})\omega_{lh}$ will be absorbed by resonant electrons, and a saturation of pump wave takes place. In order to avoid any confusion, it is important to precise that, in the present case, lower-hybrid cavitons are cigar-shaped aligned with a major axis parallel to B_o, in contrast with common Langmuir cavitons which are pan-cake shaped with a minor axis parallel to B_o. Present mechanisms allow to evidence the advantage of a simultaneous energization of both, electrons and ions, which can get 10-50 keV and 20 MeV regimes, respectively. This mechanism could also explain the simultaneous observation of hard X-rays and γ-rays, as mentioned above.

5 Conclusions

In summary, one can define a two-component scenario (Figure 7) for the formation of high-energetic particles by collisionless shock waves: one related to the so-called "macroscopic" processes, involving the direct interactions of particles with macroscopic fields at the shock front; these interactions lead to the formation of free-energy sources through the shock. The other part of the scenario includes all the different microinstabilities excited by these free-energy sources: these are "microscopic" processes. Both processes require the inclusion of kinetic effects. However, the relative importance of macroscopic and microcopic processes strongly depends on the Mach regime, geometrical effects (propagation angle), β-plasma conditions, and needs to be studied in detail. Numerical simulations reveal to be quite appropriate tools for such a challenge. At this level, the numerical code of concern needs to be chosen with a particular care:

a) MHD codes are excluded in the present case where formation of energetic particles is of main interest.

b) Hybrid codes (including ions as particles and electrons as a massless fluid) are quite appropriate to study the overall dynamics of the shock in regimes where ions play a dominant role. All time and space scalelength of electrons are neglected since $m_e=0$; then, the dynamics of the shock can be easily followed

1) Scenario :

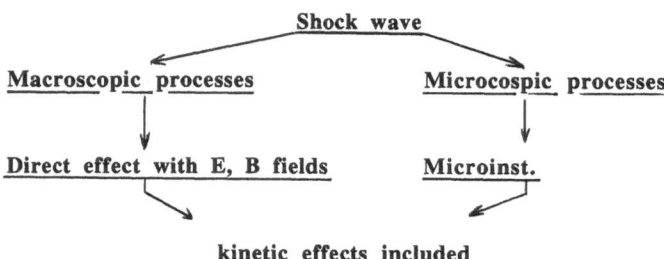

2) What numerical codes ? :

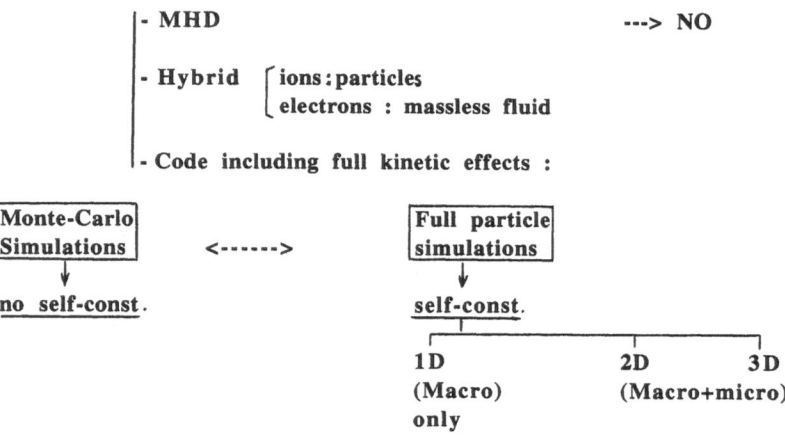

Fig. 7 Summary of the scenario including the two main types of wave-particle inter-actions, and of the simulation codes commonly used at the present time.

during long time runs covering tens of ion gyroperiods. However, such types of codes will fail for accounting for any phenomena related to electron kinetic effects.

c) Particle codes are quite appropriate to follow simultaneously ions and electrons. Two approaches are complementary:

The first approach, corresponding to Monte-Carlo simulations, follows the dynamics of particles within prescribed profiles of E and B fields. The self-consistency is lost but such a code provides helpful informations at reasonable costs, on the mechanisms for particle energization and their location through the shock profile. In addition, it is possible to switch on or off some effects (as the particle diffusion as in Veltri et al.(1990); this procedure allows to estimate quantitatively the relative impact of a given effect on the shock itself.

The other approach, corresponding to full particle codes, is the most appropriate solution but can be quite costly. At this level, a careful approach is again necessary. One-dimensional codes will exclude any resistive effects supported by cross-field instabilities (a second dimension is necessary); however, these will be quite adapted to study the first component of the scenario quoted above (interactions with macroscopic fields) in a self-consistent way. In addition, these interactions can be analyzed in details via parametric studies in θ, β, M_A etc.... Two-dimensional codes are adapted to include instabilities by a self-consistent way as illustrated by a few examples in Section 4. Parametric studies based on the orientation of the upstream magnetic field, the size of the simulation box (adapted in order to eliminate/to include certain instabilities), and variations in plasma parameters are quite rich procedures for gathering informations. However, the analysis of different instabilities within the full frame of a collisionless shock and through a self-consistent approach is still a very challenging task; on the same, studies focussed on curvature effects of collisionless shocks through a self-consistent approach have not been resolved yet. Computer runs with 3-D full particles codes are now possible, thanks to the rapid increase in computers performances both on vectorized and multi-parallel machines. Although they still stay in very limited number and are restricted to relatively small sizes of simulation boxes at the present time, these codes represent very promising tools for further analysis.

References

1. Akimoto K., K. Papadopoulos and D. Winscke, J., Plasma Phys., 34, 445, 1985.
2. Amstrong, T.P., M.E. Pesses and R.B. Decker, in Collisionless Shocks in the Heliosphere: Reviews of Current Reserach, ed. by B.T. Tsurutani and R.G. Stone , in Geophys. Monog. Ser., 35, 271, 1985.
3. Anderson, R.R., G.K. Parks, T.E. Eastman, D.A. Gurnett, L.A. Frank, J. Geophys. Res., 86, 4493, 1981.
4. Balikin M., M. Gedalin, and D. Lominadze, Adv. Space Res., 9, 135, 1989.
5. Balikin M., M. Gedalin, and A. Petrukovich, Phys. Re. Lett., 70, 1259, 1993.
6. Balikin M. and M. Gedalin, Geophys. Res. Lett., 21, 844.
7. Bingham R., J.J. Su, J.M. Dawson and K.Mc. Clements, Proceedings of the Cargese Work., Nov. 1991; also UCLA Report PPG-1383 Phys. Dept. 1991.
8. Decker R.B.and L. Vlahos, J. Geophys. Res.,90 ,47, 1985.
9. Decker R.B.and L. Vlahos, The Astrophys. J., 306, 710, 1986.
10. Drury, L. O C, Report. Progr. Phys., 46, 973, 1983.
11. Dum C.T., R. Chodura, and D. Biskamp, Phys. Rev. Lett., 32, 1231, 1974.
12. Eastman T.E., R.R. Anderson, L.A. Frank, and G.K. Parks, J. Geophys. Res., 86, 4379, 1981.
13. Feldman W.C., R.C. Anderson, S.J. Balme, S.P. Gary, J.T. Gosling, D.J. Mc Comas, M.F. Thomsen, G. Paschman, and M.M. Hoppe, J. Geophys. Res., 88, 96, 1983a.

14. Feldman W.C., R.C. Anderson, S.J. Bame, J.T. Gosling, and R.D. Zwickl, J. Geophys. Res., 88, 9949, 1983b.
15. Forrest D.J. and E.L. Chupp, Nature, 305, 291, 1983.
16. Forslund and Shonk, Phys. Rev. Lett., 25, 1699, 1970.
17. Ginzburg, V.L. and V.V. Zheleznyakov, Astron. Zh. 35, 364, 1958.
18. Greenstadt E.W. and R.W. Fredricks, in Solar System Plasma Physics III, ed. by C.K. Kennel, L. J. Lanzerotti and E.N. Parker, 1979, North Holland, Amsterdam, 3.
19. Kane, S.R., K. Kai, T. Kosugi, S. Enome, P.B. Landecker, and D.L. Mc Kenzie, Ap. J., 271, 376, 1983.
20. Kane, S.R., E.L. Chupp, D.J. Forrest, G.H. Share and E. Rieger, Ap. J.(letters), 300, L95, 1986
21. Krasnosel'skikh V.V., E.N. Kruchina, G. Thejappa and A.S. Volokitin, Astron. Astrophys., 149, 323-329, 1985.
22. Krasnoselskikh V.V., M. Balikin, M. Gedalin, and B. Lembege, Advances in Space Scien., Cospar meeting, Hambourg, in press, 1994
23. Lampe M. and K. Papadopoulos, Astrophys. J., 212, 886, 1977.
24. Lembege B., in Physical Processes in Hot Cosmic Plasmas, Eds. W. Brinkmann et al., Kluwer Academic Publishers, Netherlands, 81-139, 1990.
25. Lembege B. and J.M. Dawson, Phys. Fluids, 30, 1110, 1987.
26. Lembege B. and J.M. Dawson, Phys. Rev. Lett., 62, 2683, 1989a.
27. Lembege B. and J.M. Dawson, Phys. Fluids, 1, 1001, 1989b.
28. Lembege B. and J.M. Dawson, Report DT/CRPE/1173, 1989c
29. Lembege B. and P. Savoini, Phys. Fluids, 4(11), 3533, 1992
30. Liewer P., V.K. Decyk, J.M. Dawson, and B. Lembege, J. Geophys. Res., 96, 9455, 1991.
31. Maxwell A., and M. Dryer, Solar Phys., 73, 313, 1981.
32. Maxwell A., and A.R. Thompson, Astrophys. J., 135, 138, 1962.
33. Mc Lean D.J., in Radio Physics of the Sun, eds. M.R. Kundu, T.E. Gergely, Reidel, Dordrecht, 223, 1980.
34. Mc Lean, D.J. and D.J. Nelson, Izv. Vyssh. Uchebn. Zaved. Radiofizika 20, 1359, 1977.
35. Mikhailovskii, Theory of Plasmas instabilities, Vol. 1, Consultants bureau N.Y., 1974
36. Nelson G.J. and R.D. Robinson, Proc. Astron. Soc. Australia, 2, 370, 1975.
37. Ohsawa Y. and J.I. Sakai, The Astrophys. J., 332, 439, 1988.
38. Pantellini, Phd. Thesis, 1992.
39. Papadopoulos K., in Plasma Astrophysics, eds. T.D. Guyenne, G. Levy, ESA SP-161, 313, 1981.
40. Papadopoulos K., in Collisionless Shocks in the Heliosphere, Reviews of Current Research, Geophys. Monograph. Ser., Vol. 35, 59, Ed. B.T. Tsurutani and R.C. Stone, AGU, Washington DC, 1985.
41. Pikelner, S.B. and M.A. Gintzburg, Astron. Zh., 40, 842, 1963.
42. Savoini P., Phd. Thesis, University Paris VI, 1991.
43. Savoini P. and B. Lembege, J, Geophys, Res., Vol. 99, A4, 6609, 1994.
44. Tanaka, K., Publ. Astr. Soc. Japan, 39, 1, 1987.
45. Tanaka M., and K. Papadopoulos, Phys. Fluids, 26, 1697, 1983.
46. Thomsen M.F., H.C. Barr, S.P. Gary, W.C. Feldman, and T.E. Cole, J. Geophys. Res., 88, 3035, 1983.
47. Tokar R.L., D.A. Gurnett, and W.C. Feldman, J. Geophys. Res., 89, 105, 1984.

48. Tokar R.L., C.H. Aldrich, D.W. Forslund and K.B. Quest, Phys. Rev. Lett., 56, 1059, 1986.
49. Tsurutani, B. T., and R.P. Lin, J. Geophys. Res., 90, 1, 1985.
50. Vaisberg O.L., A.A. Galeev, G.N. Zastenker, G.N. Klimov, S.I. Nozdrachev, M.N. Sagdeev, R.Z. Sokolov, A. Yu, and V.D. Shapiro, JETP, 85, 1232, 1983.
51. Veltri P.A., A. Mangeney and J.D. Scudder, J. Geophys. Res. 90, 14939, 1990.
52. Wild J.P. and S.F, Smerd, Ann. Res. Astrom. Astrophys. 10, 159, 1972.
53. Winske D., M. Tanaka, C.S. Wu, and K.B. Quest, J. Geophys, Res., 90, 123, 1985.
54. Wu C.S., D. Winske, Y.M. Zhou, S.T. Tsai, P. Rodriguez, M. Tanaka, K. Papadopoulos, K. Akimoto, C.S. Lin, M.M. Leroy, and C.C. Goodrich, Space Sci. Rev., 37, 63, 1984.
55. Zaitsev, V.V., Izv. Vyssh. Uchebn. Zaved., Radiofisika 20, 1379, 1977.

Surprises in the Radio Signatures of CMEs

N. Gopalswamy and M. R. Kundu

Department of Astronomy, University of Maryland, College Park, MD 20742 USA

Abstract. We discuss several results regarding the relationship between coronal mass ejections (CMEs) and metric radio emissions which have changed our understanding of these phenomena considerably. Imaging observations of metric radio emissions along with coronagraph observations have been used to obtain these results. We consider the following: (i) Why slow CMEs are associated with metric type IV radio emission contrary to the earlier belief, (ii) Why shocks piston driven by the CMEs are not seen in the solar corona, (iii) Thermal radio emission from the CMEs and their implications to CME-flare relationship and (iv) Radio signatures of coronal disconnection events.

1 Introduction

Coronal mass ejections (CMEs) are large scale magnetic structures hurled from the Sun into the corona and eventually into the interplanetary medium. The CMEs are associated with a wide range of phenomena during their initiation and subsequent propagation: Flares, prominence eruptions, metric type II and type IV radio bursts, interplanetary shocks and kilometric type II bursts. In this review we are concerned with the metric radio emissions and their relationship with the CMEs. The metric radio bursts of type II and type IV are thought to be produced by nonthermal particles. The association of a CME with a type IV radio burst can be very complex, involving a wide variety of physical processes such as particle acceleration, coronal dynamics and radio emission mechanisms. We present observational evidence which seem to support the view that the origin of nonthermal particles responsible for type IV bursts is not linked to the CME speed as has been thought previously. In the case of type II bursts the nonthermal particles are believed to be produced in the front of a shock wave moving through the corona. Since the range of shock speeds inferred from radio observations is close to the range of CME speeds, the type II shocks were thought to be driven by CMEs. However, solar flares can also give rise to blast waves which can produce type II radio bursts. Since flares are often associated with CMEs, it is hard to decide whether the shock is produced by the CME or by the flare from temporal associations alone. Imaging radio observations play a crucial part in deciding whether the shock is generated by a flare or by a CME. We shall address this controversy in detail.

We have found a new way of studying the flare-CME relation from radio observations alone, when a CME is observed in the free-free radiation from the main body of the CME. The continuous coverage due to high time resolution of radio observations enables us to make accurate conclusions about flare-CME timing. A problem closely related to the moving type IV bursts is the radio signature of a coronal disconnection event. Coronal disconnection events represent detachment of coronal magnetic structures from the sun, resulting in a pair of moving and a stationary structures. We describe an observation that shows radio signatures of a coronal disconnection event which is also a strong evidence for magnetic field reconnection in the solar corona. The paper is organized as follows: Section 2 discusses radio signatures of slow CMEs; section 3 presents the controversial origin of type II shocks; section 4 explains how CME-flare relationship can be studied from radio observations and the final section 5 discusses the coronal disconnection events.

2 Radio Signatures of Slow CMEs

During the *Skylab* era, Gosling et al. [1] reported that in a majority of cases, radio bursts accompanied only fast ($> 500~kms^{-1}$) CMEs. Since the coronal Alfven speed is $\sim 500~kms^{-1}$, it was thought that only fast CMEs produce MHD shock waves which accelerate charged particles resulting in type II and type IV radio emission. The CME speed was thought to decide the energy of the nonthermal electrons responsible for radio emission (Dulk [2]): slow transients ($< 400~kms^{-1}$) produce no nonthermal particles, intermediate-speed transients (400-$1000~kms^{-1}$) produce mildly relativistic electrons (~ 100 keV) and fast transients ($\geq 1000 kms^{-1}$) produce relativistic electrons (≥ 1 MeV). When new data came in from the *Solwind* and SMM-C/P instruments, radio bursts were found in association with slow CMEs and in several cases, no radio bursts were observed in association with fast CMEs. Table 1 shows a list of slow CMEs (85-350 kms^{-1}) associated with type II and/or type IV bursts. Association of metric radio bursts with slow CMEs can be attributed to the following reasons:

Table 1. Slow CMEs associated with Metric Radio Bursts

Date	Radio Burst	CME speed (kms^{-1})	Reference
Jun 27 1984	Type II, Type IV	350	Gopalswamy & Kundu [3]
Nov 12 1984	Type II	150	St. Cyr & Webb [4]
Feb 17 1985	Type II, Type IV	240	Kundu et al.[5]
Feb 22 1985	Type II	160	St. Cyr & Webb [4]
Feb 2 1986	Type IV	140	Gopalswamy & Kundu [6]
Nov 6 1986	Type IV	85	Gopalswamy & Kundu [7]

(i) Shock waves may not be the source of nonthermal particles responsible for type IV. It is known that only the advancing front-type IV bursts are associated with high speed ($> 1000 \ kms^{-1}$) shocks. In the case of expanding arches and isolated plasmoids, which are more common, there is no such association. Thus there is no basis for expecting these moving type IV bursts to be associated with shocks. Isolated plasmoids are associated with erupting filament [6] which is known to move behind the CME frontal loop. If these plasmoids detach from the filament through a reconnection process, then the same process can produce non-thermal particles that get trapped in the plasmoid to generate the moving type IV emission. Non-thermal particles may also be produced during an associated flare which get trapped in moving magnetic structures during a CME event, resulting in the expanding arch variety of type IV bursts.

(ii) CMEs that are not super-Alfvenic can still produce slow and intermediate shocks ahead of them. Kundu et al. [5] showed that the moving type IV burst can be produced by electrons accelerated in the front of a slow shock. They found that lower hybrid turbulence could be generated in the shock front and in the downstream which can accelerate electrons to high energies (up to 300 keV). Slow mode shocks have speeds above sound speed but below Alfven speed. Since the sound speed in the corona is $\sim 170 \ kms^{-1}$, one may not expect any shock formation ahead of the CMEs moving at speeds less than the sound speed. This seems to be consistent with the fact that no activity was associated with extremely slow CMEs (speed $< 50 \ kms^{-1}$) observed by the SMM-C/P.

(iii) The type II burst may be produced by a blast wave from the flare site, independent of the CME in which case, the CME speed is of no consequence. Cane and Reames [9] found that the type IV bursts occur mostly in intense long duration flares. Long duration flares are known to be associated with CMEs and thus they concluded that the type IV bursts occur only in flares accompanying CMEs. Thus, one has to search for the origin of nonthermal particles responsible for the moving type IV bursts in flares rather than in CMEs. The CME speed is important only if they are directly responsible for particle acceleration.

3 Origin of Type II Shocks: Flares or CMEs?

When the CMEs were first detected by the OSO-7 coronographs the association between CMEs and type II radio bursts were studied in three events. Stewart et al. [10, 11] reported two events that had the estimated positions of type II bursts above and near the CME leading edge. These observations were considered as evidence for shocks "piston-driven" by the CME. However, during the third event (Dec. 14, 1971), Kosugi [12] found that the type II burst was too far ahead of the moving type IV burst to have any connection to the white light cloud (CME); rather, it might have originated in a flare a blast wave [13]. Thus, the origin of shocks responsible for coronal type II bursts became controversial right from the time the CMEs were first detected. The piston-driven shock gained support during the *skylab* era due to two studies: (i) Type II/type IV radio bursts were associated only with high speed CMEs, and (ii) All radio bursts that occurred within 45 deg of the solar limb were associated with CMEs [14]. Using the *Solwind* coronagraph data, Sheeley et al [15] found that about a third of all metric type II bursts were not associated with CMEs, suggesting a different origin of type II shocks. Furthermore, the type II bursts without CMEs were found to be associated with short duration (impulsive) X-ray events confirming flare origin for these type IIs. In the SMM era, Wagner and MacQueen [16] argued that even when a CME is associated with a type II burst, the shock wave could still be flare blast wave which eventually propagates through the CME material. Is it possible to distinguish between the two types of shocks from observations?

One can think of the following characteristics for the piston-driven shocks: (i) if the shock is fast mode, the speed of the shock can exceed the CME speed only by a factor less than 2 (see e.g., [17]). (ii) the type II burst must be located at the stand-off distance from the CME leading edge. (iii) Geometrical compatibility and position angle coincidence between the shock and the CME. To test these characteristics, Gopalswamy and Kundu [17] collected all the CME events in the literature for which simultaneous imaging observations exist for type II radio bursts. They could find only a dozen events for the entire period since the *Skylab* era. Positional analysis showed that in 8 out of 12 events, the type II burst position was well behind the CME leading edge. In two of the 4 remaining cases, the type II position was above the leading edge of the CME. However, the shock and CME speeds were not compatible in these two cases: the shock speeds were greater than the CME speed by 7 and 3 times for the October 27, 1973 ([18]) and the June 27, 1984 ([3]) events respectively. During the latter event, the type II shock had an extent at least 20 times larger than the lateral size of the CME. In the remaining two cases, one had a type II that ended even before the CME onset (Feb. 10, 1986 event - Kundu and Gopalswamy [19]); in the other, the CME and the shock had the same speed while the projected onset time of the flare better agreed with the type II onset. Clearly, the available evidence seems to support the view that the type II shocks originate only in flares, consistent with the well known statistical result that about 90% of all type II bursts can be associated with flares [20]. We would like to emphasize that the number of events is not large enough to make a rigorous statistical analysis.

Why don't we, then, observe the CME-driven shocks? This question is also relevant to Kahler et al's [21] result that about a third of fast CMEs are not associated with type II bursts. Sheeley et al. [15] suggested two reasons: (i) some of the fast CMEs did not attain super Alfvenic speeds, and (ii) some piston-driven shocks could not excite type II bursts in the lower corona. A more likely reason could be that the shock formed at heights inaccessible to the ground based radio instruments ([22]). The pre-eruption state of a CME is a coronal helmet streamer which is usually located at a height $\sim 1.5~R_\odot$ ([23]). To this, we need to add the distance over which the CME accelerates to become super-Alfvenic and the stand-off distance. Fig.1 shows an SMM CME with the pre-eruption streamer at a height of $3R_\odot$. If this CME drives a shock, it should start from a height $> 3R_\odot$ and hence could not have been detected by any radio instrument since

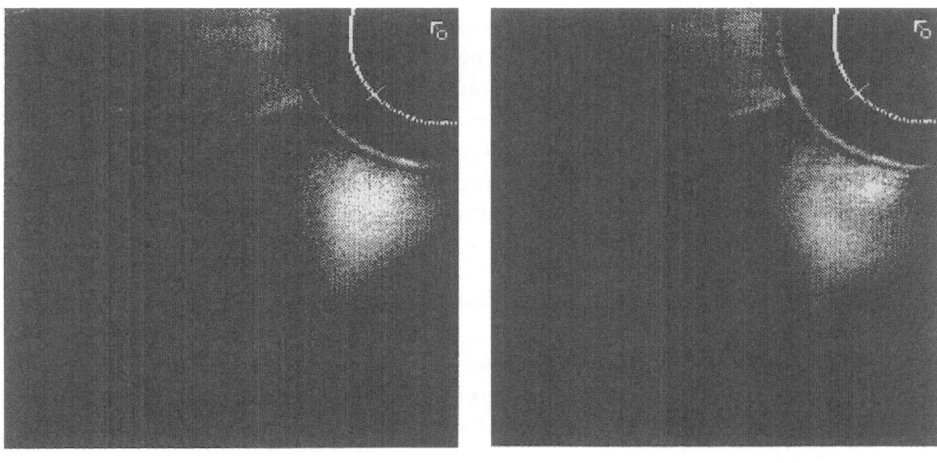

AUG 18, 1980   11:43   11:54

Fig. 1. A time sequence of SMM-C/P images showing a CME forming out of a streamer in the high corona (courtesy: A. Hundhausen)

the coronal plasma frequency at this height is less than the ionospheric cut-off frequency.

In practice, there may be a range of heights at which the shock is formed. If the initial height of the streamer is smaller and the CME accelerates faster, it may form a shock in the upper corona. Robinson et al. [24] found that 75% of the metric type II bursts associated with interplanetary type II bursts had a starting frequency < 45 MHz. Thus, there may be a small number of metric type II bursts which may be due to CME shocks and continue into the interplanetary medium producing interplanetary type II bursts. Robinson et al. [25] obtained another interesting result: about 20% of all type II events contain pairs of type II bursts. One may speculate that in these cases both a flare blast wave and a piston-driven shock are formed within the corona and hence are observed by the radio instruments. Cane and Stone [27] and Robinson et al. [26] found that the CME-associated type II bursts had a lower starting frequencies which could be taken to imply piston driven shocks. However, a detailed statistical analysis by Cane and Reames [9] suggests that the starting frequency might tell us about the impulsiveness of the flare, rather than suggesting that type II shocks are driven by CMEs. In fact this analysis is consistent with the theoretical and numerical analysis result that the distance from the site of energy release at which a shock forms has an inverse dependence to the rate of energy input (see [28] for a review).

To observe a CME driven shock, we need observations of type II bursts in the frequency range 2-20 MHz, which is an unexplored region between the coronal observations (above ionospheric cut-off) and the interplanetary observations (below 2 MHz). The SORS radio spectrometer on board CORONAS-I operates in the frequency range 0.1-30 MHz. A combination of SOHO coronagraphs and radio instruments in WAVES experiment of WIND has the capability of clarifying some of the unsolved issues in the relation between radio bursts and CMEs. In order to increase the number of events with simultaneous imaging in radio and white light, SOHO coronagraphs and the Giant Meterwave Radio Telescope (GMRT) will be of help in the near future.

4 CME-Flare Relationship from Radio Observations

During early days of CME studies, it was believed that CME was driven either by flares or by prominence eruptions because of a significant association of these two near surface activities with CMEs. CMEs without prominence cores and the fact that the cavity and the frontal loop travel faster than the prominence, cannot be explained by a prominence driver. There are three arguments against a flare driver: (i) Scale sizes: the size of CME is an order of magnitude larger than the scale of flares ([31]. This can be understood from the fact that CMEs form out of large scale streamer structures while flares occur in active region magnetic fields. (ii) "Flare Poor" CMEs: statistically significant number of CMEs are known to occur without any flares [29]; these CMEs seem to form out of streamers overlying high latitude prominences ([30]). (iii) Relative Timing: detailed comparison of the height-time history of CMEs and X-ray flare light curves shows that the CME onset precedes the flares onset by tens of minutes ([30, 31]). The CME measurements are made in the high corona (above the occulting disk of the coronagraph) as compared to near surface measurement of flares, and hence there is a criticism that the onset comparisons may not be accurate. This objection has partly been overcome, thanks to the powerful combination of SMM-C/P and Mauna Loa K-coronameter data which provided CME measurements at heights as low as 1.2 $R_\odot$. Since the white light images are generally of poor cadence, there is usually a large uncertainty in CME-flare timing. It is with reference to the flare-CME-timing that a surprise came from radio observations. We had an opportunity to observe the CME and radio bursts continuously and simultaneously and hence we could establish the flare-CME relationship from radio observations alone ([32, 33]). In the following we review the results from this observation.

The CME of January 16, 1986 observed by the SMM-C/P was also imaged by the Clark Lake radioheliograph at three frequencies (73.8, 50 and 38.5 MHz). The CME was accompanied by three isolated type III bursts and a nonthermal continuum at meter-decameter wavelengths. Figure 2 (left) shows the radio CME (extended contours) at three frequencies and the location of radio bursts. White light CME had roughly the same lateral extent as the radio CME. At the position angle of the radio bursts, but behind the limb (within 20 deg) was active region 4713. Thus any flare manifestations at other wavelengths might have been occulted. However, the nonthermal continuum is a good indicator of flare and

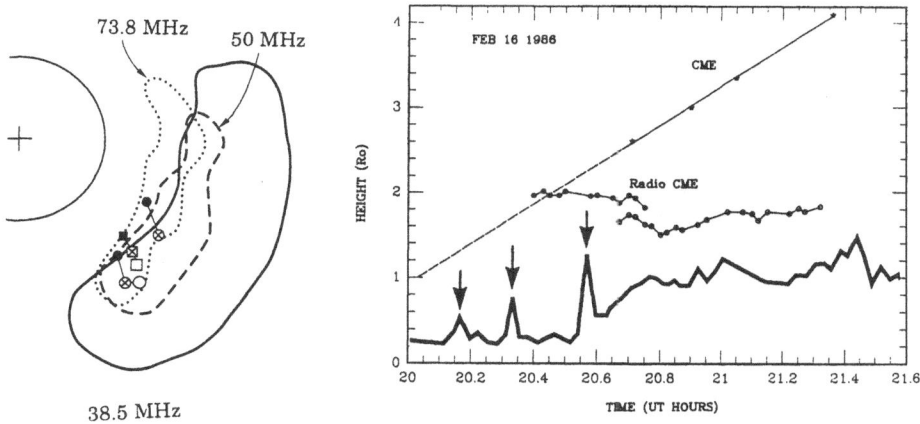

Fig. 2. left: Radio CME (extended contours) and nonthermal bursts. The filled, crossed and the open circles represent centroids of type III bursts at 73.8,50 and 38.5 MHz. The filled, crossed and open squares represent the centroids of the nonthermal continuum at 73.8,50 and 38.5 MHz respectively. Right: Height-time plots of white light (marked CME) and radio CME (curve with filled circles) and the nonthermal continuum (curve with open circles). The brightness temperature variation is shown at the bottom (thick curve). The arrows point to type III bursts on this curve and the brightness increase after the type IIIs is the nonthermal continuum. Note that the start of the radio CME coincides with the extrapolated trajectory of the white light CME.

occurs at the flash phase of the flare. Fig.2 (right) shows the height-time plot of the white light CME, the radio CME and the nonthermal continuum along with the brightness temperature of the nonthermal continuum. Note that the onset time and height of the radio CME coincides with the extrapolated trajectory of the white light CME. We see that the nonthermal continuum starts ∼ 15 min after the start of radio CME at 73.8 MHz demonstrating that flare follows the CME and hence could not have driven the CME. Here, we have compared the onset of radio CME at 73.8 MHz (corresponding to a height of 1.7 $R_\odot$). However, if the CME had started from a lower height the time delay between the CME and flare will be greater; e.g., if the CME departed from the solar surface, the delay will be ∼ 30 min. In practice, of course, we do not expect the CME to start from the surface owing to the large scale nature of the pre-eruption structures (viz., streamers). Moreover, if we allow for an initial acceleration of the CME before reaching the constant speed, we arrive at a larger delay between the onset times the CME and flare. This is the first time, the relationship between the flare and the CME onset could be studied from imaging radio observations alone, and the results are consistent with white light observations. Further details of this event could be found in [32, 33].

5 Coronal Disconnection Events

There are four possible magnetic structures that could be observed in the corona and interplanetary medium as shown in Fig. 3 (McComas et al. [34]). The familiar examples of ejected closed field structures (C) is the plasmoid observed as moving type IV radio bursts observations ([6, ?]). The U-shaped structure (D) is given the name 'disconnection event' since it appears to be 'pinched-off' from pre-existing coronal structures. The disconnection were thought to be very rare: there are only three coronagraph events reported in the literature ([35, 36, 37]) In addition, there is a report based on old eclipse observation by Cliver [38]. However, a recent search in the ground based and space-borne coronagraph data has revealed that $\sim 10\%$ of CME events may be classified as coronal disconnection events (Webb, 1993 private communication) and hence it is important to understand their nature.

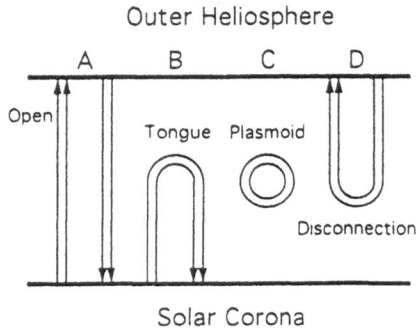

Fig. 3. Schematic diagram showing the four fundamental magnetic structures expected in the corona and interplanetary medium (after McComas et al.)

From the radio point of view, the detached structures have to be inferred from the radio emission produced by nonthermal particles trapped in these structures. The coronal disconnection event of 1986 February 13 seem to correspond to the case of B & D in Fig. 3 and the Clark Lake radioheliograph observed a nonthermal continuum as expected. This is the first detection of radio signature of a coronal disconnection event (Gopalswamy & Kundu [37]).

Figure 4 shows a Clark Lake radioheliogram at 73.8 MHz at 18:58 UT on February 13, 1986. The bright contours over the west limb is the radio continuum. The weaker, inner source close to the limb is a noise storm and is not related to the disconnection event. The locations of the U-shaped structure as

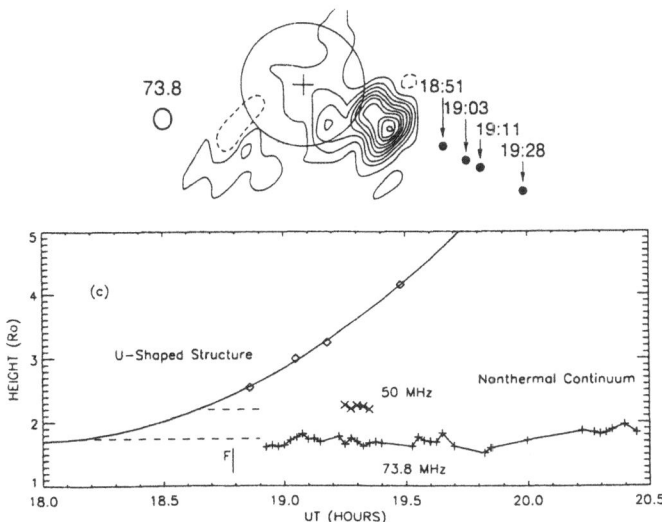

Fig. 4. top: Relative location of the U-structure (filled circles) and the continuum (contour). Bottom: Height-time plot of the U-structure and the continuum

observed by the SMM-C/P instrument are shown by black dots at the same position angle as the centroid of radio continuum. The nonthermal continuum was bursty and lasted for at least $\sim$ 2 hrs. Fig. 4 shows the height-time plot of the U-shaped structure and the nonthermal continuum at 73.8 and 50 MHz. The observed heights of the U-shaped structure (denoted by diamond symbols in Fig. 4) can be fit to a parabola which gives a starting height of $\sim$ 1.7 $R_\odot$. The second derivative of the trajectory gives an acceleration of $\sim$ 0.11 kms^{-2} which is about twice the value reported by McComas et al. [36]. The speed of the U-shaped structure is $\sim$ 195 and 220 kms^{-1} at the height of 73.8 and 50 MHz continuum sources. The coincidence of the height of 73.8 MHz radio source and the starting height of the U-shaped feature suggests that the radio emission started with the disconnection. In this scenario, the reconnection has to be an ongoing process with the reconnection region moving to larger heights at later times, somewhat similar to the formation of post-flare loops. Numerical experiments designed to simulate the disconnection events do suggest outward motion of the reconnection region ([39]). From the transient time structure of the continuum and the observed source parameters, the acceleration rate seems to be $\sim$ 1.3 $\times 10^{24}$ $ergs^{-1}$, somewhat smaller than the typical energy release in solar flares.

References

[1]   Gosling, J. T. et al: Solar Phys. **48** (1976) 389
[2]   Dulk, G. A. : in *Radio Physics of the Sun, (eds.) M. Kundu & T. E. Gergely* (1980) 419

[3]   Gopalswamy, N. and Kundu, M. R.: Solar Phys. **111** (1987) 347
[4]   St.Cyr, O. C. and Webb, D. E.: Solar Phys. **136** (1991) 379
[5]   Kundu, M. R. et al.: ApJ **347** (1989) 505
[6]   Gopalswamy, N. and Kundu, M. R.: Solar Phys. **122** (1989) 91
[7]   Gopalswamy, N. and Kundu, M. R.: in *Max91/SMM Solar Flares: Observations and Theory* (eds.) R. M. Winglee and A. L. Kiplinger (1990) 139
[8]   Gopalswamy, N. and Kundu, M. R.: ApJ (Lett.) **365** (1990) L31
[9]   Cane, H. V. and Reames, D. V.: ApJ **325** (1988) 895
[10] Stewart, R. T. et al. : Solar Phys. **36** (1974) 219
[11] Stewart, R. T. et al. : Solar Phys. **36** (1974) 203
[12] Kosugi, T.: Solar Phys. **48** (1976) 339
[13] Solar Phys. **28** (1973) 495
[14] Munro, R. H. et al.: Solar Phys. **61** (1979) 201
[15] Sheeley, N. R. Jr. et al.: ApJ **279** (1984) 839
[16] Wagner, W. J. and MacQueen, R. M.: Astron. Astrophys. **120** (1983) 136
[17] Gopalswamy, N. and Kundu, M. R.: in *Particle Acceleration in Cosmic Plasmas* AIP conference proceedings 264 (eds.) G. P. Zank and T. K. Gaisser (1992) 257
[18] Gergely, T. E. Kundu, M. R. and Hildner, E.: ApJ **268** (1983) 403
[19] Kundu, M. R. and Gopalswamy, N.: in *Eruptive Solar Flares* (eds.) Z. Svestka, B. V. Jackson and M. E. Machado Springer Verlag (1992) 269
[20] Dodge, J. C.: Solar Phys. **42** (1975) 445
[21] Kahler, S. et al.: Solar Phys. **93** (1984) 133
[22] Bougeret, J.-L.: in *Collisionless Shocks in the Heliosphere: reviews of current research* (eds.) B. T. Tsurutani and R. G. Stone (1985) 13
[23] Koutchmy, S. and Livshits, M.: Space Sci. Rev. **61** (1992) 393
[24] Robinson, R. D. Stewart, R. T. and Cane, H. V.: Solar Phys. **91** (1984) 159
[25] Robinson, R. D. et al.: Solar Phys. **97** (1985) 145
[26] Robinson, R. D. et al.: Solar Phys. **105** (1986) 149
[27] Cane, H. V. and Stone, R. G.: ApJ **282** (1984) 339
[28] Steinolfson, R. S.: in *Collisionless Shocks in the Heliosphere: reviews of current research* (eds.) B. T. Tsurutani and R. G. Stone (1985) 1
[29] Wagner, W. J.: in *Proc. Kunming Workshop on Solar Physics and Travelling Interplanetary Phenomena* (eds.) C. de Jager and C. Biao (1985) 978
[30] Hundhausen, A. J.: in *Many faces of the Sun* (eds.) K. Strong, J. Saba and B. Haisch (1994) (in press)
[31] Harrison, R. A. and Sime, D. G.: Astron. Astrophys. **208** (1989) 274
[32] Gopalswamy, N. and Kundu, M. R.: Solar Phys. **143** (1993) 327
[33] Gopalswamy, N. and Kundu, M. R.: Adv. Space Res. **13** (1993) (9)75
[34] McComas, D. J. et al.: Geophys. Res. Lett. **21** (1994) 1751
[35] Illing, I. E. and Hundhausen, A. J.: JGR **88** (1983) 10210
[36] McComas, D. J., Phillips, J. L., Hundhausen, A. J. and Burkepile, J. T.: Geophys. Res. Lett. **18** (1991) 73
[37] Gopalswamy, N., Kundu, M. R. and St.Cyr, O. C.: ApJ Lett. **424** (1994) L135
[38] Cliver, E. W.: Solar Phys. **319** (1989) 122
[39] Linker, J. A., Van Hoven, G. and McComas, D. J.: JGR **97** (1992) 13733

The SOHO Mission

Bernhard Fleck

Space Science Department of ESA, ESTEC, P.O. Box 299, NL-2200 AG Noordwijk,
The Netherlands

Abstract: SOHO, the Solar and Heliospheric Observatory, is a joint ESA/NASA
mission to study the sun from its interior to, and including, the solar wind in
interplanetary space. It is currently scheduled for launch in 1995. In this paper
a mission overview is given, comprising scientific objectives, payload, spacecraft,
operations, and data and ground system.

1 Mission Overview

The space-based Solar and Heliospheric Observatory (SOHO) is a joint venture
of ESA and NASA within the frame of the Solar Terrestrial Science Programme
(STSP), the first "Cornerstone" of ESA's long-term programme "Space Science
— Horizon 2000". The principal scientific objectives of the SOHO mission are
a) to reach a better understanding of the structure and dynamics of the solar
interior using techniques of helioseismology, and b) to gain better insight into
the physical processes that form and heat the Sun's corona, maintain it and give
rise to its acceleration into the solar wind. To achieve these goals, SOHO carries
a payload consisting of 12 sets of complementary instruments which are briefly
described in Section 2. An assessment of the current status of the scientific
subjects to be addressed by SOHO together with a description of the science
policy evolution leading to this comprehensive observatory concept can be found
in Huber and Malinovsky-Arduini (1992).

The SOHO spacecraft (Figure 1) is scheduled for launch on an Atlas II-AS in
July 1995. After a transfer phase of ca. 4 months it will be injected into a halo
orbit around the Sun-Earth L1 Lagrangian point, about 1.5×10^6 km sunward
from the Earth, where it will be continuously pointing to Sun center with an
accuracy of 10". Pointing stability will be better than 1" over 15 min intervals.
The halo orbit will have a period of 180 days and has been chosen because,
1) it provides a smooth Sun-spacecraft velocity change throughout the orbit,
appropriate for helioseismology, 2) is permanently outside of the magnetosphere,
appropriate for the "in situ" sampling of the solar wind and particles, and 3)
allows permanent observation of the Sun, appropriate for all the investigations.
The Sun-spacecraft velocity will be measured with an accuracy better than 0.5
cm/s. SOHO is being designed for a nominal operational lifetime of two years,
but on-board consumables will be sized for up to six years. The total mass of
the spacecraft is 1850 kg, and 1150 W power will be provided by the solar panels.
The payload weighs about 640 kg and will consume 450 W in orbit.

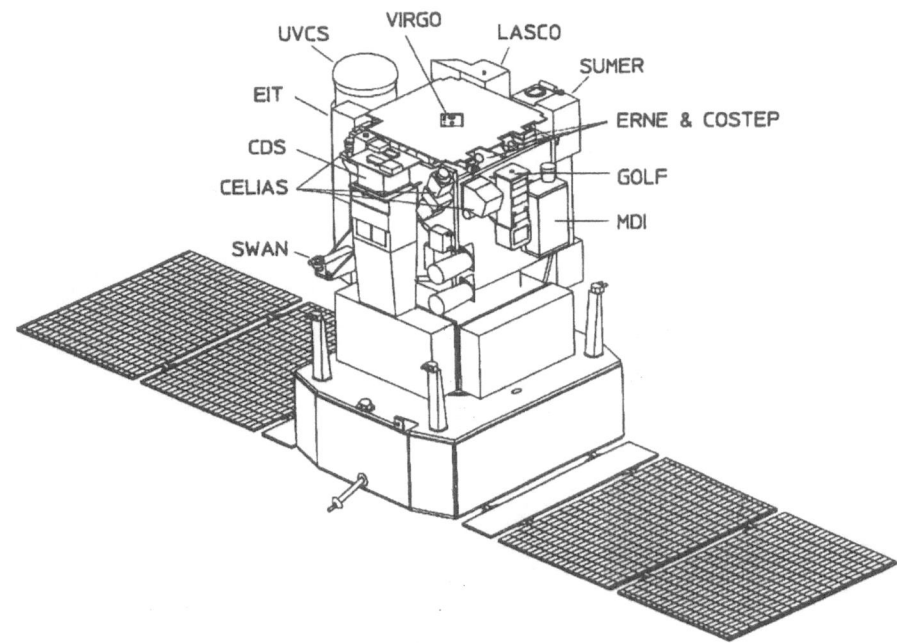

Fig. 1. SOHO spacecraft schematic view.

2 Scientific Payload

The scientific payload of SOHO comprises 12 sets of instruments (see Table 1) which can be divided into three main groups, according to their area of research: helioseismology instruments, solar corona instruments, and solar wind "in-situ" instruments. Detailed instrument descriptions by the consortia involved can be found in the ESA volumes edited by Domingo and Guyenne (1989) or Mattok (1992).

2.1 Helioseismology instruments

GOLF (Global Oscillations at Low Frequencies), using a very stable sodium vapour resonance scattering spectrometer, aims to obtain observations of the global solar velocity field ($l \leq 3$) with a sensitivity of better than 1 mm/s over the whole frequency range from $0.1\,\mu$Hz to 6 mHz (periods from 3 min to 100 days). It will also measure the long-term variations of the global average of the line-of-sight magnetic field with a precision of 1 mG.

    VIRGO (Variability of solar IRradiance and Gravity Oscillations) will perform high sensitivity observations of p- and g-mode solar intensity oscillations with a 3-channel sun-photometer measuring the spectral irradiance at 402, 500, and 862 nm, and with the 12 resolution elements Luminosity Oscillations Imager

Table 1. SOHO Payload

Investigation	Measurements	Technique	Bit Rate (kb/s)
HELIOSEISMOLOGY			
GOLF	Global Sun velocity oscillations (ℓ=0-3)	Na-vapour resonant scattering cell, Doppler shift and circular polarization	0.160
VIRGO	Low degree (ℓ=0-7) irradiance oscillations and solar constant	Global Sun and low resolution (12 pixels) imaging, active cavity radiometers	0.1
MDI/SOI	Velocity oscillations with harmonic degree up to 4500	Doppler shift and mag. field obs. with Michelson Doppler Imager	5 (+160)
SOLAR ATMOSPHERE REMOTE SENSING			
SUMER	Plasma flow characteristics (T, density, velocity) chrom. through corona	Normal incidence spectrometer, 50-160 nm, spectral res. 20000-40000, angular res. 1.5"	10.5 (+21)
CDS	Temperature and density: transition region and corona	Normal and grazing incidence spectrometers, 15-80 nm, spectr. res. 1000-10000, angular res.≈3"	12 (+22.5)
EIT	Evolution of chromospheric and coronal structures	Images (1024 x 1024 pixels in 42' x 42') in the lines of He I, Fe IX, Fe XII and Fe XV	1 (+26.2)
UVCS	Electron and ion temp., densities, velocities in corona (1.3-10 $R_\odot$)	Profiles and/or intensity of several EUV lines between 1.3 and 10 $R_\odot$	5
LASCO	Structures evolution, mass, momentum and energy transp. in corona (1.1-30 $R_\odot$)	1 internal and 2 externally occulted coronagraphs Spectrometer for 1.1-3 $R_\odot$	4.2 (+26.2)
SWAN	Solar wind mass flux anisotropies and its temporal variations	Scanning telescopes with hydrogen absorption cell for Ly-α light	0.2
SOLAR WIND 'IN SITU'			
CELIAS	Energy distrib. and comp. (mass, charge, charge state) of ions (0.1-1000 keV/e)	Electrostatic deflection, time-of-flight measurements, solid state detectors	1.5
COSTEP	Energy distribution of ions (p, He) 0.04-53 MeV/n and electrons 0.04-5 MeV	Solid state and plastic scintillator detector telescopes	1.01
ERNE	Energy distribution and isotopic composition of ions (p - Ni) 1.4-540 MeV/n and electrons 5-60 MeV	Solid state and scintillator crystal detector telescopes	1.01

(LOI) ($l \leq 7$). The relative accuracy of these data will be better than 1 ppm (for 10 s integration time). VIRGO will also measure the solar constant with an absolute accuracy of better than 0.15% using two different types of absolute radiometers (PMO6 and CROM). Both GOLF and VIRGO lay particular emphasis on the very low frequency domain of low order p- and g-modes which penetrate deep into the solar core and which are difficult to observe from the Earth because of noise effects introduced by the Earth's diurnal rotation and the transparency and seeing fluctuations of the Earth's atmosphere.

The Solar Oscillations Imager (SOI) focusses on the intermediate to very high degree p-modes. By sampling the Ni I 676.8 nm line with a wide-field tunable Michelson interferometer, SOI will provide high precision solar images (1024×1024

pixels) of the line-of-sight velocity, line intensity, continuum intensity, longitudinal magnetic field components, and limb position. It can be operated in a full
disk mode (2" equivalent pixel size) to resolve modes in the range $3 \leq l \leq 1500$ or
in a high-resolution mode (0.62" pixel size) to resolve modes as high as $l = 5000$.
SOI will run 4 different observing programs: The "structure program" provides
a continuous 5 kbits/s data stream of various spatial and temporal averages of
the full disk velocity and intensity images. It runs at *all times*. The "dynamics
program" operates during 60 consecutive days each year with continuous high
rate telemetry (+ 160 kbits/s). Each minute, a full disk velocity image and either
a full disk intensity image or a high resolution velocity image will be transmitted. The "campaign program" will be conducted during 8 consecutive hours each
day when the high rate telemetry is available. This is a very flexible operations
mode to perform a variety of more narrowly focussed scientific investigations
(e.g. studies of meso- and supergranulation, active region seismology, etc.). Finally, the "magnetic field program" will provide several real-time magnetograms
per day for planning purposes and correlative studies.

2.2 Coronal instruments

The solar atmosphere remote sensing investigations are carried out with a set
of telescopes and spectrometers that will produce the data necessary to study
the dynamic phenomena that take place in the solar atmosphere in and above
the chromosphere. The plasma will be studied by spectroscopic measurements
and high resolution images at different levels of the solar atmosphere. Plasma
diagnostics obtained with these instruments will provide temperature, density,
and velocity measurements of the material in the outer solar atmosphere.

In the past, the coronal observations with the best spatial and spectral resolution covered only limited spectral ranges, and were obtained from rockets, i.e. on
a *snapshot* basis. SOHO will provide what is now vitally needed: an extended
and concerted investigation of the physical structures, the dynamics and evolution of the transition region and corona on a *synoptic* basis. In addition, given
the capability to make both (*remote-sensing*) coronal and (*in-situ*) interplanetary measurements, SOHO will help to establish the nature of the relationship
between conditions in the regions of origin of the solar wind and the observed
flow properties at 1 AU, in particular the elusive acceleration.

SUMER (Solar Ultraviolet Measurements of Emitted Radiation) is an UV
telescope equipped with a normal incidence spectrometer to study plasma flows,
temperatures, densities, and wave motions in the upper chromosphere, transition region and lower corona with high spatial (1.5") and high time resolution
(typically 10 s) by measuring line profiles and intensities of UV lines in the range
from 500 to 1600 Å. The spectral coverage varies between 20 and 44 Å with a
spectral resolving power of $\lambda/\Delta\lambda = 18.800 - 40.000$. With SUMER, it should be
possible to measure velocity fields in the transition region and corona down to
1 km/s.

At shorter wavelengths (150 to 800 Å) CDS (Coronal Diagnostic Spectrometer), a Walter II grazing incidence telescope equipped with both a normal inci-

dence and a grazing incidence spectrometer, will measure absolute and relative intensities of selected EUV lines to determine temperatures and densities of various coronal structures. The spatial resolution of CDS is about 3", the spectral resolution varies between 2000 and 10000.

EIT (Extreme-ultraviolet Imaging Telescope) will obtain full sun high resolution EUV images in 4 emission lines (Fe IX 171 Å, Fe XII 195 Å, Fe XV 284 Å, and He II 304 Å) corresponding to 4 different temperature regimes. The wavelength separation is achieved by multilayer reflecting coatings deposited on the four quadrants of the telescope mirrors and a rotatable mask to select the quadrant illuminated by the Sun. A 1024×1024 CCD camera with an effective pixel size of 2.6" is used as detector.

UVCS (UltraViolet Coronagraph Spectrometer) is an occulted telescope equipped with high resolution spectrometers to perform spectroscopic observations of the solar corona out to 10 solar radii to locate and characterize the coronal source regions of the solar wind, to identify and understand the dominant physical processes that accelerate the solar wind, and to understand how the coronal plasma is heated. One of the gratings is optimized for line profile measurements of Ly-α, another one for line intensity measurements in the range 944 to 1070 Å.

LASCO (Large Angle and Spectrometric COronagraph) is a triple coronagraph having nested, concentric annular fields of view with progressively larger included angles. The fields of view of the three coronagraphs C1, C2 and C3 are 1.1-3, 1.5-6 and 3-30 solar radii, respectively. All three coronagraphs will use 1024×1024 CCD cameras as detectors. C1 will not only be the first spaceborne "mirror coronagraph", but it will also be the first spaceborne coronagraph with spectroscopic capabilities. It is equipped with a Fabry-Perot interferometer to perform spectroscopic measurements with a spectral resolution of $\approx 700\,\text{mÅ}$ in the lines Fe XIV 5303 Å, Fe X 6374 Å, Ca XV 5964 Å, Na D_2, and Hα. Straylight levels of C1 vary between $10^{-6}\,B_\odot$ at $1.1\,R_\odot$ and $10^{-7}\,B_\odot$ at $3\,R_\odot$. System tests of C2 produced straylight levels of about $10^{-10}\,B_\odot$ and for C3 a straylight level of about 10^{-12} was measured.

SWAN (Solar Wind ANisotropies) will measure the latitude distribution of the solar wind mass flux from the equator to pole by mapping the emissivity of the interplanetary Ly-α light.

2.3 Solar wind "in-situ" instruments

The instruments to measure "in situ" the composition of the solar wind and energetic particles will determine the elemental and isotopic abundances, the ionic charge states and velocity distributions of ions originating in the solar atmosphere. The energy ranges covered will allow the study of the processes of ion acceleration and fractionation under the various conditions that cause their acceleration from the "slow" solar wind through solar flares.

CELIAS (Charge, ELement and Isotope Analysis System) consists of three mass- and charge-discriminating sensors based on the time-of-flight technique, making use of electrostatic deflection, post-acceleration and residual energy measurements. It will measure the mass, ionic charge and energy of the low and high speed solar wind, of suprathermal ions, and of low energy flare particles. It also carries SEM (Solar Extreme-ultraviolet Monitor), a very stable photodiode spectrometer which will continuously measure the full disk solar flux in the He II 304 Å line as well as the absolute integral flux between 170 and 700 Å.

To study the energy release and particle acceleration processes in the solar atmosphere as well as particle propagation in the interplanetary medium, COSTEP (COmprehensive SupraThermal and Energetic Particle analyser) will measure energy spectra of electrons (up to 5 MeV), protons and He nuclei (up to 53 MeV/nuc). ERNE (Energetic and Relativistic Nuclei and Electron experiment), having the same scientific objectives as COSTEP, will measure energy spectra of elements in the range Z=1–30 (up to 540 MeV/nuc), abundance ratios of isotopes as well as the anisotropy of the particle flux.

3 Operations

3.1 Experiment Operations Facility (EOF)

The diagram in Figure 2 shows the basic data connections that will be present for the SOHO science operations. The SOHO Experiment Operations Facility (EOF), to be located at NASA's Goddard Space Flight Center (GSFC), will serve as the focal point for mission science planning and instrument operations. At the EOF, experiment PI representatives will receive real-time and playback flight telemetry data, process these data to determine instrument commands, and send commands to their instruments, both in near real-time and on a delayed execution basis. They will be able to perform data reduction and analysis, and have capabilities for data storage. To accomplish these ends, the appropriate experiment teams will use workstations that will be connected to an EOF Local Area Network (LAN). There will be connections from the EOF to external facilities to allow transfer of incoming data from GSFC support elements, remote investigator institutes, other solar observatories, and ESA facilities. There will also be connections for the EOF to interact with the SOHO Mission Operations Control Center (SMOCC) and other required elements at GSFC for scheduling and commanding the SOHO flight experiments. Short term and long term data storage will be either within the EOF or at an external facility with electronic communication access from the EOF.

The EOF will consist of approximately 3200 square feet of space in Building 3 of GSFC. This space is contiguous to the SOHO Mission Operations Control Center where the Flight Operations Team works. Because there is insufficient space in the EOF to house all of the personnel for the resident experiment teams, additional space in Building 26 has been provided to SOHO. In this Experiment Analysis Facility (EAF) modular workspace will be available for resident PI team members and for visiting scientists and engineers. The Building 26 space will be shared with the Solar Data Analysis Center (SDAC) and with other elements of

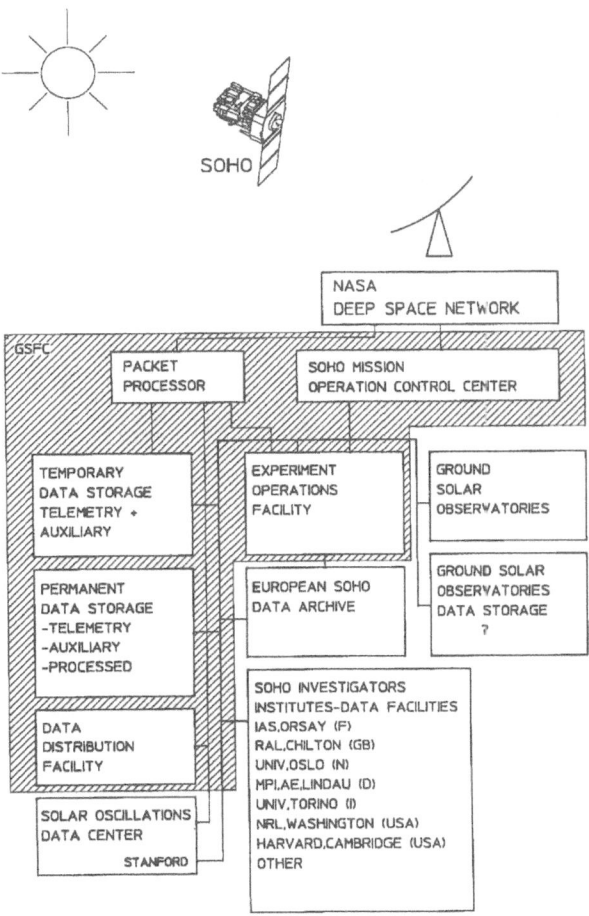

Fig. 2. SOHO ground system: basic functions for science operations

the ISTP program, and it will include a conference area. A high capacity data communications link between the EOF and the EAF is being implemented, but real-time experiment operations will not take place from the EAF.

3.2 Telemetry

The Deep Space Network (DSN) will receive S/C telemetry during three short (1.6 hrs) and one long (8 hrs) station pass per day (Figure 3). Science data acquisition during non-station pass periods will be stored on-board and played back during the short station passes. The MDI high data rate stream (160 kbits/s) will be transmitted only during the long station pass. For 2 consecutive months per year continuous data transmission, including MDI high data rate, will be supported by DSN. Whenever there is data transmission, the basic science data (40 kbits/s) will be available in near real-time at the EOF. From the EOF the SOHO investigators will control the operation of the instruments via near-real time commands. These will be verified and sent immediately to the spacecraft.

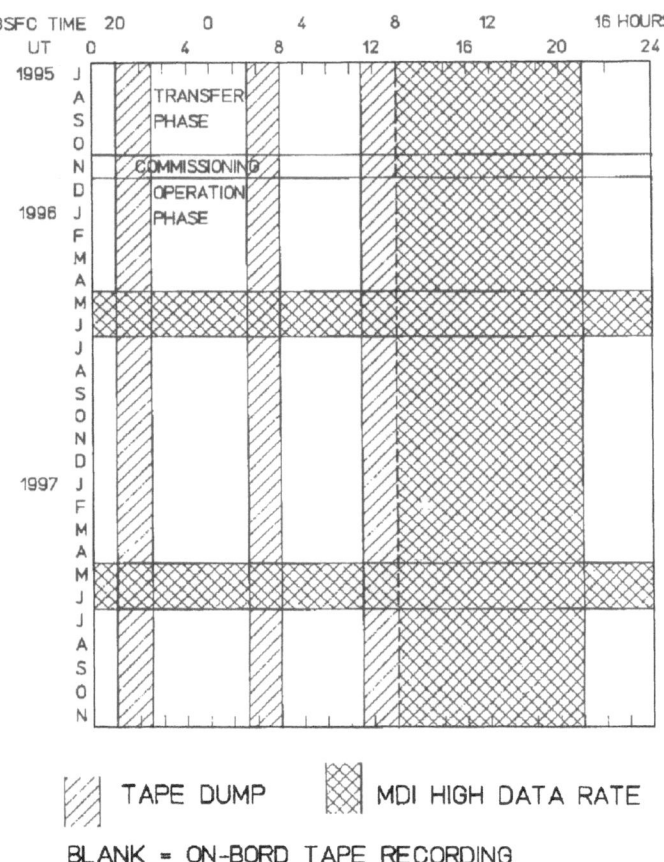

Fig. 3. SOHO telemetry and real time operation plan

The connections to the spacecraft instruments will be such that commands can be sent to the instruments for immediate execution. One will be able to decide where they want to point on the sun, what sequence of operations is desired and send those commands to the instruments.

The data, when received on the ground, will be immediately forwarded to the EOF where scientists will be able to view the current data from: their own experiment, other SOHO experiments, and certain ground based observatories.

The ability to receive and view real time data will greatly facilitate our ability to optimize the scientific return from the experiments. The electronic connections will also greatly improve our ability to do joint research with ground based observatories.

3.3 Operations policy

The Science Working Team (SWT) will set the overall science policy and direction for mission operations, set priorities, resolve conflicts and disputes, and consider Guest Investigator observing proposals. During SOHO science operations, the SWT will meet every three months to consider the quarter year starting in one month's time and form a general scientific plan. If any non-standard DSN contacts are required, the requests must be formulated at this quarterly meeting. The three-month plan will then be refined during the monthly planning meetings of the Science Operations Team (SOT), composed of those PI's or their team members at the EOF, which will allocate observing sessions to specific programs. At weekly meetings of the SOT, coordinated timelines will be produced for the instruments, together with detailed plans for spacecraft operations. Daily meetings of the SOT will optimize fine pointing targets in response to solar conditions and adjust operations if DSN anomalies occur.

3.4 Campaigns

During agreed periods one or several experiment teams and, if agreed, teams from other spacecraft or at ground observatories will run, in collaboration, observation campaigns to address specific topics. The periods of two-months continuous near real-time observation will probably be the most convenient for campaigns that require continuous observation during more than 8 hours, or that require coordination with ground observations only feasible from particular observatories around the world.

If the ground-based observatory is one that has electronic links that allow near real-time imaging transmission to and from the EOF, the coordination will be no different than if the ground based observatory were one of the SOHO experiments. If no real-time data transmission is needed or possible, the coordinated operation will be an agreed time of simultaneous observations. Generally speaking the coordinated observations with ground observatories will need a longer time lead in their planning to insure availability of the facility and coincidence of the SOHO real-time coverage.

4 Data

NASA will forward real-time science data to the EOF over the LAN to the PI workstations. Playback data will be sent to the EOF in the same manner with transmission delays from DSN (approximately 3 hours) and processing delays to turn the data around (approximately 2 hours).

4.1 Summary data

The summary data will be used both to plan observations at the EOF and to provide an overview of the observations that have been obtained from SOHO. Table 2 provides a list of the proposed contributions from each PI team. The summary data will consist of three classes of data. The first two classes will consist of a representative image from each of the imaging experiments, and key parameters from the non-imaging experiments. The third class will be a list of observing programs and start/stop times of data sequences. Together these data will provide a synopsis of solar conditions and the science programs that have been carried out by the observatory.

Table 2. Summary Data

MDI	Full disk magnetograms + full disk continuum images
EIT	Full disk images: Fe IX, Fe XII, Fe XV, and He II
UVCS	1.2-10 $R_\odot$ Ly α intensity and coronal temperature
LASCO	1.1-30 $R_\odot$ corona white light, 1.1-3 $R_\odot$ Fe XIV, Fe X, Ca XV
GOLF	Global magnetic field (6-hour mean)
VIRGO	Solar constant each day
SWAN	Heliospheric Ly α
CELIAS	Solar wind speed, heavy ion flux (5 min. averages)
CEPAC	Particle flux every 5 min
SUMER	Operation modes, time, area coverage
CDS	Operation modes, time, area coverage

The summary data will be the responsibility of the PI teams, and are to be generated as quickly as possible after receipt of data at the EOF. The data will be generated from quick-look science data and will have only preliminary calibrations performed on them. The summary data will therefore NOT be citable. An event log will be maintained in the summary data file . The event log will provide a registry of events that may be of general interest and it may also include events identified by observatories other than SOHO, but which might be relevant to the SOHO observations. Each of the instruments will have an observation program summary data file even if they have only a few entries. The observation program will have both an "as-planned" and an "as-run" data file. The EOF will hold an on-line copy of the Summary data and event time files.

The summary data to be provided by the SOHO PIs will greatly enhance the capability for joint research with scientists throughout the community. These data will be available electronically on a daily basis.

4.2 SOHO Data System

A SOHO Data System will be built which will provide access, visualisation, and analysis tools for solar data available in electronic form. This data system will have the following features:

- Provides network access to all Solar data archive centres
- Holds on-line data for SOHO planning (updated through daily ftp)
- Reads several data formats (FITS, CDF, Yohkoh, and others)
- Query: graphical interfaces (both IDL widgets and Mosaic forms) for data searches with all common parameters (time, position, wavelength, etc.) (of course, there will also be a command line where standard SQL commands can be submitted)
- Visualisation: tools for display, overlays, movies, grids, zoom, forward rotation, limb determination, image enhancement, etc.
- On-line library of SOHO campaigns (descriptions and data)
- User-library, on-line with uniform documentation

All SOHO data will be available in FITS format.

5 Guest Investigators

A SOHO Guest Investigator Programme has been envisaged from the onset of the STSP program. To ensure the maximum exploitation of the SOHO data in order to extract the best scientific output from the mission, and to attract special capabilities and expertise from outside the SOHO teams, selected Guest Investigators (GI's) will have the opportunity to acquire and/or analyze specific data sets, or, for some experiments, to become part of the PI teams.

Each approved SOHO GI will be attached to a SOHO experiment team, with whom to interact during the Guest Investigation. Two types of GI participation are foreseen, depending upon the nature of the SOHO experiment involved. For the coronal experiments (CDS, EIT, LASCO, SUMER, SWAN, and UVCS), GI participation will be of a traditional nature (like for SMM or Yohkoh): GI's will have priority rights for the analysis of certain datasets (either newly acquired, or from the archive), or priority rights for a certain type of analysis of datasets otherwise available to the whole SOHO experiment team. An example of the first is the study of a specific event, for example a CME, and an example of the latter is a statistical study, say a study of the magnitude of redshifts as a function of position on the solar disc.

The data from the helioseismology (GOLF, VIRGO, MDI), and from the particle experiments (CELIAS, COSTEP, ERNE) are of a totally different nature; they do not lend themselves to being split up into "events", or time intervals, each of which could be studied by different investigators. Hence the mode of participation of GI's attached to these instruments will be different. It is envisaged that, possibly for a limited period of time, approved GI's will be included as members of the PI team with the same rights and obligations as regular Co-I's. Approval of proposals for these SOHO experiments will depend on whether the proposed work adds to the expertise existing within the SOHO experiment team – an example could be the implementation of a statistically superior method of analyzing time series for the helioseismology instruments.

Currently, it is planned to have the first Announcement of Opportunity (AO) issued a few months before launch and the first round of GI investigations beginning about 1 year after launch. Further, it is planned that the AO for the SOHO GI programme will be renewed every year, until several years after the end of the mission, with a similar review cycle each time. The duration of a GI investigation is in principle one year, but the experiment teams are free to extend this time.

6 Coordinated Research

The SOHO payload has been conceived as an integrated package of complementary instruments, such that their measurements would produce a complementary set of data for the study of the phenomena in the solar atmosphere, in the solar wind, and in the solar interior. Therefore, to achieve the scientific aims of the mission, it is essential to operate the instruments in a coordinated programme and analyse the data in a correlative and cooperative spirit. Further, to maximize the scientific return of the mission, it is essential to have a close collaboration with ground-based solar observers (both radio and optical) and other space-based solar researchers (Yohkoh, Coronas, etc.). Finally, it should be emphasized that SOHO is part of STSP and that there is a large fraction of space science and plasma physics that may benefit from a coordinated approach between "in situ" observations such as obtained by Cluster, Ulysses, Geotail, Interball, Wind, or Polar and the remote and "in situ" measurements taken by SOHO.

References

1. Domingo, V., Guyenne, T.D. (eds.): 1989, The SOHO mission — scientific and technical aspects of the instruments, ESA SP-1104
2. Huber, M.C.E., Malinovsky-Arbuini, M.: 1992, *Space Science Reviews* **61**, 301
3. Mattok, C. (prep.): 1992, Coronal streamers, coronal loops, and coronal and solar wind composition, ESA SP-348

SUMER – Solar Ultraviolet Measurements of Emitted Radiation

Klaus Wilhelm

Max-Planck-Institut für Aeronomie, D-37189 Katlenburg-Lindau, Germany

Abstract. The experiment *Solar Ultraviolet Measurements of Emitted Radiation (SUMER)* is designed for the investigations of plasma flow characteristics, turbulence and wave motions, plasma densities and temperatures, structures and events associated with solar magnetic activity in the chromosphere, the transition zone and the corona. Specifically, SUMER will measure profiles and intensities of extreme ultraviolet (EUV) lines emitted in the solar atmosphere ranging from the upper chromosphere to the lower corona; determine line broadenings, spectral positions and Doppler shifts with high accuracy; provide stigmatic images of selected areas of the Sun in the EUV with high spatial, temporal and spectral resolution and obtain full images of the Sun and the inner corona in selectable EUV lines, corresponding to a temperature range from 10^4 to 2×10^6 K. SUMER will be flown on the ESA/NASA spacecraft Solar and Heliospheric Observatory (SOHO) to be launched in September 1995. SOHO will be positioned near the first Lagrangean point (L1) of the Sun-Earth system.

Keywords. EUV emission lines, coronal heating, solar wind acceleration

1 Scientific Objectives

Spectral imaging in EUV emission lines will allow us to study many physical parameters of the solar atmosphere: plasma density and temperature, abundances of species, velocity fields, topologies of the plasma structures and their time evolution. Such measurements are of prime importance for the following areas of science:

• Solar physics. Coronal heating. The existence of a corona requires a persistent energy input to compensate for radiative and conductive losses and for the solar wind expansion. The presence of any non-radiative mechanical heating mechanism should manifest itself by fluctuations, which may be detected by optical means in terms of line broadenings and Doppler shifts.

Solar wind acceleration. Modern concepts of high-speed solar wind acceleration hinge on the amount of wave energy available in the corona to drive the wind by wave pressure gradient forces, although it is also possible that the solar wind represents the accumulated effect of many small isolated acceleration sites.

Structure of the solar upper atmosphere. At the base of the corona small and large inhomogeneities are in a continuous state of fluctuation on time scales ranging from seconds to days and sometimes weeks.

• Stellar physics. Many stars have coronae and winds. The Sun is the only star where theoretical concepts of coronal heating and wind generation mechanisms may be tested observationally by resolving the plasma structures.

• Plasma physics. The topology of the solar plasma is defined by magnetic fields, which are rooted in the photosphere and convection zone.

• Solar-terrestrial relationship. The Earth, in common with the other planets, is immersed in the solar wind and radiation field.

2 The SUMER Instrument

The SUMER scientific objectives call for observations in the extreme ultraviolet (EUV) from 500 to 1600 Å, where emission lines of atoms and ions in the temperature range of $10^4 - 2 \times 10^6 \ K$ can be observed. We want to obtain dynamic and diagnostic information on the properties of the solar atmosphere, with a spatial scale of $\sim 1 \ arcsec$, a temporal resolution of down to 0.25 s or even 60 ms, and a spectral resolving power of $\lambda/\Delta\lambda = 17700 - 38300$ ($\Delta\lambda \stackrel{\frown}{=}$ pixel size) for examining the solar chromosphere, transition region and corona.

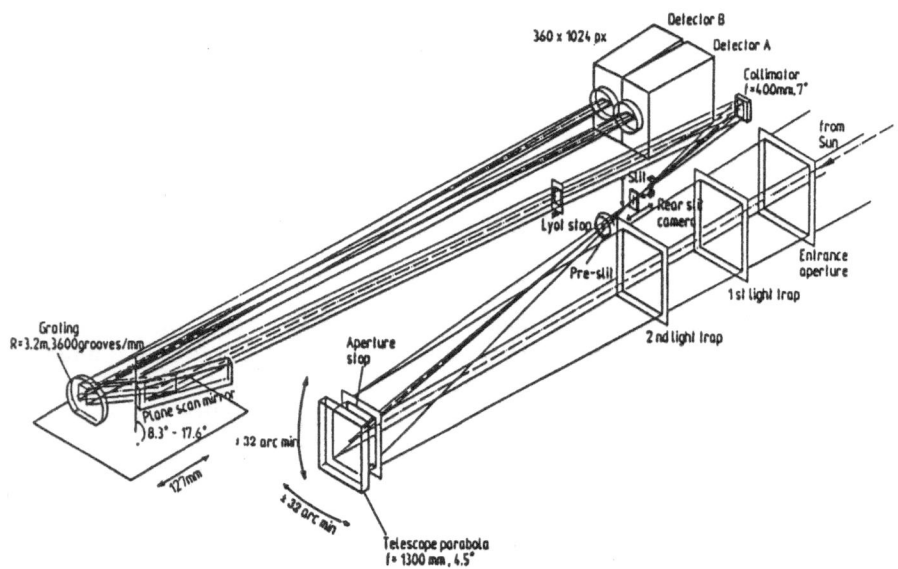

Fig. 1. Optical layout of the SUMER instrument. Control of the instrument requires the activation of up to 8 mechanisms. The slit camera is observing in visible light.

The SUMER optical design is shown in Fig. 1 and is based upon two parabolic mirrors, a plane mirror and a spherical concave grating. The first off-axis telescope parabola, which has pointing and scan capabilities of $\pm 32\ arcmin$ and step sizes of 0.38 $arcsec$ in two directions, mirrors the Sun on the spectrometer entrance slit. The second off-axis parabola collimates the beam leaving the slit. This beam is then deflected by the plane mirror onto the grating. Two 2-dimensional detectors, located in the focal plane of the grating and used alternatively, collect monochromatic images of the slit in two orders simultaneously. Coverage of the full spectral range requires a wavelength scan implemented by rotating the plane mirror. Pointing verification will be performed with the help of a rear slit camera operating in visible light. A baffle system, consisting of an entrance aperture, light traps, an aperture stop, a pre-slit and a Lyot stop completes the design, whose basic characteristics are given in Table 1.

Table 1. The optical design in numbers (1 $arcsec \mathrel{\hat{=}} 715\ km$ on the Sun).

Item	Parameter	Value
Telescope	focal length	1302.77 mm
	collecting area	$90 \times 130\ mm^2$
Slits	lengths	120 and 300 $arcsec$ (1 $arcsec \mathrel{\hat{=}} 6.315\ \mu m$)
	widths	0.3, 1 and 4 $arcsec$
Collimator	focal length	399.60 mm
Grating	groove density	3600 lines/mm
	spherical radius	3200.78 mm
Detectors	pixel size	$26.5 \times 26.5\ \mu m^2$
A + B	array size	360 spatial px · 1024 spectral px
	spatial scale	680 km/px
	spectral scale	21 $m\text{Å}/px$ (2nd order)
		42 $m\text{Å}/px$ (1st order)

SUMER will be equipped with two cross delay line detectors (XDL) provided by O.H.W. Siegmund and his team at Berkeley. The pixel size will be 26.5 × 26.5 μm and the detector format will be 360 spatial pixels × 1024 spectral pixels. A KBr photocathode will cover the central 512 spectral pixel columns, whereas the side areas will be bare. This will allow us to distinguish between 1st and 2nd order lines which will be observed simultaneously. An attenuator will cover the first and the last 30-pixel columns and will provide an attenuation of 10 for the intense Lyman-α line.

2.1 Angular and Spatial Resolutions

In the diffraction limit, 80 % of the energy would be encircled in 0.5 $arcsec$ for the telescope mirror. The actual performance is shown in Fig. 2 in terms of the point spread function (PSF) of this mirror. The PSF is characterized by a half width at half maximum (HWHM) of 0.35 *arcsec* and 80 % of the energy is within a radius of 1.3 *arcsec*. The logarithmic presentation of the PSF gives some indication on the scattering properties of the mirror. For elongated structures on the Sun, such as arcs, the telescope line spread function (LSF) could be relevant and is thus included as well. It has a HWHM of 0.45 *arcsec*.

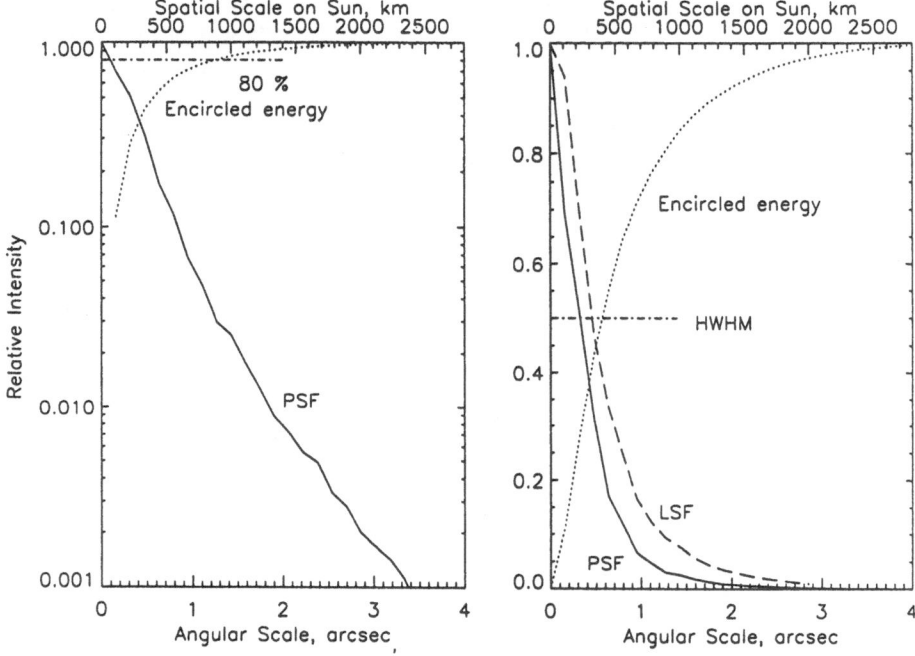

Fig. 2. SUMER telescope point spread function (PSF) and line spread function (LSF) in linear and logarithmic scales. Also shown are the encircled energy curves.

A study of the angular resolution of the instrument has to take all system parameters into account and must be performed separately for axes along and transverse to the slit. Across the slit the resolution will not be influenced by the collimator and grating system, although these elements will introduce an additional spectral blur. To evaluate the resolution, we use a modified Rayleigh criterion and require a minimum of less than 0.81 of the maximum to be detectable between two maxima. An essential feature of the instrument is its dynamic imaging capability, i.e., scanning the field-of-view in 0.38-*arcsec* steps. The pointing stability of SOHO is also of importance. The design goal is a jitter of less than 0.8 *arcsec* (2 σ). Under these conditions, we obtain the results presented in Table 2 where we assumed parallel linear solar structures of 0.3-*arcsec* widths.

Taking into account that 715 km on the Sun as seen from L1 correspond to 1 $arcsec$, the instrument will thus be able to resolve spatial elements on the Sun with dimensions down to 1000 km.

Table 2. Angular resolution of SUMER in $arcsec$ as a function of SOHO's pointing stability.

	Transverse to Slit		Parallel to Slit
SOHO jitter	Slit width, $arcsec$		Pixel size, $arcsec$
($2\,\sigma$), $arcsec$	0.3	1.0	1.0
0.4	1.2	1.5	2.0
0.8	1.4	1.7	2.3
1.6	2.1	2.2	2.6

2.2 Spectral Resolution

The spectral resolution will mainly be determined by the detector scales, but the characteristics of the observed EUV line play an important rôle, too. For bright lines with Doppler widths of $\Delta\lambda_D = 120 - 160\ m\text{Å}$, we find a line shift sensitivity of 0.1 px and a detection limit for line broadening of 0.2 px allowing us to perform Doppler measurements of plasma motions down to velocities of 1 km/s.

In laboratory measurements using a hollow cathode source with a collimator, we obtained spectra such as the one shown in Fig. 3. Although the exact line widths of the lines of the source are not known, it can be seen that the instrument performance in terms of spectral resolution will be adequate for the observations.

3 Operational Phase

The SUMER instrument will be part of the SOHO payload. SOHO (Solar and Heliospheric Observatory), an ESA/NASA spacecraft, which will be launched in 1995 and will be placed in a halo orbit around the Lagrangean point L1. The operational phase with continuous Sun-pointing will last two years with the option of extending it to a total of six years.

SUMER can be used as an observatory. Most operations will be planned several days if not weeks in advance and will require some knowledge about general conditions on the Sun, such as active region, prominence, sunspot and coronal hole locations. This information should be supplied by the various SOHO instruments and ground-based facilities as a result of synoptic solar observations, e.g. $H\alpha$, magnetic, X-ray, coronal data etc. Some pre-planned sequences will need target definitions down to one day prior to the operation. Pre-planned sequences

6 Klaus Wilhelm

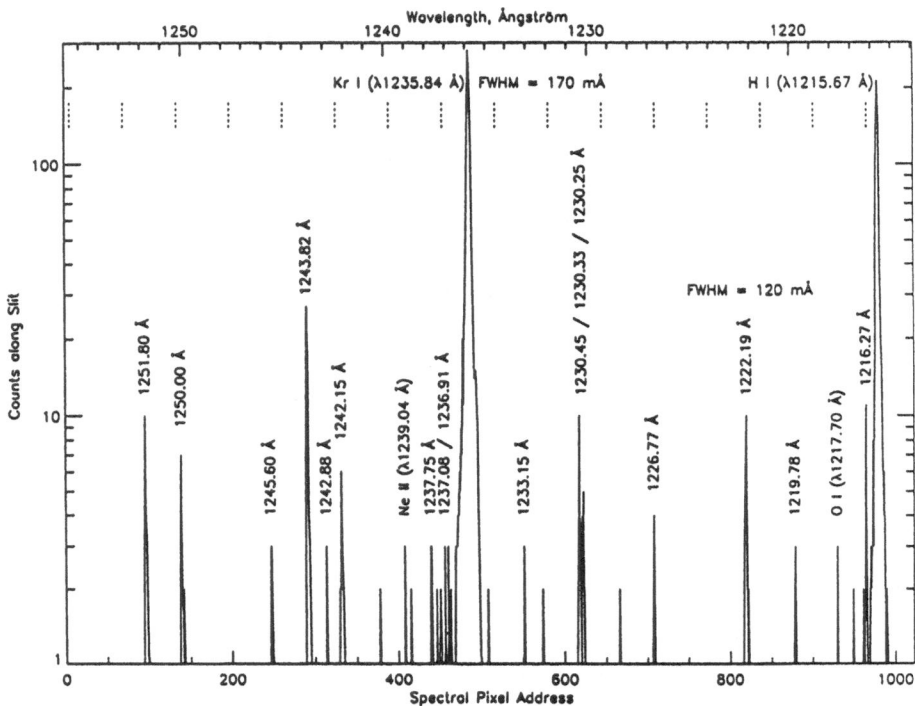

Fig. 3. Spectrum of the hollow cathode calibration source near the Kr I line at 1235.84 Å. The full detector is shown in its spectral dimension covering approximately 42 Å in 1st order.

may be interrupted by short-term planning (timescales of about a day) in response to changes in solar conditions, such as the appearance of a new active region on the east limb, the strengthening of a prominence or the variations of a coronal hole boundary. It is anticipated that the experiment workstations at the ground station will display near real-time data whenever possible and thus one may envisage situations where sudden changes in solar conditions demand a reappraisal of the planned schedule. On such occasions the SUMER team may decide to uplink a command to their experiment in real time, thus changing the operating sequence within minutes rather than hours. An essential element of the SUMER operational scheme will be co-operation with other SOHO instrument teams as well as with ground-based observers.

Acknowledgements. The SUMER project is financially supported by BMFT/ DARA, CNES, NASA and PRODEX (Swiss contribution). The instrument development has been carried out by a large team of co-investigators, associate scientists and engineers.

E I T :
The Extreme Ultraviolet Imaging Telescope
Synoptic Observations of Small and Large-Scale Coronal Structures

Frédéric Clette[1], Jean-Pierre Delaboudinière[2] (PI), Kenneth P. Dere[3], Pierre Cugnon[1] and the EIT Science Team[*]

[1] Observatoire Royal de Belgique, B–1180 Bruxelles, Belgium
[2] Institut d'Astrophysique Spatiale, F–91405 Orsay, France
[3] Naval Research Laboratory, Washington, D.C. 20375–5000, USA

Abstract. The EIT will provide wide-field images of the corona and transition region, on the solar disc and up to $1\,R_\odot$ above the limb. Its normal incidence multilayer-coated optics will select the spectral emission lines of four ions (Fe IX, 171 Å; Fe XII, 195 Å; Fe XV, 284 Å; He II, 304 Å), providing a sensitive temperature diagnostic in the range 6.10^4 to 3.10^6 K. This SOHO instrument will thus probe the coronal plasma on a global scale, as well as the underlying cooler and turbulent atmosphere. The EIT's characteristics and performances are presented, and prospects for coordinated observations with ground-based radio observatories are outlined.

1 Scientific Objectives

With the exception of the LASCO coronagraph working in visible light, the EIT will be the only instrument on SOHO providing wide-field images of the solar corona (Delaboudinière 1989). Working at extreme ultraviolet (EUV) wavelengths, it will show the on-disc corona. As such, the EIT will take over and continue the legacy of the SKYLAB and YOHKOH/SXT instruments (Eddy 1979; Ogawara et al. 1992).

The primary scientific objective of the EIT is to study the dynamics and evolution of coronal structures over a wide range of timescales and temperatures, to bring new insight in the various mechanisms responsible for the coronal heating and the solar wind acceleration.

The EIT's science program will address the following basic topics :

– the long term behaviour of global structures in the corona (holes, sector boundaries, streamers, etc.) over a significant part of a solar activity cycle, in response to the evolution of the photospheric magnetic field (e.g. Wang and Sheeley 1993).

---

[*] A.H. Gabriel, G.E. Artzner, R. Howard, D. Michels, R. Catura, J. Lemen, R. Stern, J. Gurman, W. Neupert, E.L. Van Dessel, C. Jamar, P. Rochus, A. Maucherat.

- an extensive survey of the temporal and spatial distribution of X-ray bright points, and of their connection with the underlying chromospheric and photospheric phenomena (Webb et al. 1993; Parnell et al. 1994).
- the mapping and morphological analysis of candidate sources of the solar wind (e.g. polar plumes : Walker et al. 1993).
- the evolution of the chromospheric network : flows and oscillations propagating in the chromosphere and transition region (Bocchialini et al. 1994; Cargill et al. 1994).
- the dynamics of a wide range of energetic phenomena, which are all the consequence of energy release by magnetic reconnection but at very different scales : microflares, explosive events, flaring loops, jets, coronal mass ejections (CME) and transient brightenings (Shimizu et al. 1994).

The SXT experiment brought the new concept of a dynamical corona, with continuous destruction and emergence of magnetic flux lines (magnetic restructuring, Tsuneta et al. 1992). Thanks to its sensitivity to plasma temperature, the EIT is able to improve the picture by distinguishing for example flux tubes (magnetic fibres) belonging to the same magnetic region but with very different plasma temperatures (Kjeldseth-Moe 1992). Providing a good image contrast, the EIT will also allow the observation of faint features in coronal holes, like thin magnetic arches connecting bright points in holes to distant active regions. Only a few examples of such elusive features, connecting widely spaced sectors of the corona, have been observed so far (Strong et al. 1992). By using the rotation of optically thin plasma, three-dimensional tomographic or stereoscopic reconstruction of magnetic field configurations will also be possible (Davila 1994; Aschwanden and Bastian 1994).

2 Instrument Description

2.1 Characteristics

The EIT is based on normal incidence optics, with two mirrors in a Ritchey-Chrétien configuration (Fig. 1). The four quadrants of the mirrors are coated with different EUV reflecting multilayers (Chauvineau et al. 1992), providing narrow band images at four selected wavelengths with a spectral resolution between 10 and 100 (Fig. 2). By placing a rotating mask in front of the telescope, only one sector is illuminated at a time.

The image sensor is a UV sensitive 1024 × 1024-pixel backside thinned CCD (Moses et al. 1993). The CCD readout is digitized to 14 bit resolution. Three successive aluminium filters, prevent visible and UV light from reaching the CCD and keep the straylight level below 10^{-5}. They are located respectively at the entrance aperture, on a filter wheel near the focus and in front of the CCD chip at the focus. The main telescope body consists in a cylindrical vacuum vessel with an air-tight front door, in order to protect the internal filters against acoustic vibrations during storage and launch. The only external element is the radiator which will passively cool the CCD down to $-80\,^{\circ}$C. Table 1 summarizes the main instrument characteristics and Fig. 3 shows its general layout.

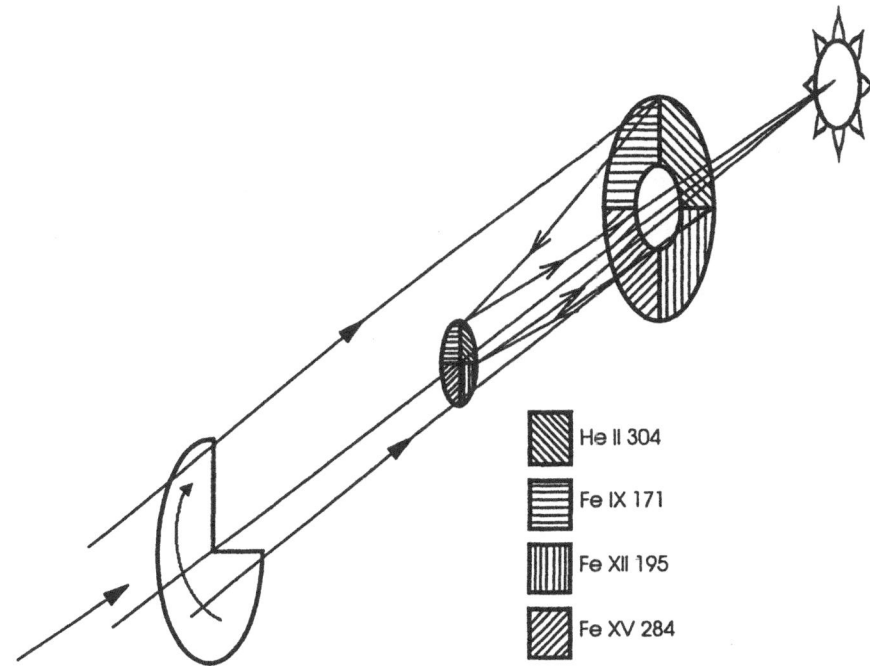

Fig. 1. Optical principle.

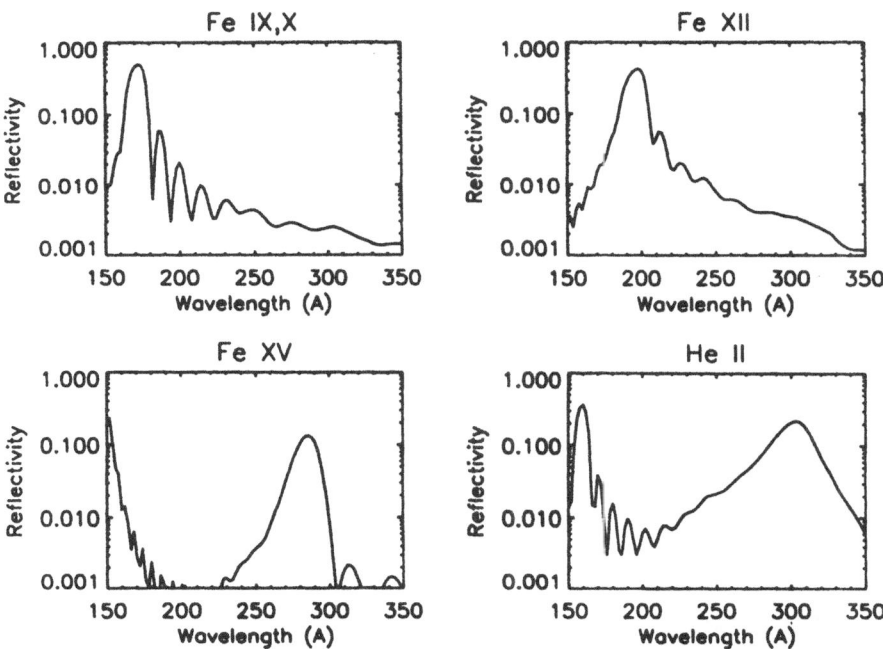

Fig. 2. Reflectivities of the four multilayer coatings (single reflection).

Table 1. EIT : Instrument Characteristics.

TELESCOPE
Mirrors : Zerodur, Hyperbolic shape
Reflectivity : 5 to 25 % (two reflections)
Aperture : 115.5 mm
Focal length : 1650 mm
Optical resolution : 1 arcsec
Collecting area : 13.3 cm^2
Straylight : 10^{-12}
Wavelengths : 171 Å, 195 Å, 284 Å, 304 Å
Spectral resolution : 10 to 100

CCD CAMERA
Sensor type : TEKTRONIX, thinned, back-side illuminated
Sensitive Area : 1024 × 1024 pixels ($21\mu m \times 21\mu m$)
Field of view : 45 × 45 arcmin
Pixel size : 2.6 arcsec
Oper. temperature : $-80\,°$C
Dynamic range : 2.10^5
Quantum efficiency : 0.24 at 304 Å, 0.45 at 171 Å

GENERAL
Total mass : 17.5 kg
Telemetry rate :   1 Kbits/s (continuous)
: 26.2 Kbits/s (selected periods)

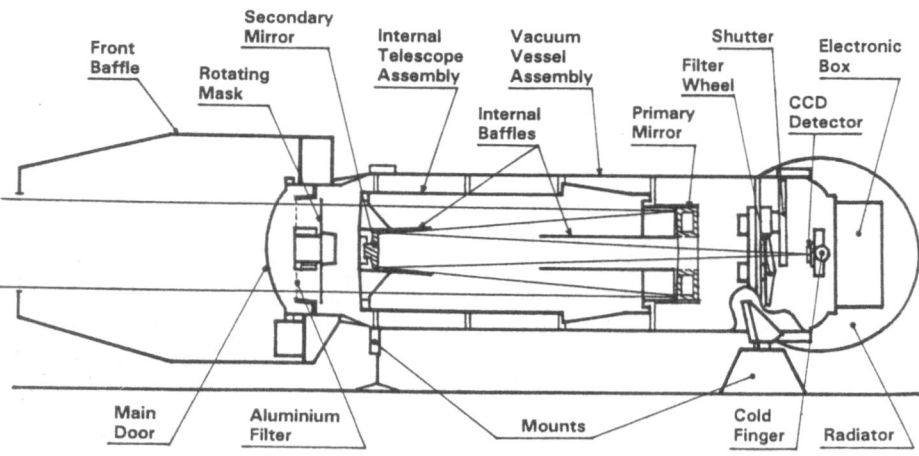

Fig. 3. Instrument layout.

2.2 Instrument Capabilities and Performances

While the angular pixel size of the CCD is 2.6 arcsec, the intrinsic optical resolution of the telescope is 1 arcsec. A slight increase in image quality is thus expected relative to the SXT operating on YOHKOH (Tsuneta et al. 1991). Overall, the image contrast will be high thanks to ultrasmooth mirrors.

The EIT's four spectral bands are centred respectively on three coronal lines (Fe IX, 171 Å; Fe XII, 195 Å; Fe XV, 284 Å) and the He II line (304 Å) produced in the upper-chromosphere and transition region. The spectral responses to the quiet Sun spectrum exhibit a good off-band rejection for all mirror sectors (Fig. 4). The respective temperature responses, computed either for the selected line alone or for the real solar spectrum, are compared in Fig. 5. A slight degree of contamination by lines formed at other temperatures is apparent. This is a consequence of the limited spectral resolution.

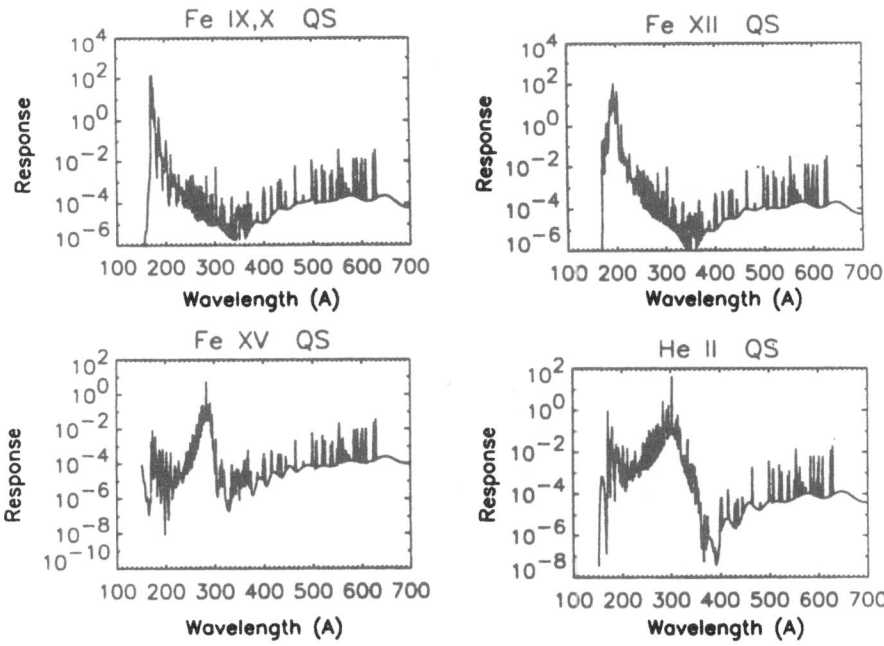

Fig. 4. Response to the average quiet Sun spectrum for each mirror quadrant, expressed in electrons/(s×pixel×Å). The average solar spectrum was synthesized using plasma parameters taken from Dere and Mason (1993).

Each emission line marks a restricted temperature interval within a wide temperature range, from 4.10^4 K in the transition region to 3.10^6 K, a typical temperature for the hot plasma in active regions. By computing the intensity ratio between images made at different wavelengths, a very sensitive temperature diagnostic will thus be produced on a global scale for the cool transition region

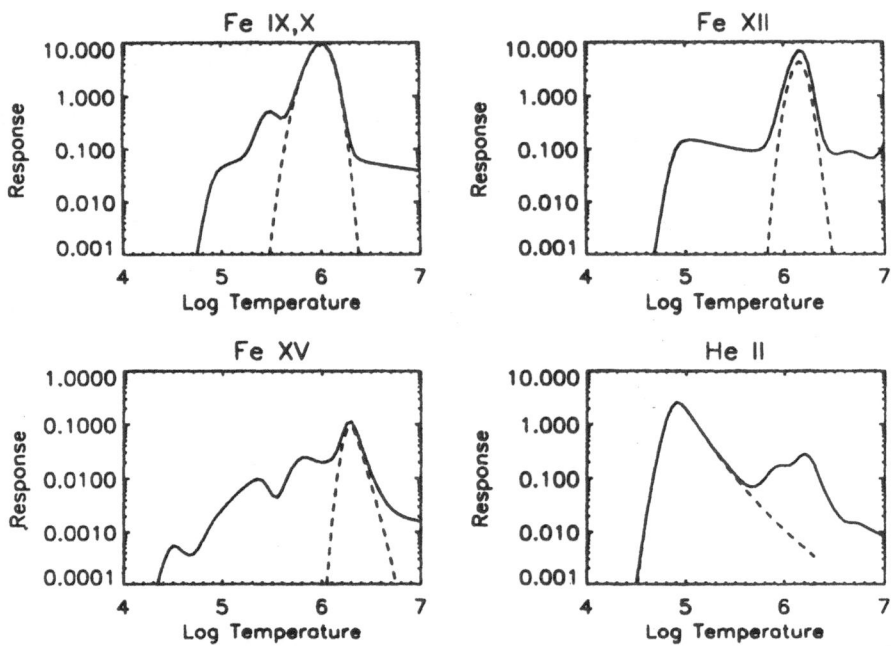

Fig. 5. Temperature response, in electrons/(s×pixel), for each observing wavelength : using the selected emission line only (dashed line) or taking into account the complete solar spectrum (solid line).

and most of the corona. These temperature measurements will then be refined for selected regions using the spectroscopic instruments (CDS, SUMER). Moreover, an indirect pressure diagnostic might also be obtained from observed properties of coronal loops, as suggested by Peres et al. (1994). Overall, the EIT differs from "flare" instruments imaging the Soft X-ray emission of super-hot plasma above 5.10^6 K, and having a limited diagnostic capability at average coronal temperatures. By its design, the EIT is perfectly adapted to the study of steady state processes which contribute to a significant part of the energy input and dissipation in the corona.

2.3 Observational Program

The full field of view (FOV) is 45 by 45 arcmin and oriented parallel to the solar rotation axis. Depending on the operating mode, the FOV may be reduced electronically to a half-field, to a North-South or East-West slab or to a square window of e.g. 128 × 128 pixels. Due to the limited telemetry rate, two modes of image compression will be used : the Rice algorithm (lossless, factor $\simeq$ 3) and the Adaptive Discrete Cosine Transform (lossy, factor $\simeq$ 10). EIT's main imaging modes are summarized in Table 2.

Table 2. EIT : Imaging Modes. Transmission times are given for the minimum teleme-try rate of 1 kb/s and with a 1÷10 compression. The telemetry rate will be multiplied by ≈10 during restricted periods.

Imaging Mode	FOV (pixels)	Pixel size (")	Readout time (s)	Transmission time
Full Sun (HR)	1024^2	2.5	21	6 m
Full Sun (LR)	512^2	5.0	5	90 s
Quarter Sun	512^2	2.5	10	90 s
Meridian slab	128×1024	2.5	2.5	45 s
Window	128^2	2.5	0.7	6 s

The EIT will be used in three main observing modes :

- Synoptic mode : up to three sets of full-Sun high-resolution images in all four wavelengths, transmitted every day.
- Survey mode : medium-resolution partial image mode.
- Event triggered mode : fast sequences of partial high-resolution images covering a small area of interest (i.e. a flaring region), down to 4 × 4 arcsec.

The choice of operating mode will be mainly driven by the available telemetry rate. Most of the time, this rate will be 1kb/s on average, but a higher rate (≈ 10 kb/s) will be available during a period of 30 minutes everyday and continuously for two consecutive months each year. Everyday, several sets of full-disc synoptic images at all four wavelengths will be transmitted. Fast-rate time series of up to 100 sub-images will be possible in the context of coordinated campaigns or event driven observations of active phenomena. The maximum time resolution is 5 seconds. On average, the daily data production will amount to 50 MBytes.

As the EIT is the only payload instrument on SOHO able to monitor the corona over the whole solar disc and beyond the limb, it will provide the global coronal context for SOHO science planning and also to aid in the interpretation of phenomena observed by other SOHO or ground-based instruments.

3 Coordination with Ground-Based Radio Observatories

In order to complete the information provided by the EIT and other SOHO experiments in EUV emission lines, simultaneous observations made from space and from the ground in other spectral regions will be necessary. Radio observations will complement the diagnostics derived from ion emission lines detected in EUV by measuring the electron component of the coronal plasma and the coronal magnetic field. Both techniques have in common the ability to observe the corona on the disc, and in both cases, the contrast of flaring regions is high.

As the EIT provides wide-field solar images, two-dimensional radio observations will be the most useful for direct correlative studies with EIT's data. Observations made from microwaves to decimeter wavelengths will be necessary to sample the solar atmosphere from the chromosphere-transition region up to $2\,R_\odot$ in the corona, i.e. the wide altitude range spanned by EIT's images.

Fundamental science topics relevant to radioastronomy are :

- the mapping of the sector magnetic structure as a function of time, to analyze the coronal rotation and trace back the sources of fast solar wind (Hewish and Bravo 1986; Alurkar et al. 1993).
- the morphological study of active region *magnetospheres* (Lang et al. 1993) and helmet streamers (Schmahl et al. 1994), incorporating the magnetic field configuration and strength, as well as tracing electron flows (Gopalswamy et al. 1987).
- the search for precursors of coronal disturbances (shock fronts, CME's, radio bursts). Local changes in the magnetic connectivity associated with these events will be monitored, to define how the magnetic reordering takes place during such transients.
- the determination of the EUV transient propagation in relation with radio bursts (source position, spectra, e.g. Gopalswamy and Kundu 1993).
- the identification of the source of radio noise storms. The emergence of magnetic flux, which seems to be correlated with the onset of storms, is revealed by the appearance and expansion of EUV loop systems (Stewart et al. 1986; Raulin and Klein 1994).

A prospective survey of ground-based radio facilities was published by the Working Group 6 of the Joint Organisation for Solar Observations (JOSO) (Fleck 1993). A majority of these facilities provide spectroscopic or polarimetric data but lack a two-dimensional imaging capability. Moreover, many instruments are not dedicated to solar observation and will only be available for short periods during campaigns. As a consequence, only a few dedicated radioheliographs remain in operation nowadays. The participation of those few instruments is thus essential in view of joint studies and science planning with all SOHO coronal instruments and with the EIT in particular.

4 Conclusion

The EIT is a very versatile EUV wide-field instrument. It will combine imaging of the on-disc corona with diagnostic capabilities, in order to probe the inner and intermediate corona as well as the transition region over a wide range of spatial scales. It will provide the temporal and topological context of coronal phenomena and will thus aid in unifying SOHO and ground-based investigations. As such, it will offer good opportunities for cross-analysis with radio instruments.

References

Alurkar, S.K., Janardhan, P. and Hari Om Vats : Comparison of single-site interplanetary scintillation solar wind speed structure with coronal features. Solar Phys. **144** (1993) 385–397

Aschwanden, M.J. and Bastian, T.S. : VLA stereoscopy of solar active regions. I. Methods and tests. Astrophys. J. **426** (1994) 425–433

Bocchialini, K., Vial, J.-C. and Koutchmy, S. : Dynamical properties of the chromosphere in and out of the solar magnetic network. Astrophys. J. **423** (1994) L67–L70

Cargill, P.J., Chen, J. and Garren, D.A. : Oscillations and evolution of curved current-carrying loops in the solar corona. Astrophys. J. **423** (1994) 854–870

Chauvineau, J.P. et al. : Description and performance of mirrors and multilayers for the Extreme ultra-violet Imaging Telescope (EIT) of the SOHO mission. SPIE Proc. Ser. **1546** (1992) 576–586

Davila, J.M. : Solar tomography. Astrophys. J. **423** (1994) 871–877

Delaboudinière, J.-P. et al. : EIT – Solar Corona Synoptic Observations with an Extreme-Ultraviolet Imaging Telescope. in "The SOHO Mission", ed. V. Domingo, ESA **SP-1104**, ESA, Noordwijk (1989) 43–48

Dere, K.P. and Mason, E. : Nonthermal velocities in the solar transition zone observed with the High-Resolution Telescope and Spectrograph. Solar Phys. **144** (1993) 217–241

Eddy, J.A. : "A New Sun : The Solar Results From SKYLAB", Rep. **NASA-SP402**, NASA (1979)

Fleck, B. : Ground-based observatories. in Proc. of the LASCO/EIT Science Consortium Meeting (June 8-11, 1993)., ed. K.P. Dere, Part 2, Naval Research Laboratory, Washington D.C. (1993)

Gopalswamy, N., Kundu, M.R. : Thermal and nonthermal emissions during a coronal mass ejection. Solar Phys. **143** (1993) 327–343

Gopalswamy, N., Kundu, M.R. and Szabo, A. : Propagation of Electrons Emitting Weak Type III Bursts in Coronal Streamers. Solar Phys. **108** (1987) 333–345

Hewish, A. and Bravo, S. : The sources of large-scale heliospheric disturbances. Solar Phys. **106** (1986) 185–200

Kjeldseth-Moe, O. : CDS and SUMER Observations of Fine Structure and Dynamic Loops : Experience from HRTS. in Proc. of the First SOHO Workshop, ed. C. Mattok, ESA **SP-348**, ESA, Noordwijk (1992) 155–165

Lang, K.R. et al. : Magnetospheres of solar active regions inferred from spectral-polarization observations with high spatial resolution. Astrophys. J. **419** (1993) 398–417

Moses, D. et al. : Extreme ultraviolet response of a Tektronix 1024×1024 CCD. SPIE Proc. Ser. **2006** (1993) 252

Ogawara, Y. et al. : The Status of Yohkoh in Orbit : An Introduction to the Initial Scientific Results. Publ. Astron. Soc. Japan **44**, No. 5 (1992) L41–L44

Parnell, C.E., Priest, E.R. and Golub, L. : The three-dimensional structures of X-ray bright points. Solar Phys. **151** (1994) 57–74

Peres, G., Reale, F. and Golub, L. : Loop models of low coronal structures observed by the Normal Incidence X-ray Telescope (NIXT). Astrophys. J. **422** (1994) 412–415

Raulin, J.P. and Klein, K.-L. : Acceleration of electrons outside flares : evidence for coronal evolution and height-extended energy release during noise storms. Astron. Astrophys. **281** (1994) 536–550

Schmahl, E.J., Gopalswamy, N. and Kundu, M.R. : Three-dimensional coronal structures using Clark Lake observations. Solar Phys. **150** (1994) 325–337

Shimizu, T. et al. : Morphology of active region transient brightenings with the YOHKOH Soft X-ray Telescope. Astrophys. J. **422** (1994) 906–911

Stewart, R.T., Brueckner, G.E. and Dere, K.P.: Culgoora radio and SKYLAB EUV observations of emerging flux in the lower corona. Sol. Phys. **106** (1986) 107–130

Strong, K.T. et al. : Observation of the Variability of Coronal Bright Points by the Soft
    X-Ray Telescope on Yohkoh. Publ. Astron. Soc. Japan 44, No. 5 (1992) L161–L166
Tsuneta, S. et al. : The Soft X-ray Telescope for the SOLAR-A mission. Solar Phys.
    136 (1991) 37–67
Tsuneta, S. et al. : Global Restructuring of the Coronal Magnetic Fields Observed
    with the Yohkoh Soft X-Ray Telescope. Publ. Astron. Soc. Japan 44, No. 5 (1992)
    L211–L214
Walker, A.B.C., Deforest, C.E., Hoover, R.B. and Barbee Jr., T.W. : Thermal and
    Density Structure of Polar Plumes. Solar Phys. **148** (1993) 239–252
Wang, Y.-M. and Sheeley Jr., N.R. : Understanding the rotation of coronal holes.
    Astrophys. J. **414** (1993) 916–927
Webb, D.F., Martin, S.F., Moses, D. and Harvey, J.W. : The correspondence between
    X-ray bright points and evolving magnetic features in the quiet Sun. Solar Phys.
    144 (1993) 15–35

The Ultraviolet Coronagraph Spectrometer

G. Noci[1], J.L. Kohl[2], M.C.E. Huber[3], E. Antonucci[4], S. Fineschi[1], L.D. Gardner[2], G. Naletto[5], P. Nicolosi[5], J.C. Raymond[2], M. Romoli[3], D. Spadaro[6], L. Strachan[2], G. Tondello[5], A. van Ballegooijen[2]

[1]Università di Firenze
[2]Harvard-Smithsonian Center for Astrophysics
[3]European Space Agency
[4]Università di Torino
[5]Università di Padova
[6]Osservatorio Astrofisico di Catania

Abstract The Ultraviolet Coronagraph Spectrometer (UVCS) is an instrument onboard the Solar and Heliospheric (SOHO) spacecraft, a joint ESA/NASA mission to be launched in 1995. The UVCS will provide ultraviolet spectroscopic measurements to determine the primary plasma parameters of the solar corona (temperatures, densities, velocities), from its base to as high as 10 $R_\odot$. We review briefly, here, its science objectives and give an instrument description.

1 Science Objectives

The objective of the UVCS investigation is to answer certain fundamental questions concerning acceleration processes in the corona and their relationship to solar wind properties near the earth. What are the dominant plasma heating and acceleration processes in the solar corona? What is the role of MHD waves? What are the dissipation mechanisms? How are heavy ions accelerated? How is the composition of the solar wind established and what produces the abundance variations seen in the solar wind? Where is the slow speed wind generated? Are coronal holes the only sources of non-transient high speed wind? Closely related to these questions are a variety of other problems. Why do some regions emit low speed wind and others high speed wind? What are the roles of thermal pressure gradients, wave-particle interactions and suprathermal particles in accelerating solar wind from different types of regions? Why are the particle fluxes in high and low speed winds similar? How does the coronal magnetic geometry affect the outflow of the solar wind? What is the role of small-scale structures (polar plumes, spicules) in generating high speed wind? What is the thermal state of gas expelled during coronal mass ejections and what are the plasma conditions in the associated shocks?

To answer such questions, we must determine the physical conditions in coronal regions where the solar wind is accelerated. The number of measured parameters must be sufficient to significantly constrain theoretical solar wind models. Although a substantial amount of data on the electron density structure of the corona already exist, there are only isolated measurements of other critical plasma parameters, except close to the surface ($r < 1.3\ R_\odot$). The UVCS instrument is designed to measure the primary plasma parameters of the solar corona from its base to as far as $10\ R_\odot$:

Temperatures: electron temperature (T_e), effective temperatures of protons (T_p) and several minor ions ($T_{O^{5+}}$, $T_{Mg^{9+}}$, $T_{Si^{11+}}$, $T_{Fe^{11+}}$). They are determined from line profile measurements, and include turbulent and wave velocities, as well as thermal motions. The electron temperature is determined not only from the profile of the Ly-α scattered by the free electrons, but also from the fraction of neutral hydrogen.

Densities: electron and ion densities (N_e, N_{H^0}, $N_{N^{4+}}$, $N_{O^{5+}}$, $N_{Mg^{9+}}$, $N_{Si^{11+}}$, $N_{Fe^{11+}}$). Obtained from line total intensities and from the polarized radiance in the visible continuum (N_e).

Flow velocities: of the electron/proton plasma ($V_{e/p}$) and ions ($V_{O^{5+}}$, $V_{Mg^{9+}}$, $V_{Si^{11+}}$). Obtained from Doppler shifts and Doppler dimming of spectral lines.

Furthermore, the UVCS can measure the ultraviolet fluxes of several bright stars to directly compare the ultraviolet intensities of the sun and stars with a single, absolutely calibrated instrument.

2 Diagnostics

The primary diagnostic tool for UVCS will be the resonantly scattered Ly-α profile. Its width is a direct measure of the velocity distribution of neutral hydrogen atoms along the line of sight. Except at very great heights, this will be the same as the proton velocity distribution due to rapid charge transfer. The Ly-α intensity relative to the disk intensity is sensitive to the density of coronal H^0 and to the velocity of the observed gas away from the sun, which shifts the absorption profile of coronal H^0 away from the chromospheric Ly-α emission profile (Doppler dimming). Knowledge of the hydrogen neutral fraction can be obtained from the electron density measured with the UVCS White Light Channel, LASCO, or other white light coronagraph, and from the electron temperature (see below).

The next most important diagnostic will be the intensities of O VI $\lambda1032$ and $\lambda1037$ and their ratio. Close to the sun these lines are collisionally excited, their ratio is 2:1, and their intensity is proportional to $N_{OVI}N_e$. At greater heights, resonant scattering of transition region emission begins to dominate, and the line ratio and intensities change. Low solar wind speeds are sufficient to Doppler dim the lines. Because $\lambda1037$ is pumped by a nearby C II line, the line ratio is velocity-sensitive up to about 250 km/s. The Mg X and Si XII resonance doublets have similar characteristics.

Third, UVCS will measure the profile of electron-scattered Ly-α. Its width determines the electon temperature, T_e.

Since the total intensity of the resonantly scattered Ly-α is not sensitive to velocity for $v \leq 80km/sec$, the electron temperature can be obtained also by comparison of the Ly-α intensity with that of the visible continuum (from the White Light Channel).

Other spectral lines will also be important, particularly [Fe XII] $\lambda1242$. This will be combined with the other strong lines and occasionally with fainter lines such as Ly-β, He II $\lambda1085$, and O V $\lambda630$ to determine the electron temperature and elemental abundances.

3 Instrument Overview

The Ultraviolet Coronagraph Spectrometer (UVCS) consists of an occulted telescope and a high resolution spectrometer assembly (see Figure 1). Three spherical telescope mirrors focus co-registered images of the extended corona onto the entrance slits of the spectrometer assembly. The spectrometer assembly consists of three sections:

– The Ly-α section is a toroidal grating spectrograph with an entrance slit mechanism, a neutral density filter inserter, a grating mechanism, and a windowed XDL detector (the Ly-α detector). This section is optimized for line profile measurements of H I 1216 Å and it is also used for the Fe XII line at 1242 Å.
– The O VI section is a toroidal grating spectrograph with an entrance slit mechanism, a neutral density filter inserter, a grating mechanism, and an open XDL detector (the O VI detector). The O VI detector is optimized for measurements of the O VI lines at 1032 and 1037 Å (in first order), and is also used for the Si XII lines at 499 and 521 Å (in second order). The O VI section includes a mirror in the path of the Ly-α beam, to focus the Ly-α and Mg X 610/625 radiation from the grating onto the O VI detector.
– The white light channel (WLC) measures the polarized radiance of the K corona. It consists of an entrance aperture, a polarimeter assembly, and a photomultiplier tube. The polarimeter assembly consists of a rotatable half-wave plate, a fixed linear polarizer, a bandpass filter and a lens.

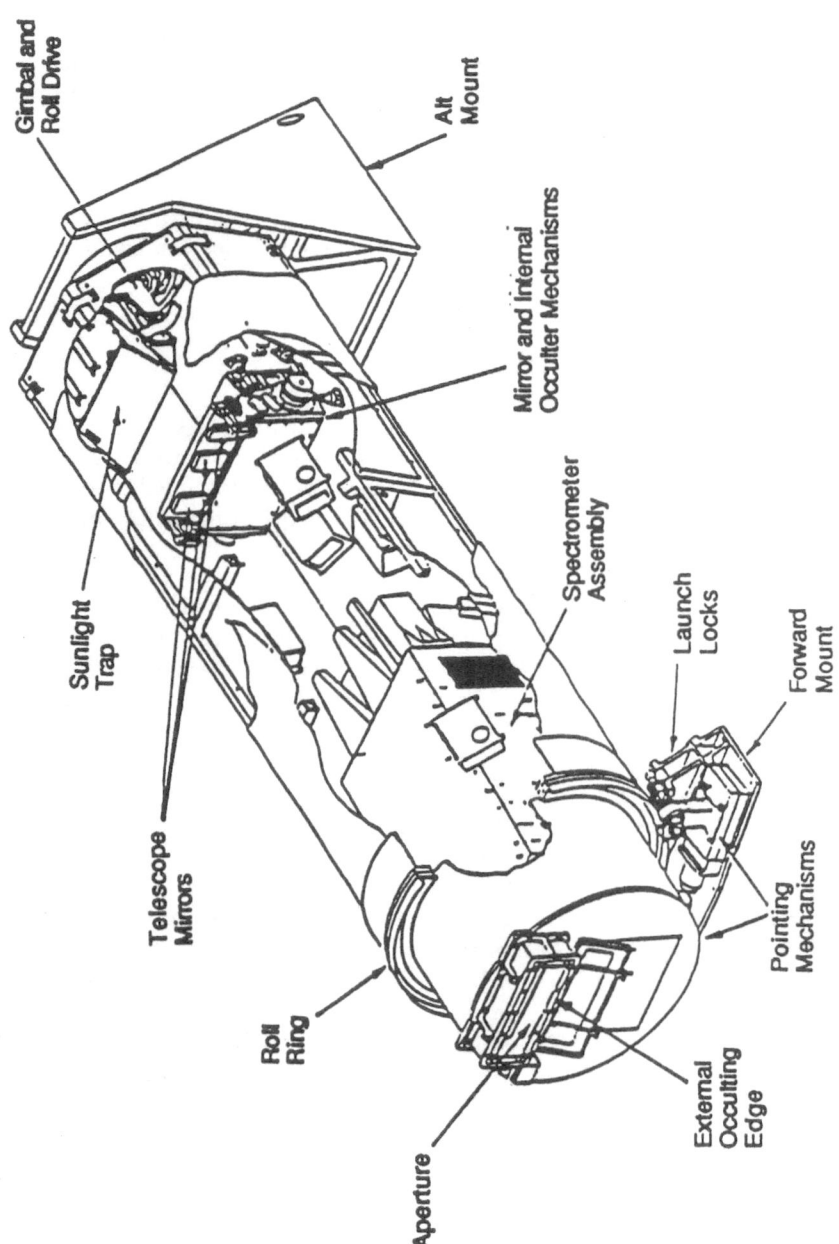

Figure 1 Diagram of the UVCS instrument.

3.1 Telescope Assembly

The following is a brief description of the occulted telescope section of the UVCS. One side of the rectangular entrance aperture acts as a linear external occulter that shields the telescope mirrors from direct sunlight; the other sides serve to limit the amount of light entering the instrument. For sun-center pointing the light from the solar disk enters a sunlight trap, while coronal radiation from 1.2 $R_\odot$ to 10 $R_\odot$ illuminates three telescope mirrors. The external occulter and the telescope mirrors are placed such that radiation from 1.2 $R_\odot$ just reaches the edges of the mirrors and radiation from 10 $R_\odot$ completely fills the mirrors. The telescope mirror/internal occulter mechanism tilts the three mirrors, allowing the coronal images to be scanned across the entrance slits of the spectrometer assembly. An internal occulter, just in front of the telescope mirrors, blocks light diffracted and scattered by the external occulter that would otherwise be imaged onto the slits and dominate the coronal radiation in many cases. This internal occulter is translated across the mirrors as they tilt, so as to cover those parts of the mirrors that would specularly reflect the unwanted light through the slits.

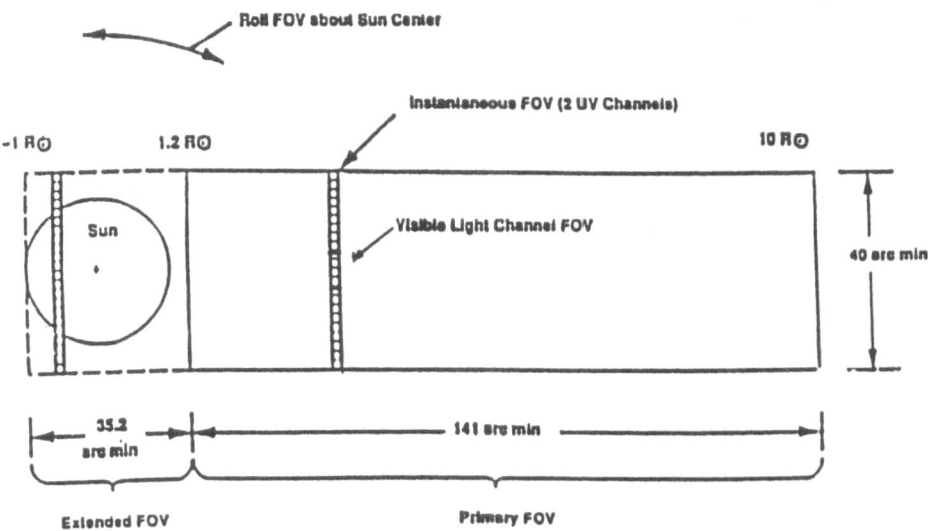

Figure 2 Field of View of the UVCS EUV channels.

The field-of-view (FOV) of the UVCS is illustrated in Figure 2. When the center line and roll axis of the instrument is pointed at solar disk center (as is done in most coronal observations), the primary FOV of UVCS extends from 1.5 to 10 $R_\odot$ in the corona. The instantaneous FOV is the region of the corona which is imaged onto the spectrograph entrance slits at a particular time. For the EUV channels the instantaneous FOV is a region at radial distance r from solar disk center, with a length of 40 arcmin and a width determined by the width of the spectrograph entrance slit. For the WLC the instantaneous FOV is a region of 10 × 10 arcsec located at the center of the 40 arcmin slice. The position r is determined by the angle of the telescope mirror. The roll mechanism allows the telescope assembly to be rotated about its axis, thus providing 360 degree coverage of the solar corona. The FOV can be extended onto the solar disk by offset pointing the instrument axis relative to sun center, using the pointing adjustment mechanism.

3.2 Spectrometer Assembly

The entrance slits of the Ly-α and O VI spectrometers have a length corresponding to an angular distance of 40 arcmin in the corona. Ly-α and O VI entrance slit mechanisms permit a range of entrance slit widths to be selected. The WLC has a fixed entrance aperture, corresponding to an instantaneous FOV of 10 × 10 arcsec located near the center of the 40 arcmin FOV of the EUV channels. All three sections include filter mechanisms, which allow neutral density filters to be inserted behind the respective entrance slits.

The Ly-α and O VI spectrometer sections use toroidal concave diffraction gratings in Rowland circle mounts that produce stigmatic images at the subject wavelengths. The Ly-α grating has 2400 lines/mm, and the O VI grating has 3600 lines/mm. The angles of the gratings are controlled by grating mechanisms. The null positions place the H I 1216 Å line on the center of the Ly-α detector and the O VI 1032 Å line on the center of the O VI detector. Grating motions allow the spectral lines to be placed at different locations on the detectors. The wavelength ranges for each detector are listed below.

The WLC has a filter mechanism which allows a neutral density filter to be inserted just behind the entrance slit. Light from the entrance slit passes through a rotating half-wave plate and a fixed linear polarizer. The waveplate can be rotated into positions 30, 0, or -30 degrees. The selected linear polarization component passes through a color filter and is focused by an achromatic lens onto a visible light detector.

Detectors

The two EUV detectors are Crossed Delay Line Detectors (XDLs). The detectors have 360 pixels in the spatial direction and 1024 pixels in the spectral direction. The Ly-α detector has a MgF$_2$ window and a KBr photocathode. The O VI detector is windowless, but has a protective cover which is to be removed before launch. Both XDLs have vacion pumps to maintain vacuum conditions inside the detector during ground operations.

Each XDL detector has an associated image processor (IP) located in the telescope assembly. Following completion of an exposure, image data is transferred from the IP to either the primary or redundant electronics in the remote electronics unit (REU).

The visible light detector (VLD) consists of a single photomultiplier tube. The device can be used in either a photon-counting mode or a photo-diode mode. Since the amount of visible light data is rather small, generally a number of measurements will be collected in a buffer before the data is telemetered.

Wavelength Coverage

The Ly-α detector has a MgF_2 window (short-wavelength cutoff at 1130 Å) and a KBr photocathode. At the null position of the grating the Ly-α spectral line (λ 1216 Å) is centered on the detector. The nominal wavelength range on the detector is 1145 to 1287 Å. With grating motions the wavelength range can be extended down to 1070 Å (detection limited by window transmission) and up to 1361 Å. The dispersion is 5.54 Å/mm, or 0.1385 Å per pixel (total width 141.8 Å). The following solar lines fall in the nominal wavelength range: H I λ 1216, Fe XII λ 1242, N V λ 1238, Si III λ 1206.

The dispersion of the O VI channel is 3.70 Å/mm or 0.0925 Å per pixel in first order, and 1.85 Å/mm or 0.0463 Å per pixel in second order. The null position of the grating is such that O VI λ 1032 is near the center of the O VI detector. The primary purpose of the O VI detector is to observe O VI λ 1032/1037 in first order. It is also used to observe Si XII λ 499/521 in second order. The wavelength range for this detector is 984 to 1080 Å (*i.e.*, 492 to 540 Å in second order). Grating motions can extend this range down to 937 and up to 1126 Å (*i.e.*, 469 to 563 Å in second order). The following solar lines fall within the full range: O VI λ 1032/1037, C III λ 977 and He II λ1085 (first order), Si XII λ 499/521 (second order). The O VI channel also includes a convex mirror which focusses the Ly-α radiation onto the O VI detector. This provides a redundant capability to measure the Ly-α profile, and also allows observations of Mg X λ 610/625 in second order. The nominal wavelength range for the redundant Lyα response is 1220 to 1124 Å (610 to 562 Å second order). Grating motions extend this range from 1268 to 1076 Å (634 to 536 Å second order).

The spectral resolution of the Ly-α and O VI channels will be adequate to resolve the profiles of the Lyman lines, but it might, depending on the line widths, be marginal for the lines of heavier ions in the standard mode. Therefore, it might be necessary to obtain several exposures at different grating positions to measure the profiles of heavier ions.

3.3 Flight Software

The UVCS flight software resides in the REU. The software performs the following functions: (1) instrument control; (2) spacecraft interface; (3) mechanism control; (4) image processing. The instrument control software includes software for all observing and standby modes. This includes "housekeeping" software to monitor instrument health and to safeguard the instrument against over-exposure of detectors by direct sunlight. The spacecraft interface software is responsible for reception of commands and telemetry of science and housekeeping data. Mechanism control software performs the functions required to safely operate instrument mechanisms. Image processing software performs the functions related to the operations of the Image Processors (loading of detector masks, taking of an exposure).

The flight software provides the user with the capability to operate the instrument interactively (only during contact periods) or in batch mode. Most UVCS observations will be performed by executing "observation sequences" stored in onboard memory. The sequences contain detailed information about UVCS instrument configuration, exposure times, etc. Occasionally new or revised sequences will need to be generated and uplinked in order to address specific scientific objectives or correct for problems with the instrument. The UVCS workstations at the experiment operation facility (EOF) contain the software tools needed to generate, uplink and verify new sequences and associated parameter data.

4 Collaborative Studies

The UVCS scientific output will be greatly enhanced by the combination of the UVCS data with those from the other instruments on SOHO and from ground based observatories. This is true, in particular, for the radio observations, which will increase the amount of information acquired on a given coronal feature or help to establish the proper boundary conditions for models of coronal structures. Examples of radio data suitable for collaborative studies are the following ones:

- Interplanetary scintillations. Observations of scintillation of radio sources provide information on the velocity of diffracting irregularities in the solar wind plasma. These will be compared with the UVCS velocity data below 10 $R_\odot$, and will permit a sort of cross-calibration; beyond this height they will complement the UVCS data to yield information on solar wind streams from the low corona to large heliocentric distances.
- Decimetric and centimetric data. Observations with the Very Large Array at dm and cm wavelengths are expected to yield electron temperature, electron density, and magnetic field strength, at the coronal base. These will be used to define the physical conditions at the inner boundary of the UVCS field of view.

− Electron temperatures. For selected coronal features (streamers, coronal holes) the direct measurements of electron temperature made by the UVCS, by observing the profile of the electron scattered Ly-α, will be compared with the radio determinations based on the brightness temperature.

5 References

An extended description of the coronal and solar wind science that can be addressed with the UVCS instrument is contained in the following literature:

Esser, R., Leer, E., Habbal, S.R. and Withbroe, G.L., A Two-fluid solar wind model with Alfvén waves: Parameter study and application to observations, *J. Geophys. Res.*, **91**, 2950, 1986.

Esser, R., and Withbroe, G.L., Line–of–sight effects on spectroscopic measurements in the inner solar wind region, *J. Geophys. Res.*, **94**, 6886, 1989.

Esser, R., Broadening of the resonantly scattered H I Lyman alpha line caused by Alfvén waves, *J. Geophys. Res.*, **95**, 10261, 1990.

Esser, R., and Leer, E., Flow of oxygen ions in the solar wind acceleration region, *J. Geophys. Res.*, **95**, 10269, 1990.

Gabriel, A. H., Measurements of the Lyman α corona, *Sol. Phys.*, **21**, 392, 1971.

Kohl, J.L., Weiser, H., Withbroe, G.L., Noyes, R.W., Parkinson, W.H., Reeves, E.M., MacQueen, R.M., and Munro, R.H., Measurements of coronal kinetic temperatures from 1.5 to 3 solar radii, *Astrophys. J.*, **241**, L117, 1980.

Kohl, J.L., and Withbroe, G.L., EUV spectroscopic plasma diagnostics for the solar wind acceleration region, *Astrophys. J.*, **256**, 263, 1982.

Kohl, J.L., Gardner, L.D., Strachan, L., and Hassler, D.M., Ultraviolet spectroscopy of the extended solar corona during the Spartan 201-1 mission, *Space Science Rev.*, (in press) 1994.

Noci, G., Kohl, J.L, and Withbroe, G.L., Solar wind diagnostics from Doppler-enhanced scattering, *Astrophys. J.*, **315**, 706, 1987.

Spadaro, D., and Ventura, R., Spectral lines from source regions of the solar wind: The O VI resonance doublet, *Astron. Astrophys.*, (in press), 1994.

Strachan, L., Measurements of outflow velocities in the solar corona, Ph. D. thesis, Harvard Univ., May 1990.

Strachan, L., Kohl, J.L., Weiser, H., Withbroe, G.L., and Munro, R.H., A Doppler dimming determination of coronal outflow velocity, *Astrophys. J.*, **412**, 410, 1993.

Strachan, L., Gardner, L.D., Hassler, D.M., and Kohl, J.L., Preliminary results from 201-1: Coronal streamer observations, *Space Science Rev.*, (in press), 1994.

Withbroe, G.L., in *Activity and outer atmosphere of the sun and stars*, 11th course, SAAS-Fee, 1, eds. Benz, A.O., Chmielewski, Y., Huber, M.C.E., and Nussbaumer, H., Observatoire dé Geneve, 1981.

Withbroe, G.L., Kohl, J.L., Weiser, H., Noci, G., and Munro, R.H., Analysis of coronal HI Lyman alpha measurements from a rocket flight on 13 April 1979, *Astrophys. J.*, **254**, 361, 1982.

Withbroe, G.L., Kohl, J.L., Weiser, H., and Munro, R.H. Probing the solar wind acceleration region using spectroscopic techniques, *Space Science Rev.*, **33**, 17, 1982.

Withbroe, G.L. and Raymond, J.C., Plasma diagnostics for the outer solar corona, UV and XUV Fe XII lines, *Astrophys. J.*, **285**, 347, 1984.

Withbroe, G.L., Kohl, J.L., Weiser, H., and Munro, R.H. Coronal temperatures, heating, and energy flow in a polar region of the sun at solar maximum, *Astrophys. J.*, **297**, 324, 1985.

Withbroe, G.L., Kohl, J.L., and Weiser, H., Analysis of coronal H I Lyman-alpha measurements in a polar region of the sun observed in 1979, *Astrophys. J.*, **307**, 381, 1986.

Withbroe, G.L., The temperature structure, mass and energy flow in the corona and inner solar wind, *Astrophys. J.*, **325**, 442, 1988.

The Charge, Element, and Isotope Analysis System CELIAS on SOHO

D. Hovestadt[1], P. Bochsler[2], H. Grünwaldt[3], F. Gliem[4], M. Hilchenbach[1], F. M. Ipavich[5], D. L. Judge[6], W. I. Axford[3], H. Balsiger[2], A. Bürgi[1], M. Coplan[5], A. B. Galvin[5], J. Geiss[2], G. Gloeckler[5], K. C. Hsieh[8], R. Kallenbach[2], B. Klecker[1], M. A. Lee[9], S. Livi[3], G. G. Managadze[7], E. Marsch[3], E. Möbius[9], M. Neugebauer[10], K.-U. Reiche[4], M. Scholer[1], M. I. Verigin[7], D. Wilken[3], and P. Wurz[2]

[1] Max–Planck–Institut für Extraterrestrische Physik, D–85740 Garching, Germany
[2] Physikalisches Institut der Universität Bern, CH–3012 Bern Switzerland
[3] Max–Planck–Institut für Aeronomie, D–37189 Katlenburg–Lindau Germany
[4] Institut für Datenverarbeitung, Technische Universität Braunschweig, D–38023 Braunschweig, Germany
[5] Department of Physics and Astronomy and IPST, University of Maryland, College Park, MD 20742, USA
[6] Space Science Center, University of Southern California, Los Angeles, CA 90089, USA
[7] Institute for Space Physics, Moscow, Russia
[8] Department of Physics, University of Arizona, Tucson, AZ 85721, USA
[9] EOS, University of New Hampshire, Durham, NH 03824, USA
[10] Jet Propulsion Laboratory, Pasadena, CA 91103, USA

1 Introduction

The CELIAS instrument is designed to study the composition of the solar wind (SW) and of solar and interplanetary energetic particles on SOHO. It consists of three different sensors with associated electronics, which are optimized each for a particular aspect of ion composition. These aspects are the elemental, isotopic, and ionic charge composition of SW or energetic ions emanating from the Sun. In addition, the Solar EUV Monitor (SEM) has been included into CELIAS for monitoring the absolute EUV flux from the Sun.

2 Scientific Objectives

Particle abundance observations of the solar wind and of solar energetic particles (SEP) by itself, and in close correlation with optically observable phenomena on the Sun, will allow to tackle basically unsolved questions in solar physics such as: (1) the steady heating process in the corona, (2) SW acceleration processes, (3) dynamic heating phenomena driven by magnetic fields or waves, e.g. heating in coronal mass ejections, filament eruptions, flares, (4) SEP acceleration processes,

(5) processes leading to variations in abundances (elemental, ionic charge, iso-topic) and (6) the relation between composition and events/regions in the solar atmosphere.

Further scientific goals of CELIAS are to study the composition and dynamics of interplanetary pick–up ions in correlation with the solar EUV flux. For this purpose a solar EUV flux monitor (SEM) for the wavelength 17 to 70 nm has been included into CELIAS as a sub–unit of the STOF sensor. The SEM serves also to monitor the total absolute EUV flux for the large EUV telescopes on SOHO.

Evidently, the understanding of the phenomena outlined above requires col-laboration with many other astrophysical disciplines. Among them are ground based observations including radio astronomical data (e.g. the well known Type III kilometric radio bursts which are related to the strong enrichment of certain species in SEP's) or the interpretation of solar wind abundances in terms of solar abundances as derived from photospheric spectra, meteoritic composition, and from solar oscillation data.

3 Experiment Overview

The CELIAS instrument consists of three different sensor units (CTOF, MTOF, and STOF) coupled to a Digital Processing Unit (DPU). All three sensors em-ploy electrostatic deflection systems in combination with time–of–flight (TOF) measurements of the impinging particles. The CTOF and STOF sensors employ

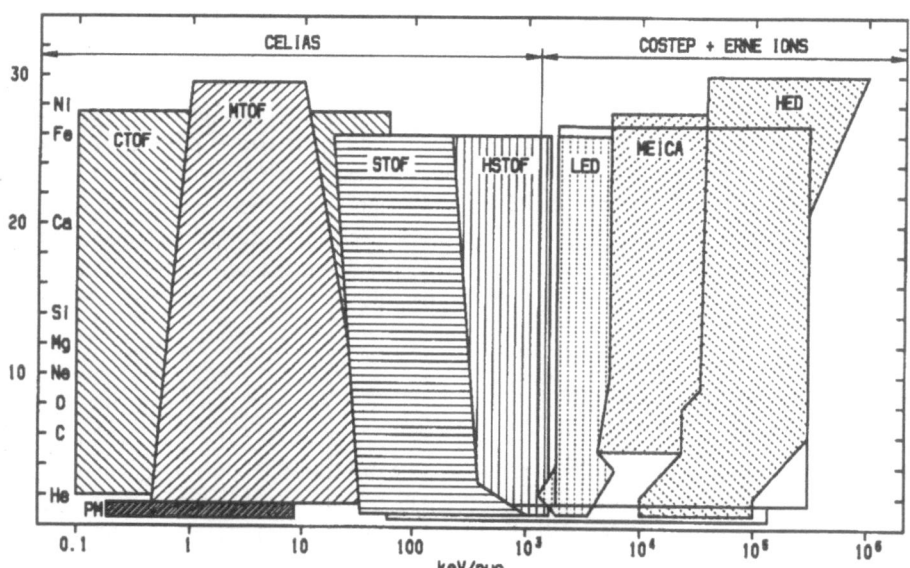

Fig. 1. Energy ranges for ions for the SOHO particle instruments.

also silicon solid state detectors to determine the residual energy of particles. Thus for these two sensors the quantities energy per charge (E/Q), time–of–flight (TOF), and energy (E) are determined for the incoming particles. CTOF and STOF are aimed at different energy ranges, CTOF being devoted to solar wind and low energy suprathermal ions, while STOF is devoted to higher energy suprathermal and low energy solar energetic particles. The MTOF sensor is a high resolution retarding potential mass analyzer with a quadrupol electric field configuration, which will be able not only to measure the composition of the less abundant elements in the solar wind, but also the isotopic composition of the more abundant heavy ions. The covered nuclear mass and energy ranges of the CELIAS ion sensors along with the planned coverage of the COSTEP/ERNE energetic particle experiments are summarized in Fig. 1.

The solar EUV monitor, SEM, is structurally connected with STOF. SEM consists basically of a diffraction grating in front of a set of three absolutely calibrated silicon light diodes, which are covered with an aluminum filter. The diodes are placed at the zero – and first order diffraction image of the Sun in order to isolate the 30.4 nm He II line.

The digital processing unit, DPU, serves all three sensors by fully handling the communications with the spacecraft. It receives rate and pulse-height data from each sensor, compresses, stores and formats it for conveying the data to the S/C telemetry. It also handles the experiment control through the commanding system and surveys the CELIAS housekeeping data.

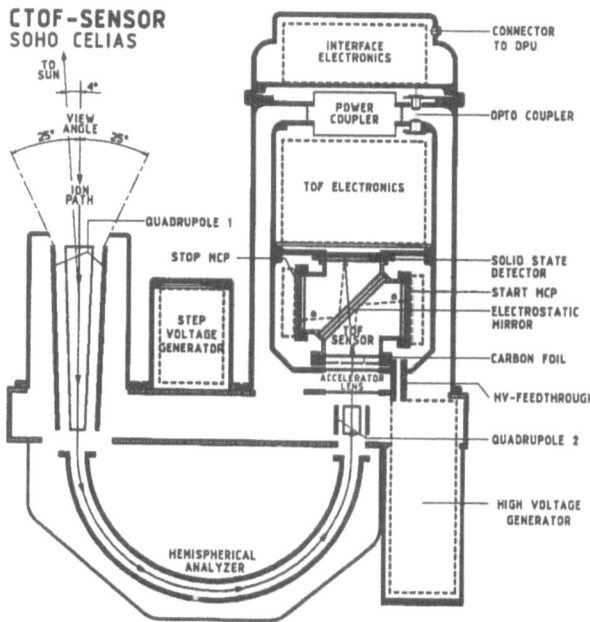

Fig. 2. Schematic view of CTOF sensor.

4 The CTOF Sensor

The solar wind Charge TOF (CTOF) sensor covers the energy per charge (E/Q) range 0.1 to 55 keV/charge, which fully includes the solar wind range. The sensor will determine the composition, charge state distribution, kinetic temperature, and speed of the more abundant solar wind ions (e.g. He, C, N, O, Ne, Mg, Si, and Fe). The sensor is designed to accept ions up to 55 keV/e corresponding to a bulk speed of about 1000 km/s for, e.g., Fe^{8+}. To cover a wide solid angle and also to allow the measurement of reasonable fraction of the thermal width of the ion distribution, the solar wind enters the instrument through an ion-optic system which accepts the solar wind within a cone of 50 deg and which is matched with an electrostatic analyzer. After penetrating through the analyzer and prior to their entrance into the TOF detector, the ions are accelerated by up to 35 keV/charge to raise their energy above the threshold of the solid state detector (SSD) and to improve the charge state resolution. Fig. 2 shows the principle of operation of this instrument.

The identification of solar wind ions is obtained by measuring the energy/charge E/Q, TOF τ, and the residual energy E_{SSD} deposited in the solid state detector. From this we can determine the mass (M), and velocity s/τ of each ion according to $M = 2(\tau/s)^2 \times E_{SSD}/a_1$ where s is the TOF path length, and a_1 takes into account the pulse-height defect in the SSD. The mass/charge ratio (M/Q) will be derived by combining the accurate measurement of E/Q in the electrostatic deflection system and TOF τ:

$$M/Q = 2(\tau/s)^2 \times (U_{acc} + a_2 E/Q),$$

U_{acc} is the acceleration voltage, and a_2 accounts for the energy loss in the carbon foil. Thus from the measurement of M and M/Q the ionic charge Q, and finally from Q and E/Q the initial energy E can be derived.

5 The MTOF Sensor

The solar wind Mass TOF (MTOF) sensor is a high resolution TOF mass spectrometer $(M/\Delta M > 100)$ which will allow over a wide range of solar wind bulk speeds to measure the elemental and isotopic composition of the solar wind. The instrument employs a wide band electrostatic preselecting deflection system (WAVE), followed by a high-resolution retarding potential TOF analyzer. The principle of operation of the mass analyzer is to measure the TOF of mostly singly ionized particles in a static electric square-well potential $(V \propto x^2)$. In such a potential – of a harmonic oscillator – the TOF depends only on M/Q:

$$\tau \propto \sqrt{M/Q}$$

Solar wind ions within the passband of the front section WAVE of the instrument pass through a thin carbon foil. In passing through the foil the solar wind ions lose a fraction of their energy, are scattered (typically by 10 deg), and

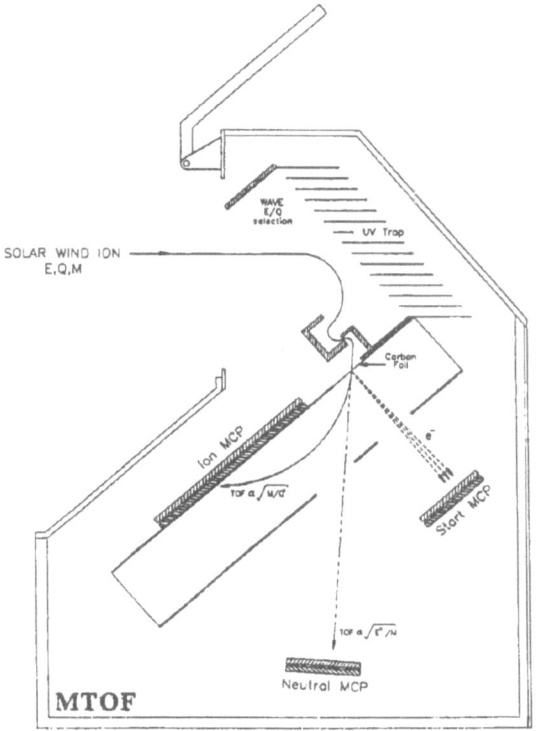

SOLAR WIND ION
E,Q,M

MTOF

Fig. 3. Schematic view of MTOF sensor.

emerge either as neutrals or singly ionized. Secondary electrons emitted from the foil will be recorded in a micro–channel–plate (MCP), while the electrostatically reflected ions are recorded in the stop–MCP. From the TOF immediately the mass of the ion can be derived. Fig. 3 shows a schematic view of the MTOF sensor system.

6 The STOF Sensor

The Suprathermal TOF (STOF) sensor is an ion telescope with a large geometrical factor of 0.1 cm^{2}sr for the measurement of the energy distribution of individual charge states of various elements of solar energetic particles. It covers the energy range from 20 to 3000 keV/charge by employing two separate sections of the electrostatic deflection system, one with curved (STOF) and one with flat deflection plates (HSTOF). Similar to the CTOF sensor it combines the selection of incoming particles according to E/Q by electrostatic deflection in a multi–gap system with a subsequent TOF analysis and a final energy measurement in a solid state detector. Only the postacceleration is not required because the suprathermal particles carry sufficient energy to respond in the solid state detector. Fig. 4 shows a schematic view of the sensor.

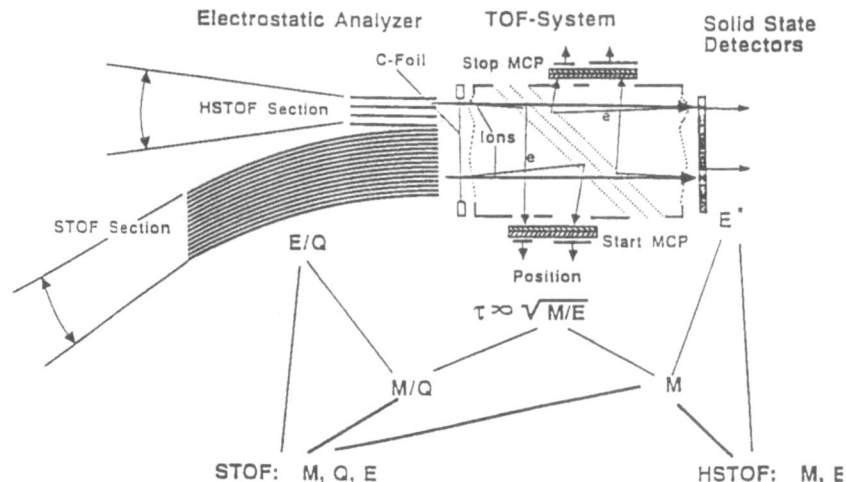

Fig. 4. Schematic view of STOF sensor.

7 The Solar EUV Monitor

The Solar Extreme–Ultra–Violet Monitor, SEM, is a highly stable photodiode spectrometer that will continuously measure the full solar disk absolute photon flux at the prominent and scientifically important He II 30.4 nm line, as well as the absolute integral flux between 15 and 70 nm. Fig. 5 shows a schematic diagram of the optical arrangement of the EUV monitor system. The entrance

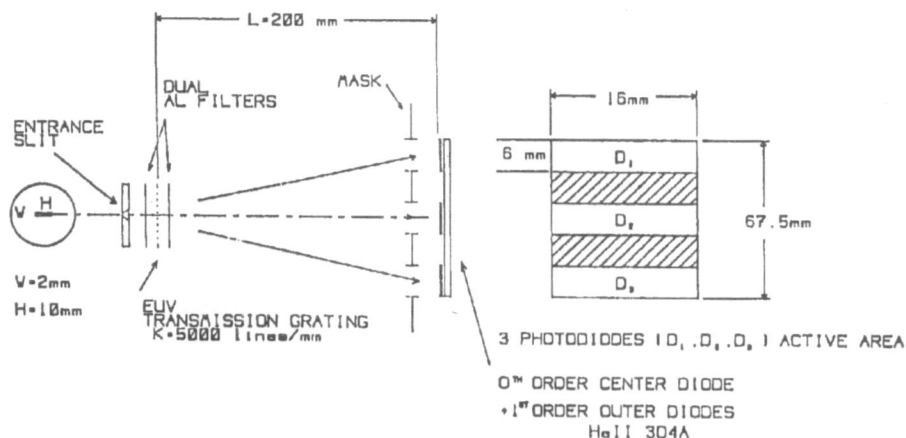

Fig. 5. Schematic view of the Solar EUV Monitor, SEM.

system consists of a 5000 line/mm transmission grating, placed between two aluminum filters (total 150 nm thick). The Al filter limits the radiation to the Al bandpass (17 – 70 nm). Any potential effect due to solar wind on the Al filters will be eliminated by the "solar wind deflector". The prominent He II 30.4 nm line will be isolated by the two side diodes while the center diode responds to the direct light (0–order). The diode current is monitored for all three diodes separately. Due to the extended source of the Sun (1/2 deg) and finite size of the entrance silt (2 mm × 10 mm) the spectrometer will have a bandwidth of ± 4.0 nm at the 30.4 nm line.

8 The Digital Processing Unit

The Digital Processing Unit, DPU, performs all tasks necessary to communicate commands and data between the sensors and the spacecraft, which in turn is linked through telemetry to ground. The main tasks are (1) to control the CELIAS sensors by DPU generated and S/C routed ground commands, (2) to record the house keeping data, (3) to record and format the counting rates of the sensors, and (4) to record, classify, and format pulse–height data. Its output data are routed through the S/C telemetry. Fig. 6 shows the basic structure of the DPU.

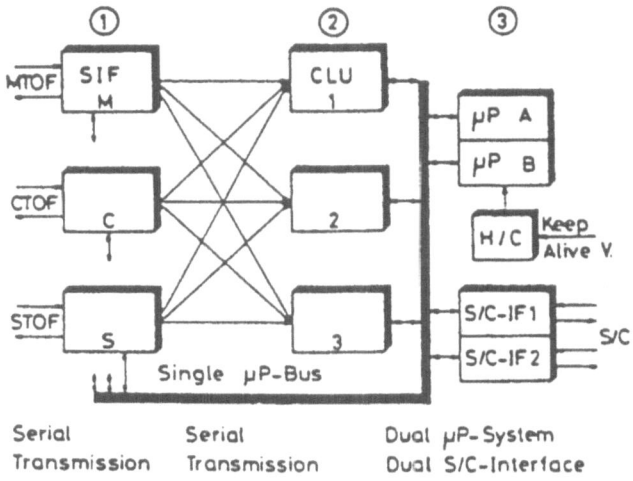

Fig. 6. Principle of operation of the CELIAS DPU: SIF sensor interface, CLU = classification unit, H/C = hard core.

Acknowledgments The CELIAS experiment is produced in a joint effort at the following hardware-providing institutions:

Max–Plank–Institut für Extraterrestrische Physik, Garching, Germany; Max-Plank–Institut für Aeronomie, Katlenburg–Lindau Germany; Physikalisches Institut der Universität Bern, Switzerland; University of Maryland, College Park, MD, USA; Institut für Datenverarbeitung, Technische Universität Braunschweig, Germany; Space Science Center, University of Southern California, Los Angeles, USA.

The work is supported in part by DARA, Germany, NASA, USA, by the Swiss National Science Foundation, and the PRODEX program of ESA.

Participants

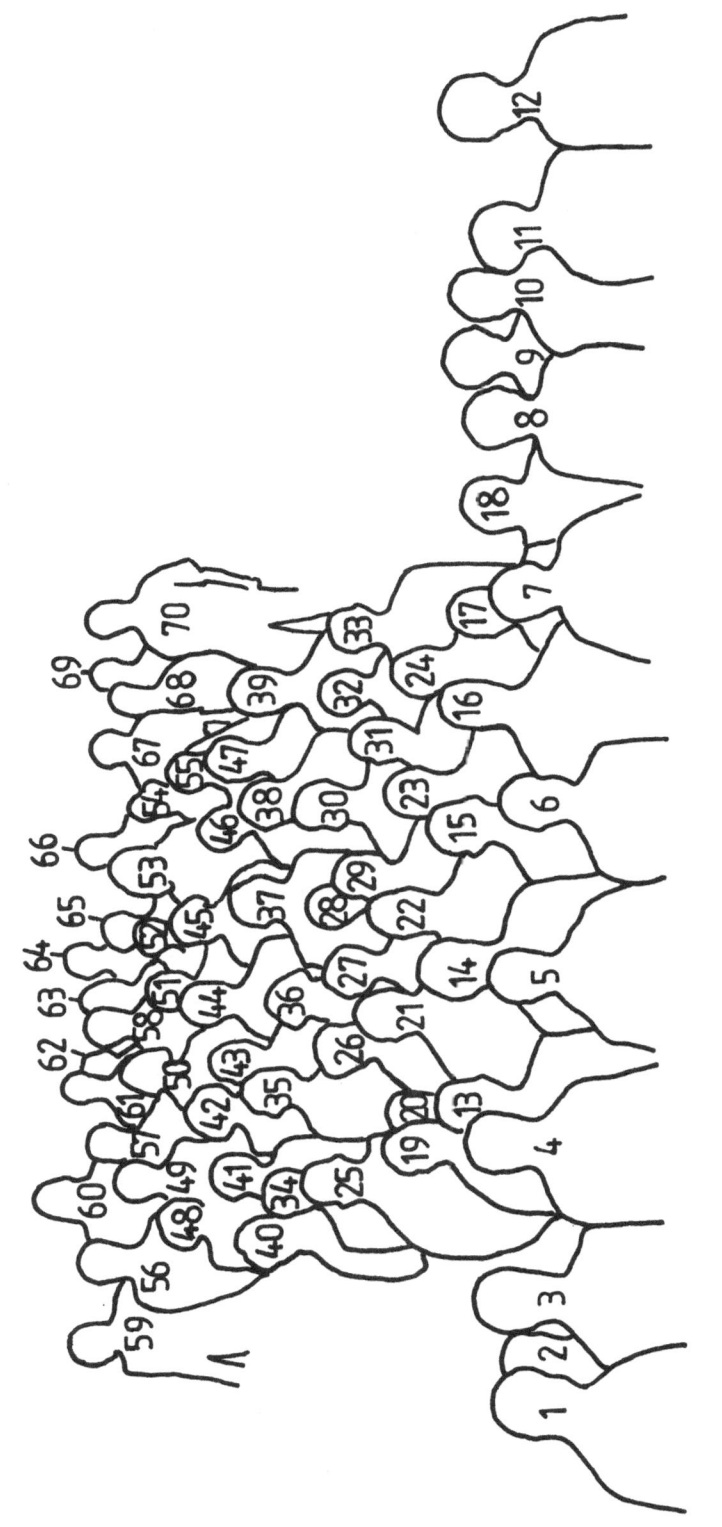

List of Participants

Apushkinskij, G. (59)	Astron. Observ., St. Petersb. Univ., Russia
Aschwanden, M. (37)	University of Maryland, College Park, USA
Aurass, H.	Astrophysikalisches Institut Potsdam, Germany
Benz, A.O. (5)	Institut für Astronomie, ETH Zürich, Switzerland
Bochsler, P.	Physikalisches Institut Univ. Bern, Switzerland
Bogod, V. (10)	SAO, St. Petersburg Branch, Russia
Borovik, V. (27)	AO Pulkovo, St. Petersburg, Russia
Chernov, G. (9)	IZMIRAN, Troitsk, Russia
Clette, F. (25)	Observatoire Royal de Belgique, Brussel, Belgium
Crosby, N.B. (44)	DASOP, Observ. de Paris-Meudon, France
Csillaghy, A. (57)	Inst. of Astronomie, ETH Zürich, Switzerland
Drago, F.	University of Florence, Italy
Enome, S. (40)	Nobeyama Radio Observatory, Japan
Fichtner, H. (54)	MPI für Radioastronomie, Bonn, Germany
Fleck, B. (22)	ESA Space Science Dept., Noordwijk, Netherlands
Fleishman, G. (56)	Ioffe Inst. f. Phys. Techn., St. Petersb., Russia
Fletcher, L. (50)	Astronomical Inst., Utrecht, Netherlands
Fu, Q.	Beijing Astronomical Observatory, China
Gelfreikh, G. (29)	AO Pulkovo, St. Petersburg, Russia
Golubchina, O. (45)	AO Pulkovo, St. Petersburg, Russia
Gopalswamy, N. (4)	University of Maryland, College Park, USA
Grebinskij, A. (52)	SAO, St. Petersburg Branch, Russia
Hackenberg, P.	Astrophysikalisches Institut Potsdam, Germany
Hanschur, U.	Astrophysikalisches Institut Potsdam, Germany
Hewish, A. (8)	Cavendish Laboratory, Cambridge, UK
Hildebrandt, J. (61)	Astrophysikalisches Institut Potsdam, Germany
Hofmann, A. (70)	Astrophysikalisches Institut Potsdam, Germany
Jamitzky, F. (21)	MPI für Extraterr. Physik, Garching, Germany
Jansen, F. (6)	Astrophysikalisches Institut Potsdam, Germany
Jiřička, K. (39)	Astronom. Institute, Ondřejov, Czech Rep.
Karlický, M. (32)	Astronom. Institute, Ondřejov, Czech Rep.
Kerdraon, A. (16)	DASOP Observ. de Paris-Meudon, France
Kirchner, T.M. (24)	Deut. Wetterdienst, Meteorologischer Dienst, Germany
Klassen, A. (64)	Astrophysikalisches Institut Potsdam, Germany
Klein, K.-L. (30)	DASOP Observ. de Paris-Meudon, France
Kliem, B. (47)	Astrophysikalisches Institut Potsdam, Germany
Korzhavin, A. (17)	SAO, St. Petersburg Branch, Russia
Kosugi, T. (15)	Nation. Astronomical Observatory, Mitaka, Japan
Krucker, S. (60)	Institut für Astronomie, ETH Zürich, Switzerland
Krüger, A. (7)	Astrophysikalisches Institut Potsdam, Germany
Krülls, W.M. (53)	Observ. de Paris, Paris-Meudon, France
Kuijpers, J. (43)	Astronomical Institute, Utrecht, Netherlands

Kundu, M.R. (34)	University of Maryland, College Park, USA
Kurths, J. (38)	MPG Universität Potsdam, Germany
Lembege, B. (35)	C.N.E.T./C.E.T.P., Issy-les-Moulineaux, France
Magun, A.	Institut of Appied Physics, Bern, Switzerland
Mann, G.	Astrophysikalisches Institut Potsdam, Germany
Meister, C.-V.	Universität Potsdam, Germany
Melnikov, V.	Radiophysical Research Inst., N. Novgorod, Russia
Mercier, C.	DASOP/URA Lasolmire, Meudon, France
Messerotti, M. (33)	Astronomical Observatory, Trieste, Italy
Murawski, K. (46)	Center for Plasma Astrophysics, Leuven, Belgium
Nagnibeda, V. (28)	Astronom. Inst., St. Petersburg Univ., Russia
Nefedjev, V. (67)	Inst. of Solar-Terrestrial Physics, Irkutsk, Russia
Noci, G.	Depart. di Astronomia, Firenze, Italy
Oraevsky, V.N. (1)	IZMIRAN, Troitsk, Russia
Orzaru, C.-M. (11)	Institute of Astronomy, Romania
Ostrowski, M. (55)	Astronomical Observatory, Krakow, Poland
Paschke, J.	Astrophysikalisches Institut Potsdam, Germany
Pick, M. (36)	DASOP Observ. de Paris-Meudon, France
Pissarenko, N. (2)	Space Research Institute, Moscow, Russia
Poquerusse, M. (13)	Observ. de Paris-Meudon, France
Rosenraukh, Y. (23)	Inst. of Solar-Terrestrial Physics, Irkutsk, Russia
Ruždjak, V. (69)	Hvar Observatory, Zagreb, Croatia
Ryabov, M. (51)	Odessa Observatory URAN-4, Ukraine
Schumacher, J. (62)	Astrophysikalisches Institut Potsdam, Germany
Schwarz, U.	MPG Universität Potsdam, Germany
Schwenn, R.	MPI für Aeronomie, K.-Lindau, Germany
Sheiner, O. (42)	Radiophysical Research Inst., N. Novgorod, Russia
Smolkov, G. (12)	Institute of Solar-Terr. Physics, Irkutsk, Russia
Tlamicha, A. (3)	Astronomical Observatory, Ondřejov Czech Rep.
Treumann, R. (26)	MPI für Extraterr. Physik, Garching, Germany
Urbarz, H. (18)	Astronomisches Institut, Univ. Tübingen, Germany
Vilmer, N., (48)	DASOP Observ. de Paris-Meudon, France
Vlahos, L. (19)	University of Thessaloniki, Greece
Vršnak, B. (68)	Hvar Observatory, Zagreb, Croatia
Wilhelm, K. (14)	MPI für Aeronomie, K.-Lindau, Germany
Yurovsky, Y. (49)	Crimean Astrophysical Observatory, Ukraine
Zaitsev, V. (66)	Inst. of Applied Physics, N.Novgorod, Russia
Zheleznyakov, V.V. (58)	Inst. of Applied Physics, N. Novgorod, Russia
Zlobec, P. (31)	Astronomical Observatory Trieste, Italy
Zlotnik, E.Ya. (41)	Inst. of Applied Physics, N. Novgorod, Russia

Numbers in parentheses refer to the group photo: On the photo are further the LOC-assistents L. Kurth (65), A. Trettin (63), and D. Scholtz (20).

Lecture Notes in Physics

For information about Vols. 1–407
please contact your bookseller or Springer-Verlag

New Series m: Monographs